Charles H. Townes
May, '93

Lecture Notes in Physics

The Editorial Policy for Proceedings

The series Lecture Notes in Physics reports new developments in physical research and teaching – quickly, informally, and at a high level. The proceedings to be considered for publication in this series should be limited to only a few areas of research, and these should be closely related to each other. The contributions should be of a high standard and should avoid lengthy redraftings of papers already published or about to be published elsewhere. As a whole, the proceedings should aim for a balanced presentation of the theme of the conference including a description of the techniques used and enough motivation for a broad readership. It should not be assumed that the published proceedings must reflect the conference in its entirety. (A listing or abstracts of papers presented at the meeting but not included in the proceedings could be added as an appendix.)

When applying for publication in the series Lecture Notes in Physics the volume's editor(s) should submit sufficient material to enable the series editors and their referees to make a fairly accurate evaluation (e.g. a complete list of speakers and titles of papers to be presented and abstracts). If, based on this information, the proceedings are (tentatively) accepted, the volume's editor(s), whose name(s) will appear on the title pages, should select the papers suitable for publication and have them refereed (as for a journal) when appropriate. As a rule discussions will not be accepted. The series editors and Springer-Verlag will normally not interfere with the detailed editing except in fairly obvious cases or on technical matters.

Final acceptance is expressed by the series editor in charge, in consultation with Springer-Verlag only after receiving the complete manuscript. It might help to send a copy of the authors' manuscripts in advance to the editor in charge to discuss possible revisions with him. As a general rule, the series editor will confirm his tentative acceptance if the final manuscript corresponds to the original concept discussed, if the quality of the contribution meets the requirements of the series, and if the final size of the manuscript does not greatly exceed the number of pages originally agreed upon.

The manuscript should be forwarded to Springer-Verlag shortly after the meeting. In cases of extreme delay (more than six months after the conference) the series editors will check once more the timeliness of the papers. Therefore, the volume's editor(s) should establish strict deadlines, or collect the articles during the conference and have them revised on the spot. If a delay is unavoidable, one should encourage the authors to update their contributions if appropriate. The editors of proceedings are strongly advised to inform contributors about these points at an early stage.

The final manuscript should contain a table of contents and an informative introduction accessible also to readers not particularly familiar with the topic of the conference. The contributions should be in English. The volume's editor(s) should check the contributions for the correct use of language. At Springer-Verlag only the prefaces will be checked by a copy-editor for language and style. Grave linguistic or technical shortcomings may lead to the rejection of contributions by the series editors.

A conference report should not exceed a total of 500 pages. Keeping the size within this bound should be achieved by a stricter selection of articles and not by imposing an upper limit to the length of the individual papers.

Editors receive jointly 30 complimentary copies of their book. They are entitled to purchase further copies of their book at a reduced rate. As a rule no reprints of individual contributions can be supplied. No royalty is paid on Lecture Notes in Physics volumes. Commitment to publish is made by letter of interest rather than by signing a formal contract. Springer-Verlag secures the copyright for each volume.

The Production Process

The books are hardbound, and the publisher will select quality paper appropriate to the needs of the author(s). Publication time is about ten weeks. More than twenty years of experience guarantee authors the best possible service. To reach the goal of rapid publication at a low price the technique of photographic reproduction from a camera-ready manuscript was chosen. This process shifts the main responsibility for the technical quality considerably from the publisher to the authors. We therefore urge all authors and editors of proceedings to observe very carefully the essentials for the preparation of camera-ready manuscripts, which we will supply on request. This applies especially to the quality of figures and halftones submitted for publication. In addition, it might be useful to look at some of the volumes already published. As a special service, we offer free of charge LATEX and TEX macro packages to format the text according to Springer-Verlag's quality requirements. We strongly recommend that you make use of this offer, since the result will be a book of considerably improved technical quality. To avoid mistakes and time-consuming correspondence during the production period the conference editors should request special instructions from the publisher well before the beginning of the conference. Manuscripts not meeting the technical standard of the series will have to be returned for improvement.

For further information please contact Springer-Verlag, Physics Editorial Department V, Tiergartenstrasse 17, W-6900 Heidelberg, FRG

Andrew W. Clegg   Gerald E. Nedoluha (Eds.)

Astrophysical Masers

Proceedings of a Conference
Held in Arlington, Virginia, USA
9-11 March 1992

Springer-Verlag

Berlin Heidelberg New York
London Paris Tokyo
Hong Kong Barcelona
Budapest

Editors

Andrew W. Clegg
Gerald E. Nedoluha
Remote Sensing Division, Code 7200
Naval Research Laboratory
Washington, DC 20375-5000, USA

ISBN 3-540-56343-1 Springer-Verlag Berlin Heidelberg New York
ISBN 0-387-56343-1 Springer-Verlag New York Berlin Heidelberg

© Springer-Verlag Berlin Heidelberg 1993
Printed in Germany

Typesetting: Camera ready by author/editor
58/3140-543210 - Printed on acid-free paper

PREFACE

The first conference devoted exclusively to astrophysical masers was held in Arlington, Virginia, 9—11 March 1992. The 113 participants included scientists from Australia, Brazil, Canada, England, Finland, France, Germany, Greece, Italy, Japan, Russia, South Africa, Spain, Sweden, and the United States. More than 100 talks and posters were presented during the three-day conference.

In March 1991, when conference planning began, we hoped to attract as many as 50 astronomers with an interest in masers. It was a pleasant surprise when, in the last few months before the conference, it became apparent that the attendance would far exceed our expectations. The large number of conference participants shows that there is currently an extraordinarily high level of interest within the astronomical community in the study of masers. One reason, perhaps, is the wide range of astrophysical phenomena with which masers are associated: extragalactic astronomy, star formation, stellar evolution, the interstellar medium, and solar system physics, to name a few. The papers in this volume include contributions on all of these topics.

This book serves at least two purposes. First, it is a review of research on astrophysical masers since their discovery more than 27 years ago. The early years of maser research are recalled by Charles Townes in a transcript of his after-dinner talk. Additional review papers cover each major field of study. Second, it is a record of the current state of maser research, presented in 94 articles covering the latest observational and theoretical results. The quality and volume of the contributions insure that these proceedings will be an important addition to the astronomical literature.

We would like to express appreciation to several people for assistance in organizing the conference. The scientific organizing committee, comprised of R. Booth, R. J. Cohen, S. Deguchi, M. Elitzur, R. Genzel, K. J. Johnston, J. Moran, R. Norris, V. S. Strelnitski, C. H. Townes, W. D. Watson, and T. Wilson, provided valuable advice and, through their contacts, made possible a truly international gathering of scientists. Ken Johnston and Phil Schwartz were especially helpful in all stages of conference planning, and were responsible for attracting financial support. Others who helped insure that the meeting ran smoothly include Kristin Clegg and Jo-Ann Clegg, two experts in conference organization; and Al Fey

v

and Schuyler Van Dyk, who provided scientific and technical assistance. Kathy Cowan and Pat Tetrault furnished administrative support at NRL. We are particularly grateful to Amahl Drake for assistance in travel arrangements for the Russian participants—no small feat considering the Soviet Union existed when we began planning the meeting, but had disappeared by the time the meeting was held. The conference was sponsored by the Naval Research Laboratory and by NASA. We are grateful to both organizations for their support. We also acknowledge support from the National Research Council through the NRC/NRL Cooperative Research Associateship program. Lastly, we thank all of the participants for providing the subject matter for a very stimulating conference.

Naval Research Laboratory   Andrew W. Clegg
Washington, D.C.   Gerald E. Nedoluha
September 1992

CONTENTS

6. OH MASERS IN STAR-FORMING REGIONS

7. WATER MASERS IN STAR-FORMING REGIONS

8. METHANOL MASERS IN STAR FORMING REGIONS

9. PROPER MOTION

10. SCATTERING

11. VARIABILITY

12. CIRCUMSTELLAR MASERS (GENERAL)

13. CIRCUMSTELLAR OH MASERS

LIST OF PARTICIPANTS

Zulema Abraham	Universidade de São Paulo
J. Alcolea	Center for Astrophysics
Nels Anderson	Max-Planck-Institut für Extraterrest. Physik
Alice Argon	Center for Astrophysics
Willem Baan	Arecibo Observatory
Leo Blitz	University of Maryland
Eric Bloemhof	Center for Astrophysics
David Boboltz	Virginia Tech
Roy Booth	Onsala Space Observatory
Phil Bowers	SFA, Inc.
V. Bujarrabal	Centro Astronómico de Yebes
Paola Caselli	Universita di Bologna
Tim Cawthorne	Center for Astrophysics
R. Cesaroni	Universita di Firenze
Jessica Chapman	Australia Telescope National Facility
Mark Claussen	Naval Research Laboratory
Andrew Clegg	Naval Research Laboratory
R. J. Cohen	University of Manchester
Alan Collison	Naval Research Laboratory
Francisco Colomer	Onsala Space Observatory
Jim Cordes	Cornell University
Shuji Deguchi	Nobeyama Radio Observatory
Brian Dennison	Virginia Tech
Phil Diamond	National Radio Astronomy Observatory
Mara-Lana Djurhuus	University of Maryland
Amahl Drake	University of California, Santa Cruz
Wolfgang Duschl	Universität Heidelberg
Moshe Elitzur	University of Kentucky
Dieter Engels	Hamburger Sternwarte
Paul A. Feldman	Herzberg Institute of Astrophysics
Alan Fey	Naval Research Laboratory
Ralph Fiedler	Naval Research Laboratory
D. Field	University of Bristol

Jacqueline Fischer	Naval Research Laboratory
J. R. Forster	Hat Creek Radio Observatory
Roger Foster	Naval Research Laboratory
D. A. Frail	National Radio Astronomy Observatory
John Galt	Dominion Radio Astrophysical Observatory
Ralph Gaume	Naval Research Laboratory
Reinhard Genzel	Max-Planck-Institut für Extraterrest. Physik
Eric Gérard	Observatoire de Paris - Meudon
Mark A. Gordon	National Radio Astronomy Observatory
M. D. Gray	University of Bristol
Lincoln Greenhill	University of California, Berkeley
Helene Guertler	NASA Goddard
Carl Gwinn	University of California, Santa Barbara
Lauri Haikala	Helsinki University Obs.
Arsen Hajian	Cornell University
Peter Hofner	University of Wisconsin
David Hollenbach	NASA Ames
Ken Johnston	Naval Research Laboratory
Menas Kafatos	George Mason University
S. V. Kalenskii	P. N. Lebedev Physical Institute
Michael Kaufman	Johns Hopkins University
Frank Kerr	USRA/University of Maryland
William Klemperer	Harvard University
S. Knowles	Naval Research Laboratory
Hidayuki Kobayashi	Inst. of Space and Astronautical Sciences
Nikolaos Kylafis	University of Crete
Geoff Lawrence	University of Minnesota
Murray Lewis	Arecibo Observatory
S. Likhachev	P. N. Lebedev Physical Institute
Tarja Liljeström	Helsinki University Observatory
Irene Little-Marenin	University of Colorado
Philip Lockett	Centre College
Gordon MacLeod	Hartebeesthoek Radio Astronomy Observatory
Mordecai-Mark Mac Low	NASA Ames
Loris Magnani	University of Georgia

M. R. W. Masheder	University of Bristol
Demetrios N. Matsakis	U.S. Naval Observatory
L. I. Matveyenko	Russian Academy of Sciences
Gary Melnick	Center for Astrophysics
Karl Menten	Center for Astrophysics
Victor Migenes	University of Manchester
Makoto Miyoshi	Nobeyama Radio Observatory
Jim Moran	Center for Astrophysics
Michael Mumma	NASA Goddard
Robert Mutel	University of Iowa
Gerald Nedoluha	Naval Research Laboratory
David Neufeld	Johns Hopkins University
Colin Norman	Space Telescope Science Institute
Ray Norris	Australia Telescope National Facility
Sten Odenwald	BDM International
G. M. Pacheco	Instituto Nacional de Pesquisas Espaciais
Nimesh Patel	Five College Radio Astronomy Observatory
Tom Pauls	Naval Research Laboratory
Ruisheng Peng	University of Illinois
Preethi Pratap	Center for Astrophysics
Andreas Quirrenbach	U.S. Naval Observatory
Mark Reid	Center for Astrophysics
Jane Russell	Naval Research Laboratory
Phil Schwartz	Naval Research Laboratory
John H. Simonetti	Virginia Tech
V. I. Slysh	P. N. Lebedev Physical Institute
Howard Smith	Smithsonian Institution
John Spencer	Naval Research Laboratory
Robert Stencel	University of Colorado
V. S. Strelnitskij	Astronomical Council Academy of Sciences
Hiroshi Takaba	Kashima Space Research Center
Gianni Tofani	Osservatorio Astrofisico di Arcetri
Charles Townes	University of California, Berkeley
Jeffrey A. Uphoff	Virginia Tech
Irina Val'tts	Lebedev Physical Institute

Schuyler Van Dyk	Naval Research Laboratory
H. J. van Langevelde	Sterrewacht Leiden
Jürgen Vogt	Universität Ulm
William Watson	University of Illinois
Dan Weedman	NASA
Tom Wilson	Max-Planck-Inst. für Radioastronomie
Anders Winnberg	Onsala Space Observatory
Douglas Wood	National Radio Astronomy Observatory
Alwyn Wootten	National Radio Astronomy Observatory
Jeremy Yates	NRAL/Jodrell Bank

1. HISTORY

The Early Years of Research on Astronomical Masers

of

CHARLES H. TOWNES

University of California at Berkeley

All the beautiful work presented at this meeting is most impressive! I've been a bit out of touch with the field for a while; it's nice to get reacquainted and see all these important things happening.

I was asked to comment on the early days of astronomical masers and give a history, but reacted that history takes some study and can be dull. The response was well, then just give some anecdotes, and my talk is going to be more like that. History, you can read about; hopefully it's recorded somewhere. In itself, a published scientific paper tends to tell you what was actually done; what's behind it is another question and sometimes more important. So perhaps I'll illustrate, but not really expound history. But don't think that I will have it all completely straight, because everyone sees such things with their own eyes, and memories fade.

In thinking over the development of this field, I was impressed with a general proposition I have noticed before in other fields - the great tendency we all have to latch on to certain concepts and not give them up. We're often slow to deal freely with ideas. Perhaps scientists agree with each other too much, reinforcing each other and thus affirming ideas which become too fixed.

I'm reminded of the case of Michelson and his measurement of light. Michelson was a superb experimentalist. When I was in school he seemed to be almost the only great American physicist besides Millikan. He had measured the speed of light and gave its value within a certain probable error. That measurement was doctrine. As time went on, people continued to analyze this data and cut down his probable error more and more. And there were others who re-measured the speed of light, and found approximately his value. The precision of the speed of light thus seemed to get better and better over the decades, with always the same value -- Michelson's. Suddenly, there was a British radar engineer who, playing with radar during and after the second world war, reflected radar signals off objects at known distances and timed them. He didn't get Michelson's answer and was bold enough to publish this. It was far outside of the published probable error, considerably outside the error even Michelson had published. Other people started measuring and, yes, they all agreed. For the previous measurements, it turned out, people sometimes didn't agree with Michelson, but they either said nothing or just did it again until they did agree.

Such effects we all need to watch, ponder, and be humble about. It's both our blind ignorance and the rigidity with which we stick to certain conceptions about which we need humility, all of us. One might wonder at this moment what are the big mistakes, the important oversights we are making. Or for the venturesome, the question may be which of all our way-out suppositions are right. And the foolish experiments we'd like to do, the measurements we'd like to try — which ones should we really try; which are really going to pay off?

Let me start talking about apparently foolish experiments. About 30 years ago a friend of mine, head of an important department, called me up. "I'm very worried," he said, "about Alan Barrett. You recommended him to us, but the only thing he seems to want to do is this OH. But is that never going to get anywhere .... He's already failed once. What do you think?"
I said, "Well, it is risky, but I think it's a reasonable thing to try." But he replied, "I'm worried about his career and future."

In fact that was a hazardous thing for a young postdoc to undertake: an experiment that most people didn't think would work. It clearly seemed difficult, and had already failed once. But fortunately, not so long after that Alan's career suddenly was safe and looked bright -- he had stuck to it and, with helpful colleagues, had found OH.

As I understand it, shortly after the discovery Alan came to Washington for a committee meeting -- as many of us are wont to do. Gart Westerhout had brought the radio astronomers together for dinner at his house, and Alan's account of having just found OH was the talk of the dinner party. Harold Weaver was there, and the next morning he phoned his own team back in Berkeley. Hat Creek already had a suitable antenna, and he wanted the Berkeley group to immediately look for OH. Sure enough it was soon confirmed by groups in the U.S. and Australia. Alan found the first OH absorption against radiation from the supernova Cas A; confirmation come from the Galactic Center. Both Weaver and Gundermann at Harvard started looking through Gart Westerhout's catalog of bright radio sources because one needs a bright source to obtain absorption lines. About three o'clock one morning Harold had a call from the technician at Hat Creek in northern California who was doing this. The technician called up to say, "There's something wrong -- I just can't figure it out. Everything seems to be working all right, we've got some signals at the right frequency, but of the wrong sign - the power goes up instead of down. Could you suggest what I might be doing wrong?"

That, of course, was the first detection of an astronomical maser. Gundermann found similar things as she worked her way through Westerhout's list, and they were part of her thesis.

Harold Weaver's group published a paper on something that he called "mysterium." Why? The lines were all chopped up, there were many separate lines; obviously this couldn't be just a simple pair of molecular lines. He doubted it was OH at all: perhaps the apparent discovery of OH was even a mistake. He compared the new unexplained lines with astronomer's experience with nebulium. Nebulium, you may remember, was thought to probably be a new element found in nebulae and giving lines in the optical region. Ike Bowen finally explained them as due to metastable states of common atoms. So, he called it mysterium - something new and mysterious. Weaver was right that it seemed mysterious. I'm sure that didn't shake Alan Barrett's view --it had to be OH, regardlessly of the strange peculiarities observed. What was going on, however --to produce so many different and sharp lines, at high intensity, and in emission?

The first suggestion I can remember that it might be a maser came from Shklovskii. This was just in informal conversation with me. We talked about it, and the possibility certainly seemed reasonable. Tommy Gold was also suggesting quite early that it might be a maser. He and his associates produced a theory of masers excited by ultraviolet radiation. Nevertheless, it was not immediately taken very seriously. The idea seemed to grow on people that maybe this really could be it. Marvin Litvak soon began to also produce theories of how to make an astronomical maser. While people began to accept the idea more and more, what really clinched it were the experiments involving somewhat higher resolution. I understand Bernie Burke said, "Let's build an interferometer." Rogers, working with him, dashed out to Haystack, put together the two antennas there, and pretty quickly made the interferometer. A student named Jim Moran also participated. This interferometer demonstrated that the intense radiation was in fact coming from a source of rather restricted size. The temperature had to go way up, which I think pretty much clinched the maser conclusion in people's minds. There was also something rather similar done out at CalTech. The excitement was getting around, and many things were being done fairly rapidly, in parallel and by small groups.

How important was this going to be? I was an administrator at MIT when Alan and his colleagues made the discovery. Alan and I had always been close, so I was asked to chair a news-release program to announce this nice discover and tried to expound to the media how important it was, and what it would do for the future. Some reporter asked me whether it would be as important as discovery of the 21 cm hydrogen line. I had to say that I thought it was very important but doubted it could be that important. Today, I might have to change my mind. How difficult it is to foresee our scientific future! OH was, after all, somewhat like CN, CH and CH+, which were the three molecules that were previously known. They are all unstable molecules and free radicals. Many people thought that finding OH was great. It would be pursued further with radio astronomy, but would probably be the end of the road.

Opinions had to be rather negative about looking for other molecules because, while OH was worked on, for some years nobody looked for anything more. When I got out of administration and moved to Berkeley in late 1967, one of the things I decided was that, well, nobody else is going to do it, so maybe I should try. Fortunately Jack Welsh was interested, and we put together a small group -- a postdoc Dave Rank and a student Al Cheung working with me, and Doug Thornton, an engineer working with Jack Welsh. We went after ammonia in particular, which seemed to me the most obvious choice in the light of available technology. Over in the physics department, with help and advice from the Radio Astronomy group, we built a suitable amplifier and it was put on the Hat Creek antenna. It was indeed time to see if there were stable molecules in insterstellar space; NH_3 turned out to be there. After we found it, a friend of mine came to me (Professor A, I'll call him) and said, "You know, Professor B prevented me from discovering ammonia first." I asked him how was that. He replied, "I wanted to look for ammonia, and had a student who was going to do it, but Professor B told him it would be a waste of his time; it couldn't possibly de detected. Because, he said, even if there is ammonia, the low density in interstellar space means it's going to come into equilibrium with the blackbody radiation, and so there wouldn't be any net intensity -- you can't detect anything." The student refused to do it, and poor Professor A had to give up. Fortunately, I wasn't a student at that time and could feel a little more independent. And if I wasted a year or two it wasn't so serious.

Common perception at that time meant that people were pretty convinced that the density of interstellar space and interstellar clouds was quite low. There were well-known dust clouds around, but observations at 21 cm had seemed to show there was no hydrogen in the dust clouds. Actually, I found a couple of papers which speculated on whether the hydrogen could possibly be molecular in these clouds. But people weren't taking that very seriously. Furthermore, normal theory says that there must be an equilibrium of pressure everywhere so that if low density is measured in most places it's also going to be low everywhere, including in the dust clouds. That was a rather general presumption. George Field has written in a semi-popular book not long ago, that in addition he tried to talk us out of the search by explaining that ammonia couldn't exist in interstellar clouds because there was no way of forming ammonia.

Let me point out that all the above reservations about looking for ammonia were rather logical. Professor B's story was completely logical: he knew the accepted density in interstellar space; he could calculate that there aren't enough collisions to give a net absorption or emission, even if ammonia is there. George Field's argument was completely logical: there was no reasonable mechanism that anyone could invent to produce ammonia. We still haven't one.

Our lack of realism in science is not usually because we do not apply logic; it's because of the boundary conditions we put on -- a background of limiting assumptions. The logic usually is not too bad, but the model may be wrong. Of course, it's a great temptation when doing theory to make boundary conditions as simple as possible. Theoretical models are thus always over-simplified. Frequently that's a great strength, but on the other hand there may easily be a too-rigid set of oversimplifications which overlook important complications.

One of the things I've always admired about Shklovskii is that he would challenge anything, It's a pity he can't be with us here -- he's such an enjoyable and wonderful person. You should all read his book which has now come out (in English) entitled, "Six Billion Vodka Bottles to the Moon." I won't try to explain to you just what he was doing with those vodka bottles, but that's the title of the book. Now that he's left this planet, we can assume he's somewhere out there looking at interstellar clouds where there's enough ethyl alcohol to fill about 10^{25} vodka bottles ... somewhere out there still imaging and discovering.

Another way interesting experiments sometimes get done is that we just don't know any better. In this case, we just went ahead and did it, and there was ammonia. Why did we pick ammonia? For one thing, ammonia has a good strong absorption spectrum. For another, there were low-lying states of ammonia which are metastable. But I'm also sure that I was somewhat emotionally connected with ammonia because I had used ammonia in almost everything else I had done. It's lines fell at the right frequency -- we had good systems and local oscillators that could easily get into that range. We couldn't work in the millimeter range very easily at that point.

Whether or not it could be made in interstellar space, it was at least made of common atoms and hence, we reasoned, might be there.

I'm not sure just when, but at about the same time two young postdocs at NRAO – Lew Snyder and Dave Buhl -- proposed to look for water and asked for time on NRAO's antenna. The proposal went through the usual committee, which said that's a foolish waste of antenna time. These two young guys would not be allowed to use the antenna for such a misguided search. Let me point out again that this decision was completely logical. Water, even if it were there, would have to be in a high state of excitation -- 456 wave numbers above ground -- to produce the 1.35 cm line proposed, and this state had a fairly rapid decay to ground. There's no reason it would be there. I myself didn't initially plan to look for water, I too thought logic was against it. Nevertheless Snyder and Buhl turned out to be right; the committee which denied them the antenna was dead wrong.

I won't try to guess whether these two young men knew much about what we were doing; we certainly didn't know what they were doing at the time. But our group was adventuresome enough to look for water and we had an antenna. We had found ammonia, and while water seemed highly unlikely in that particular state, on the other hand it was right in the same frequency range, so why not look? The whole project was going to be Al Cheung's thesis. He'd already found ammonia, and if he found another molecule that would be a very nice addition to his thesis.

So we gave it a try on Sgr B2 which was our favorite source, and sure enough there was a tiny little line of water in the Galactic center. That was just great, but also shocking and surprising. I couldn't see why in the world it would be there. I told Al to go ahead and look in all the other sources, just to see what might be there. It was the Christmas season. We'd already sent the work on ammonia off to a journal. I was having a party for all my research group and we were at my house, having a great time. Poor Al was up there at Hat Creek slaving away all night on the antenna. But in the middle of the party Al telephoned and said, "It must be raining in Orion!" There were big water lines - bigger than anything we'd seen or imagined. So we at the party all celebrated for Al. But these strong lines were still more puzzling -- what in the world was going on? We went on to W49, which of course was still more spectacular. The astronomy - physics group published a paper on H_2O, saying we don't understand why it would be this intense, or why water would be in this state. It clearly had to be in a different place than ammonia because it was so highly excited and very different. It might be a maser, but on the other hand we couldn't quite tell -- maybe there's some other production mechanism. We were also worried —was it really water? There was an ammonia line, due to the 3,1 state, within about 10 km/s doppler shift from the water line. Could it be ammonia? We looked at all the same sources in all the other ammonia lines, and didn't see anything else at all comparable in strength. We had to say it's probably water, and might even be a maser.

At that point we moved to Maryland Point, collaborating with NRL personnel. The reason was that we had a 20-foot antenna in California, while they had an 80-foot antenna. Clearly these sources (of water at least) were much smaller than our antenna pattern, and we wanted to get better resolution. So we moved to NRL and worked with Connie Mayer and Steve Knowles. Sure enough, the intensity went way up. Intensity of the H_2O line was effectively about 50,000 K in that antenna. And in about a week's time we found time variations in W49, which meant it had to be still smaller — not more than a light-week in size. The temperature shot up to about 10^9 K or more, and then it was very clear. At that point it wasn't hard to convince colleagues that it had to be masing; water at a temperature of 10^9K could not reasonably exist.

In the meantime, there had also been a lot further work on OH. OH had been pursued at Harvard, CalTech, Berkeley, and also in England and Australia. Another important part of the story is the further resolution. As I mentioned, MIT had built an interferometer to show that OH had to be quite small, which pretty much convinced people that it was a maser. But there was still the question as to how small were these masers, and what are they? I've just been told that there was at least a rumor that I had said the sources had to be so small they would never be resolved. Of course, since that was a mistake I don't remember saying it. But anyhow, Jim Moran pursued the question at MIT and collaborated with Harvard in making a 14 KM baseline. As is now well known, the OH source broke up into spots. The same thing had been done in Great Britain by Davies, also working with interferometry, and such interferometry has now become the beautiful and interesting field we see today. I understand OH sources were also independently broken up into spots at CalTech with their interferometer, but it was the Harvard/MIT group which pursued the subject most vigorously at that time.

Why didn't somebody suggest there might be masers in interstellar space long before this? That gap is characteristic, I suggest, of our thinking problems. Once we know something then it becomes pretty obvious. Before, we somehow don't think of it. But after all, where else but in interstellar space would one easily get non-equilibrium situations and, at least for microwave frequencies, with time constants between collisions which are as short or shorter than radiation relaxation times? That's just what you want for producing population inversions. Astronomical sources are obviously places to expect that. But this simple reasoning just wasn't around initially. Shortly after masers were identified in astronomical objects, we could produce the reasoning and say to ourselves it is all pretty obvious.

There is still more to the story of Snyder and Buhl. They had been refused the opportunity to discover water at NRAO because it seemed quite illogical for it to be there. But after we at Berkeley found water, they were of course immediately given time on the antenna. They teamed up with a couple of young people from Harvard, Ben Zuckerman and Pat Palmer, and decided to look for still another molecule, formaldehyde, at a somewhat longer wavelength and one quite handy for their equipment. They found it, and before long they were finding still other molecules. Of course now we have about a hundred different molecules identified in interstellar space, and about fifty different maser transitions, but so far as I know formaldehyde is unique in that its line absorbs the 3 K background radiation. Perhaps its frequency represents the coldest radiation field outside the Earth in that it gets down below 3 K. That too is due to a non-equilibrium distribution, and so far is perhaps the only extraterrestrial case of a "DASAR" -- Darkness Amplification by Stimulated Absorption of Radiation.

I think I've been rather dazed by all the things that have been said today and all the beautiful work that's been shown. And we look forward to more. But there is one final observation I would like to make by way of perspective.

Masers and lasers have of course been of economic and technical importance, as well as of scientific interest. I sometimes point out to politicians and other administrative types that if radio astronomy had been sponsored more strongly in the United States we probably would have had masers and lasers sooner. Masers could have been detected long ago in the sky --probably as early as the 1930s, and certainly immediately after or during World War II they could easily have been detected, but nobody was looking. The United States as a whole didn't do very well by radio astronomy shortly after the war, as compared to the British and the Australians. They generally did a wonderful job, but still not quite enough.

Suppose someone had indeed looked, and seen this peculiar maser radiation. What would they have thought if they detected such radiation before masers were thought of in the lab?

Would they have attributed it to Little Green Men, just as Little Green Men was initially used as the origin of pulsars because nobody understood them. The maser radiation would have been seen to be modulated, of all things. A very successful SETI operation? No such phenomena could happen naturally according to thinking of that time. However, I'm sure that someone would have been figured it out before long and concluded yes, that's what's happening — the molecular states are inverted, and we get amplification. That would probably have suggested trying out such an idea in the laboratory. So masers and lasers -- highly applied technology, in a sense -- could well have come out of astronomy.

That's something that we should always carry along with us, that the science that we do — learning things, discovering new things -- will have consequences for our society's future which no one is able to predict. Both we and the politicians must be humble about predicting the future, the presently unknown. But we do know that we can have fun working intensively to learn new things and to understand, with considerable assurance that what we learn in almost any field of science has a good chance of being useful as well as fascinating.

2. THEORY

ON THE THEORY OF NATURAL MASERS

Vladimir S. Strelnitski
Institute for Astronomy, Russia Academy of Sciences
48 Pjatnitskaja Street
Moscow 109017, RUSSIA

ABSTRACT

A short review of some selected topics of the theory of astrophysical masers is presented. After some critical remarks about the two commonly used simplified descriptions of kinetics of the maser level populations, two nonlinear effects studied with these descriptions are considered: the cross-relaxation mechanisms of Goldreich-Kwan and of Elitzur, and the maser modes interaction in disk-like masers. The results of a qualitative analysis and a numerical simulation of possible gas-dynamical effects due to "negative" radiative pressure in strong masers are also mentioned.

1. INTRODUCTION

This paper is by no means a complete review of theory of astrophysical masers. Little, if anything, could be mentioned, for several recent years, as a new principal achievement in the natural maser theory. Yet the "classic" aspects of the theory have many times been reviewed. They can be found, for example, together with references to previous reviews and original papers, in the recent monograph by M. Elitzur (Elitzur, 1992).

This is rather a short review of some new results obtained by my colleagues and me during the recent 1-2 years. The paper touches several aspects of the astrophysical maser theory — from such well-known things as the "simplified" representation of the maser level populations, which we still try to simplify further (Sect. 2), or the Goldreich-Kwan cross-relaxation effect, for which we discuss the limits of application in Sect. 3, to less known (and, may be, vain) attempts to "make" the gas in strong masers move under the influence of their radiation pressure (Sect. 5). In Sect. 4 we discuss numerical simulations of maser amplification in a Keplerian disk, which are used for interpretation of the triplet and doublet spectra observed in several maser sources.

2. SIMPLIFIED DESCRIPTIONS OF THE POPULATION KINETICS

Many important effects in astrophysical masers (saturation, line narrowing and rebroadening, polarization, cross relaxation, etc.) were pointed out using simplified representations of maser level populations. In these representations the processes controlling populations of the maser levels are generally divided into three groups: (1) radiative transitions between these levels; (2) collisional transitions between them; (3)

15

other processes (pumping), whose net effect should be a continuous restoration of the population inversion, which the two former processes tend to cancel.

There exist two forms of simplified representation. They differ by the description of pumping, more precisely — by the way the dependence of pump rates on maser level populations is taken into account.

In the first approach four quantities describe actually the pumping: λ_1, λ_2 $(\text{cm}^{-3}\text{s}^{-1})$ — the rates of income of populations to the signal levels from other levels, and Γ_1, Γ_2 (s^{-1}) — the rate coefficients for the outlay of populations to other levels. Unfortunately, the traditional names of the λ's ("pump rates") and Γ's ("decay rates") are misleading: they may give rise to an illusion that *only* λ's are responsible for the creation of inversion, although actually the both processes — income and outlay — are equally important for that.

This illusion is only strengthened by the traditional assumption $\Gamma_1 = \Gamma_2 \equiv \Gamma$ — to simplify the solution of the statistical equilibrium equations. The population difference turns then out to be:

$$\Delta n \equiv n_2 - n_1 = \frac{\lambda_2 - \lambda_1}{\Gamma + 2C + 2R}, \tag{1}$$

where C and R are the rate coefficients for collisionally and radiatively induced transitions between maser levels (spontaneous emission, unimportant in strong masers is usually ignored). Artificiality of the separation, in Eq.(1), of the "pump" coefficients λ from the "decay" coefficient Γ becomes evident, when Eq.(1) is compared with our solution for the general case, $\Gamma_1 \neq \Gamma_2$:

$$\Delta n = \frac{\frac{2\lambda_2}{1+\Gamma_2/\Gamma_1} - \frac{2\lambda_1}{1+\Gamma_1/\Gamma_2}}{\overline{\Gamma} + 2C + 2R}, \tag{2}$$

where $\overline{\Gamma} \equiv 2(\Gamma_1^{-1} + \Gamma_2^{-1})^{-1}$. It is seen, in Eq.(2), that λ's and Γ's are equally important in creating the population inversion. When $\lambda_1 = \lambda_2$ ("net pump rate = 0", in traditional terminology!), the inversion doesn't actually vanish, and is just defined by the difference of Γ_2 and Γ_1. The real net pump rate is not $|\lambda_2 - \lambda_1|$, but $|\lambda_2 - n_2 \cdot \Gamma_2|$, or $|\lambda_1 - n_1 \cdot \Gamma_1|$. It is readily deduced from the statistical equilibrium equations that in steady state these quantities are equal in fact, and that they maintain the inversion by just compensating the summary relaxation rate, $(R + C) \cdot \Delta n$.

Note that the aim of this division of the pumping processes into λ- and Γ-processes — to introduce, in a general form, the dependence of the pump rate on maser level populations — is hardly achieved, because in many real pump schemes the λ's *will* depend on n_1, n_2. The less is the number of levels involved in the pumping, the stronger is this dependence, the worst case being a 3-level pumping scheme.

In the second approach the phenomenological probabilities, P_1, P_2, are introduced for the population exchange between the maser levels via other levels. This representation *forces* to admit (without any evidence): $n_1 + n_2 \equiv n = const$. Denoting $P \equiv P_1 + P_2$, $\Delta P \equiv P_1 - P_2$, one obtains:

$$\Delta n = \frac{n \cdot \Delta P}{P + 2C + 2R}. \tag{3}$$

The pump rate is here $|n_1 \cdot P_1 - n_2 \cdot P_2|$. Thus, the dependence of the pump rate on maser level populations turns out to be direct and total, which is another extreme: processing of populations in complicated cycles through other levels would hardly allow to represent them in such a simple form. As already mentioned, the forced assumption $n_1 + n_2 = const$ (which means: independent of P_1, P_2, C, R) doesn't seem realistic either.

We come to the conclusion that neither of the two approaches accounts correctly for the dependence of the pump rate on maser level populations. Actually they only introduce some time scale for pumping, which is to be compared with collisional and radiative relaxation time scales.

It seems, therefore, reasonable to further simplify the description by introducing the *net* rate of pumping, λ (cm^{-3}s^{-1}), just to define the pumping time scale. In stationary conditions the net rate of pumping should be equal, by its absolute value, for both maser levels (though the signs are opposite), so we need only *one* such parameter. The solution of the statistical equilibrium equations gives then:

$$\Delta n = \frac{\lambda}{C + R}. \tag{4}$$

Being simpler than "λ, Γ" and "P_1, P_2" approaches (one pump parameter instead of four in the "λ, Γ" and two in the "P_1, P_2"), this approach is more realistic (no illusionary "insight"!), and it doesn't seem less informative in no sense.

For more details see (Sumin *et al.*, 1992).

3. CROSS-RELAXATION AND LINE WIDTH

There are 3 wide-spread mistakes in people's attitude to the Goldreich-Kwan (Goldreich and Kwan, 1974, hereafter — GK) mechanism of population cross-relaxation preventing maser lines from rebroadening after saturation:
(1) that the mechanism doesn't work in principle;
(2) that it works and is efficient enough to provide line narrowing by several times (as compared with the thermal line width) under saturation;
(3) that it can explain the often observed strong $\Delta\nu(I)$ dependence in H$_2$O masers: $\Delta\nu \propto I^{-0.5}$.

Let me briefly comment on these three statements.
(1) Though it has never been claimed in open publications that GK cross-relaxation is an erroneous idea, this has many times been declared in intense non-official discussions

among specialists. These discussions have, however, persuaded even the most persistent skeptics that trapping of the IR photons, or another process reshuffling populations between the center and the wings of the local line profile (e.g. elastic collisions, Elitzur (1990)) can, in principle, prevent maser lines from rebroadening after saturation sets in.

(2) Cross-relaxation can only work until the maser emission rate remains lower than the cross-relaxation rate. Since we want, on the other hand, that the cross-relaxation work after saturation has established, i.e. when the maser emission rate exceeds the rate of inelastic collisions between the maser levels, the range of maser intensities, where this mechanism of line narrowing is active, turns out to be limited both from below and from above. And for both, IR trapping and elastic collisions, this range turns out to be less than 1 order of magnitude in intensity (Likhachev et al., 1990). This means that in its pure form (line narrowing at full saturation) this mechanism can not work, because the very transition from unsaturated to saturated regime occupies at least one order of magnitude in intensity increase. GK showed that due to cross-relaxation the line under saturation narrows by $[\log(I_2/I_1)]^{0.5}$, as under unsaturated regime. Thus after saturation begins, the line can at most be narrowed by $(\log 10)^{0.5} \approx 1.5$ times, much less than the lines in the strong masers are believed to be narrowed (by a factor of 5-6).

The problem was numerically studied with the use of "P_1, P_2" approach (Likhachev et al., 1990). One of the results is reproduced in Fig. 1. It is seen, in fact, that "switching on" the cross-relaxation can hardly conserve the narrowness of the line, acquired during unsaturated amplification, up to essentially higher intensities.

(3) The third erroneous statement is simply due to the ambiguity with the notion "gain". Some people mean gain is τ_0 , the unsaturated optical length of the maser, others - that it is I_2/I_1, or $log(I_2/I_1)$. GK implicitly assumed the latter, and then the

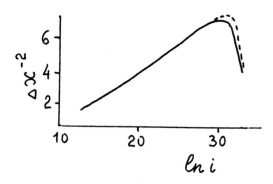

Fig. 1. Inverse squared dimensionless linewidth as a function of logarithm of dimensionless intensity at the output of a model bidirectional H_2O maser. Solid line: no cross-relaxation; dashed line: cross-relaxation of a deliberately overestimated rate (10^3 s^{-1}) is included. (From: Likhachev et al., 1990.)

dependence of $\Delta\nu$ on I is very weak (see the precedent paragraph). However, several authors erroneously interpreted the GK conclusions, having ascribed to GK the former of the above definitions of the gain. Their reasoning then (that, because under saturation $I \propto \tau_0$ (correct) and "according to GK" under cross-relaxation $\Delta\nu \propto \tau_0^{-0.5}$ (not correct), it should be $\Delta\nu \propto I^{-0.5}$) was erroneous, just because the latter was erroneous.

In conclusion: cross-relaxation by IR photon trapping or by elastic collisions, being, in principle, a correct mechanism of line narrowing after saturation, in practice, can not provide highly narrowed maser lines in full saturation; this mechanism is certainly not responsible for the observed strong dependence $\Delta\nu \propto I^{-0.5}$.

4. MODE INTERACTION IN DISK MASERS

The simplified description of maser level populations (Section 2) allowed also to investigate maser amplification in Keplerian disks. This problem arose when it was realized (Elmegreen and Morris, 1979) that several cases of triplet H_2O spectra observed in regions of star formation could be due to maser amplification in circumstellar disks seen edge-on.

Fig. 2. represents the curves of equal radial velocities in a Keplerian disk seen edge-on. The adjacent curves differ by one thermal width, thus the band between two adjacent curves contains the molecules that are coherent in radial velocity within approximately one thermal width. It is seen that there are three longest coherent paths for any observer: one along the line of sight passing through the center of the disk, and two other along the chords near the edges of the disk. This produces, in a general case, a triplet spectrum: the non-shifted central component and two side components shifted to the blue and to the red by about the rotation velocity at the edge of the disk.

The radial velocity pattern in Fig.2 is axially symmetric. It means that we have to imagine a superposition of similar patterns seen from different directions. We realize then that the chord modes intersect with the diameter modes of approximately perpendicular orientation and also intersect with other chord modes. If the maser is partly saturated, the intersecting modes interact - compete for population inversion.

Recently Cesaroni (1990) used the idea of the mode competition in a disk maser to explain the anti-correlation of the side components observed in the triplet H_2O spectrum of S255. Cesaroni assumed that the symmetry of the red and the blue side components is broken by some anisotropy of the pumping from the central source, which can produce some decrease of the population inversion in the region, where one of the side components is formed. This will result in a decrease of intensity of not only this side component, but also of the *diameter* modes, which intersect with the both chords giving the side components. As a result, the population inversion in the region, where the second side component is formed, will *increase* raising the intensity of this component, which thus will change in anti-phase with the first side component.

We (Lekht *et al.*, 1992) have further developed the model of Cesaroni — to explain the anti-correlations of not only the side components, but also of the central component relative to the side ones. We observed such anti-correlation in the triplet H_2O spectrum in S140 (Fig. 3).

Though the total number of maser modes in a Keplerian disk is very high, it was possible to reduce, for a homogeneous disk, the number of interacting radiation transfer equations to four. One of our results is shown in Fig. 4. It is seen that, due to the

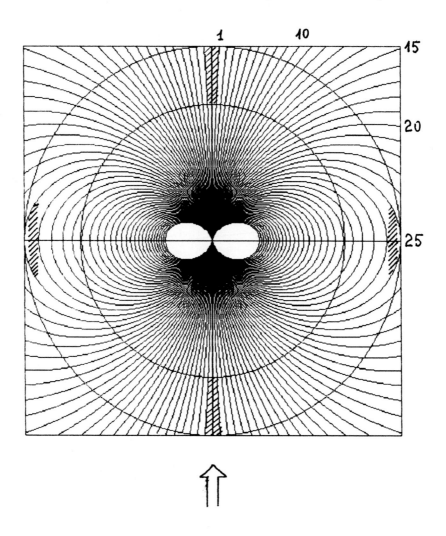

Fig. 2. Curves of equal radial velocity in the model Keplerian disk with the dimensionless (measured in thermal widths) velocity $V_0 = 25$ near the edge. The observer is at the bottom. It is supposed that the population inversion takes place within the ring between two circles. The longest amplification paths - the "diameter mode" and the "chord modes" are shaded.

diameter and chord mode interaction, the growth of pump efficiency produces first a slight correlation and then a strong anti- correlation of the output intensities of the central and the side components.

We took into account the competition for inversion not only between the chord and the diameter modes in the disk, but also between the chord modes, though indirectly. The interacting chord modes make up a ring of maser activity along the outer edge of the disk. The information about any local change of inversion (due to pump anisotropy, or any other inhomogeneity of the physical conditions within the disk) will be transferred along this ring. This can explain the anti-correlation of the side components without using the diameter modes as mediators.

Last year new observational data on the recently discovered hydrogen recombination line maser in MWC 349 appeared (Planesas *et al.*, 1991; Gordon, 1991; see also this volume; Thum *et al.*, 1992). All these data suggest that maser amplification takes place in a circumstellar Keplerian disk. The observed spectrum is a doublet, which is possible, if the active zone (between two circles in Fig. 2) is narrow. Numerical modelling allowed to restore some kinematical and physical parameters of the disk by comparing the observed and the model spectra (Smith *et al.*, 1992). It is tempting in this case also to explain the observed anti-correlations and correlations

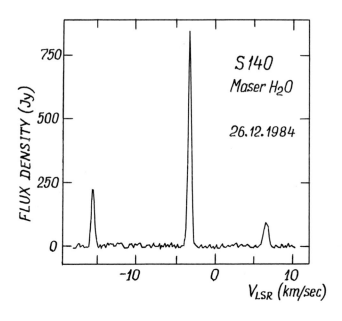

Fig. 3. A triplet spectrum of the H_2O maser in S140.

of the "red" and "blue" components of the doublet (Fig. 2 in Thum *et al.*, 1992) by the competition between the chord modes.

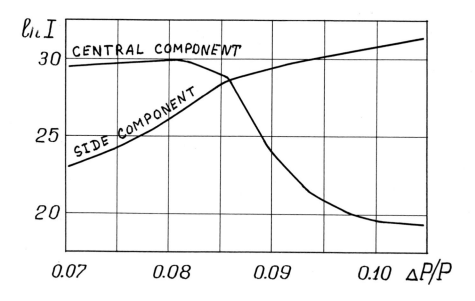

Fig. 4. Mode competition in a disk H_2O maser (after Lekht *et al.*, 1992).

5. RADIATIVE PRESSURE IN MASERS

The natural masers observed in sites of star formation, in the atmospheres of the evolved red giants and in the ISM of some active nucleus galaxies are sometimes very bright. We don't know, how large are solid angles, through which the masers radiate, but if they are not too small, one could expect some gas-dynamical effects due to the pressure of the intense maser radiation upon the amplifying gas.

It was pointed out (Strelnitski, 1972) that the radiative pressure in masers has a specificity: it accelerates the gas *inward* the source, whereas the ordinary, non-maser, radiation accelerates the absorbing gas *outward*. The anticipated *observable* dynamical effects depend on the adopted geometrical model of the maser. In the simplest, one dimensional, model the effects crucially depend on whether the maser is thought to be uni- or bidirectional. In the former case one could expect more pronounced effects for an *unsaturated* maser. This is due to the fact that the gradient of the radiative force, giving the gas a *differential* acceleration (which is needed to distort the radial velocity coherence of the amplifying molecules, thus — to produce observable consequences) is larger under unsaturated (exponential) amplification (Strelnitski, 1972). However, the local value of acceleration is proportional to the product of radiation intensity and absorption coefficient, and, at a given pump rate, this product attains its maximum when the maser saturates. Since under saturation the amplification is linear, the gradient of acceleration, in a unidirectional saturated maser, is small. In a bidirectional maser, however, the two saturated halves are accelerated toward each other, and this severely distorts the integral amplification at a given frequency producing a potentially observable change in intensity and radial velocity of the outgoing radiation (Sumin and Strelnitski, 1991; Strelnitski and Sumin, 1992).

It was found, however, that even for the most intense, H_2O, masers intensity variations with the observed time scales of less than one year require very high values of λ/n_H ratio: $\lambda/n_H \sim 1\ s^{-1}$. With the maximum rates of the hitherto proposed pump processes, $\lambda/n_m \sim 10^2\ s^{-1}$, and the upper limit for the relative number density of molecules on masering levels, $n_m/n(H_2O) < 10^{-1}$, this corresponds to the requirement $n(H_2O)/n_H > 0.1$. Though relative abundances of water vapor this high do not meet the general cosmic abundance of oxygen ($[O]\ /\ [H] \approx 7 \cdot 10^{-3}$), they can not be excluded in circumstellar environments of very young stars, because efficient processes of elements segregation should take place there (Strelnitski $et\ al.$, 1972).

It is also important that the gas-dynamical effects and the corresponding variations of the outgoing intensity are proportional to the *frequency* of the masering line . Many numerical pump models developed to explain the $6_{16} - 5_{23}$ H_2O maser predicted masering in other, higher in frequency, rotational lines as well. Our numerical solutions for radiative gas dynamics equations describing one-dimensional, bidirectional masers (see references above) have demonstrated a principal possibility of strong gas-dynamical effects (including shock waves) and strong intensity variations, if a high frequency H_2O maser (like $5_{23} - 4_{14}$, $\nu = 7.0 \cdot 10^{11}$ GHz) works in a medium highly enriched in water vapor. Future observations of time variability (possible quasi-periodic component) and search for strong high-frequency masers could shed light on whether the radiation pressure effects play any role in natural masers.

ACKNOWLEDGMENT

Most of the published and yet unpublished recent results described in this review were obtained in collaboration with my colleagues G.T.Bolgova, E.E.Lekht, S.F.Likhachev, V.O.Ponomarev, R.L.Sorochenko, A.A.Sumin. Some of them were invited to participate in this conference, but, unfortunately, were unable to come, by reasons independent of them. I consider them all as co- authors of this review too and express to them my gratitude for there help in preparation of the review.

REFERENCES

Cesaroni, R. 1990, *Astr. Ap.*, **233**, 513.

Elitzur, M. 1990, *Ap. J.*, **350**, L17.

Elitzur, M. 1992, Astrophysical Masers, Kluwer

Elmegreen, B.J., Morris, M. 1979, *Ap. J.*, **229**, 593.

Goldreich, P., Kwan, J. 1974, *Ap. J.*, **191**, 93.

Gordon, M. 1992, *Ap. J.*, **387**, 701.

Lekht, E.E., Likhachev, S.F., Sorochenko, R.L., Strelnitski, V.S. 1992, *Astron. Zh.* , submitted.

Likhachev, S.F., Strelnitski, V.S., Sumin, A.A. 1990, *Astron. Tsirk.* , **1546**, 28.

Planesas, P., Martin-Pintado, J., Serabin, E. 1991, *Ap. J.*, **386**, L23.

Smith, H.A., Ponomarev, V.O., Strelnitski, V.S. 1992, in preparation.

Strelnitski, V.S. 1972, *Astron. Zh.* , **49**, 649.

Strelnitski, V.S., Sunjaev, R.A., Varshalovich, D.A. 1972, *Comments Astrophys. Space Phys.*, **4**, 155.

Strelnitski, V.S. Sumin, A.A. 1992, *Trans. Astron. Astrophys.*, submitted.

Sumin, A.A., Strelnitski, V.S. 1991, *Astron. Tsirk.* , **1550**, 11.

Sumin, A.A., Strelnitski, V.S., Bolgova, G.T. 1992, *Trans. Astron. Astrophys.*, submitted.

Thum, C., Martin-Pintado, J., Bachiller, R. 1992, *Astr. Ap.*, **256**, 507.

LINE PROFILES IN MASERS: VELOCITY RELAXATION

Nels Anderson
Max-Planck-Institut für extraterrestrische Physik
W-8046 Garching bei München, Germany

William D. Watson
Loomis Laboratory of Physics, University of Illinois
1110 West Green Street
Urbana, IL 61801-3080, USA

ABSTRACT

Populations and velocity profiles of the rotational levels of the water molecule are calculated under conditions typical of those in astrophysical water masers. Trapped infrared radiation as well as elastic and inelastic collisions are included. The distortion of velocity profiles by a strong, narrow maser line is found with minimal assumptions and approximations. This provides the first detailed determination of the rate of velocity relaxation, accurate values of which are essential for interpreting maser line profiles in general and for inferring maser luminosities in particular. Relaxation rates for the 22-GHz maser at densities of 10^9 cm^{-3} are 2.1 s^{-1} and 5.3 s^{-1} for temperatures of 400 K and 1000 K, respectively, including reasonable elastic collision rates. Under these and other conditions, as well as for other transitions, the relaxation rate far exceeds the loss rate—usually by more than a factor of 10. Therefore, line narrowing can continue to an intensity that is at least 10 times the saturation intensity.

1. INTRODUCTION

When radiation propagates through an astrophysical maser, its spectral profile initially narrows as its intensity increases. This narrowing results from preferential amplification at the center of the spectral line, where the rate of stimulated emission and the optical depth are greatest. Since the rate of stimulated emission is greater for masing molecules at line center than for those in the line wings, the number of excited molecules at line center will be reduced from the Maxwellian value found in the absence of masing. When the intensity is sufficiently great, the reduction of excited molecules at line center is large enough that the amplification at line center falls below that in the wings. The wings grow faster than the line center from this point onward and the line rebroadens. The line ultimately expands to the thermal breadth if sufficiently high intensities are achieved.

In the absence of other processes, rebroadening begins when the maser saturates, i.e., when the rate of stimulated emission becomes comparable to the effective decay rate Γ of the maser transition. The rate γ at which molecular velocities relax toward a Maxwellian distribution may, however, exceed Γ. If so, rebroadening is postponed until the rate of stimulated emission becomes comparable to γ. Relaxation of molecular velocities occurs as a result of the trapping of infrared radiation (Goldreich & Kwan

1974, hereafter GK) and of elastic collisions (Nedoluha & Watson 1988). GK estimated the rate of relaxation due to trapped infrared radiation to be of the same order of magnitude as the Einstein A coefficients for the infrared transitions that involve the masing states. In both 18-cm OH masers and 22-GHz water masers, this rate should exceed Γ.

The foregoing relationship between spectral linebreadth and the stimulated emission rate (and hence the luminosity of the maser) has been applied to the brightest 22-GHz water masers to demonstrate that they are much more highly beamed and less luminous than previously recognized (Nedoluha & Watson 1991). The precise value of the stimulated emission rate is also important in inferring magnetic field strengths from the observed circular polarizations in water masers (Nedoluha & Watson 1992).

These properties of masers derived from line widths depend upon the relaxation rate γ. The only rate heretofore available has been the estimate of GK. The accuracy of this estimate is unclear (at least to us) because of the approximations on which it is based. We have therefore performed a calculation that avoids many of these approximations. We calculate explicitly the populations *as a function of velocity* for the lower 40 energy levels of the ortho configuration of water. The contributions of the collisions and infrared radiation to γ, Γ, and the pumping rate Λ are thus treated in a way that is easily seen to be consistent.

2. TECHNIQUE

Let $n_i(v)$ be the number density of water molecules in rotational level i per magnetic substate per unit velocity along the line of sight which have velocity v along the line of sight. Define $N_i = \int_{-\infty}^{\infty} n_i(v)\,dv$. Then in steady-state,

$$\sum_{j \neq i} \left(\frac{g_j}{g_i} C_{ji} N_j \phi(v) - C_{ij} n_i(v) \right) + \sum_{j \neq i} \mathcal{R}_{ji}(v) + \gamma_c (N_i \phi(v) - n_i(v)) = 0, \qquad (1)$$

where the sums are taken to be over the lowest 40 rotational levels of ortho-water, C_{ij} is the collisional excitation rate from level i to level j, g_i is the degeneracy of level i, $\phi(v)$ is the Gaussian of unit normalization corresponding to the one-dimensional thermal velocity profile, $\mathcal{R}_{ji}(v)$ is the net radiative flow rate from level j to level i, and γ_c accounts for elastic collisions. The collisional excitation rates C_{ij} are the rate coefficients of Palma et al. (1988b) multiplied by the total H_2 density N_{H_2}.

The term "elastic collision" in the preceding paragraph is meant to denote a collision which leaves the molecule in the same rotational level but alters its velocity. Letting q_{el} be the rate coefficient for elastic collisions, we have $\gamma_c = q_{el} N_{H_2}$. Precise coefficients are not available for H_2O-H_2. Rates for C^+-H collisions and H-H collisions, when scaled to the reduced mass appropriate for H_2O-H_2, indicate coefficients in the neighborhood of 0.7–2×10^{-10} cm^3 s^{-1} for temperatures of 100–1000 K (see Spitzer 1978). Given the similarity of the two rates despite the very different characters of the forces involved, the elastic collision rate for water is unlikely to differ substantially. Furthermore, the limited experimental measurements of H_2O elastic collision rates that are available (Palma et al. 1988a) imply values in this vicinity. Therefore, when elastic collisions are included in the following calculations, we adopt a baseline rate coefficient of $q_{el} = 10^{-10}$ cm^3 s^{-1}.

Aside from the maser transition, radiative transitions are handled with an escape probability method:

$$
\mathcal{R}_{ji} =
\begin{cases}
-A_{ij}\beta_{ij}n_i + A_{ij}(1 - \beta_{ij})\dfrac{N_j}{N_j - N_i}\left(N_i\dfrac{n_j}{N_j} - n_i\right), & j \text{ below } i \\[4mm]
\dfrac{g_j}{g_i}A_{ji}\beta_{ji}n_j + \dfrac{g_j}{g_i}A_{ji}(1 - \beta_{ji})\dfrac{N_j}{N_i - N_j}\left(N_i\dfrac{n_j}{N_j} - n_i\right), & j \text{ above } i
\end{cases}
\tag{2}
$$

where β_{ij} is an escape probability. The first terms represent radiative pumping and loss, while the second terms represent trapped radiation.

Once the profiles have been calculated, the relaxation rate γ is measured by choosing γ to give the same inversion profile in a simple phenomenological model as in the full rate equations (1). Letting u and l denote the upper and lower maser levels, respectively, the rate equations describing the phenomenological model are

$$
\begin{aligned}
\Lambda_u\phi - \Gamma n_u + Ce^{-h\nu/kT}N_l\phi - Cn_u - R\Delta n + \gamma(N_u\phi - n_u) = 0 \\
\Lambda_l\phi - \Gamma n_l + \frac{g_u}{g_l}CN_u\phi - \frac{g_u}{g_l}Ce^{-h\nu/kT}n_l + \frac{g_u}{g_l}R\Delta n + \gamma(N_l\phi - n_l) = 0
\end{aligned}
\tag{3}
$$

where $\Delta n(v) = n_u(v) - n_l(v)$, Λ_u and Λ_l are pumping rates taken from the full numerical calculation, Γ is the loss rate, also from the full numerical solution, $R(v)$ is the maser stimulated emission rate, C is the collisional transition rate from the upper maser level to the lower, and $h\nu$ is the energy of the maser transition. For $R(v)$ we use a Gaussian with a line-center rate of 1 s^{-1} and a width of one-fifth of the thermal width. We have verified, however, that rates of up to 100 s^{-1} and widths as large as the thermal width change the best-fit γ by only 0.01 s^{-1} at most.

Because these single-parameter fits are excellent (maximum deviation typically less than one part in 10^5), we conclude that our method of measuring relaxation is appropriate and accurate.

3. RESULTS

The table shows relaxation rates for a number of maser transitions for several representative temperatures and densities. For the commonly assumed conditions of 10^9 cm^{-3} and 400 K, the relaxation rate for the 22-GHz maser is 2.0 s^{-1} in the absence of elastic collisions. Relaxation rates generally increase with temperature and density; see the Table for other conditions and transitions. Elastic collisions typically contribute less than $\sim$1 s^{-1} to the relaxation rate. At high densities they can contribute more, but by then the maser begins to thermalize (although other kinds of maser pumps, e.g., two-temperature schemes, might function at higher densities where elastic collisions would be important). Given that typical Γ's are 0.01–0.1 s^{-1} while typical γ's are 0.5–5 s^{-1}, *line narrowing can continue to an intensity roughly 10 times that at which saturation sets in.*

Density	Temperature	Transition	Frequency	Loss Rate, Γ	Relaxation Rate, γ w/o EC[a]	w/EC[b]
10^8 cm^{-3}	400 K	$6_{16}-5_{23}$	22 GHz	0.013 s^{-1}	0.87 s^{-1}	0.88 s^{-1}
		$4_{14}-3_{21}$	380	0.011	0.64	
		$4_{23}-3_{30}$	448	0.013	0.55	
		$8_{27}-7_{34}$	1296	0.11	2.0	
"	1000	$4_{14}-3_{21}$	380	0.026	1.3	
		$4_{23}-3_{30}$	448	0.029	0.93	
		$8_{27}-7_{34}$	1296	0.054	2.4	
10^9 cm^{-3}	100 K	$6_{16}-5_{23}$	22 GHz	0.022 s^{-1}	0.63 s^{-1}	0.74 s^{-1}
"	400	$6_{16}-5_{23}$	22	0.10	2.0	2.1 s^{-1}
"	1000	$6_{16}-5_{23}$	22	0.23	5.1	5.3
		$4_{14}-3_{21}$	380	0.24	3.8	4.0
		$4_{23}-3_{30}$	448	0.23	3.5	
		$8_{27}-7_{34}$	1296	0.22	6.9	
$10^{9.5}$ cm^{-3}	400 K	$6_{16}-5_{23}$	22 GHz	0.32 s^{-1}	2.4 s^{-1}	2.8 s^{-1}
"	1000	$6_{16}-5_{23}$	22	0.72	6.5	7.0

[a] Without elastic collisions ($q_{el} = 0$).
[b] With elastic collisions: $q_{el} = 10^{-10}$ cm^3 s^{-1}.

REFERENCES

Goldreich, P., & Kwan, J. 1974, ApJ, 190, 27 (GK)
Nedoluha, G. E., & Watson, W. D. 1988, ApJ, 335, L19
———. 1991, ApJ, 367, L63
———. 1992, ApJ, 384, 185
Palma A., Green, S., DeFrees, D. J., & McLean, A. D. 1988a, J. Chem. Phys., 89, 140
———. 1988b, ApJS, 68, 287
Spitzer, L. 1978, Physical Processes in the Interstellar Medium (New York: Wiley), § 2.5

OH Maser Polarization

Shuji Deguchi

Nobeyama Radio Observatory, National Astronomical Observatory, Nagano 384-13, Japan

and

William D. Watson

Department of Physics, University of Illinois, Urbana, IL 61801

Abstract

The mechanism to produce circular polarization has been reinvestigated under conditions with radiative pumping and the Faraday rotation due to electrons. A new radiative pump mechanism is proposed for the OH main line masers. A parity propensity for the OH collisional excitation, which has recently been found by laboratory experiments, prohibits the collisional pumping of the main line masers. It is shown that the inversion can be created when the radiation temperature is higher than the kinetic temperature. The anisotropic radiative pump can produce unequal magnetic sublevel populations. In addition to the line overlap between the Zeeman triplet of the OH main lines, a modest amount of the Faraday rotation due to electrons can increase the degree of circular polarization of OH masers even when the angle to the magnetic filed is near $\pi/2$.

I. Introduction

OH maser emission is known to be polarized circularly and linearly. The Zeeman pairs are identified occasionally, but mostly the circular polarization is dominant and partners in pairs are often missing (Reid and Moran 1981). Very narrow polarized components have been observed in OH 1612 MHz line in the circumstellar envelopes (Cohen et al. 1987). The mechanism to produce the dominant circular polarization has been proposed by Cook (1966) and Deguchi and Watson(1986). The former mechanism requires a coherent variation of the magnetic field strength with the large scale velocity field of the cloud. While, the latter relies on the overlap of the Zeeman triplet which are separated by several thermal line width. The line overlap can create circular polarization over 90 % in most of the propagation direction except when the maser propagate perpendicularly to the magnetic field. This mechanism has been investigated in detail further by Nedoluha and Watson (1989). In these calculation, it has been assumed that the pumping is isotropic and the Faraday rotation due to electrons is neglected. In this paper, we investigate how these two effects modifies the final output of polarization.

II. Pump Mechanism

It has been recognized that a parity propensity rule exists for the collisional excitation of OH Λ-doublet states with H_2 (Andresen et al. 1984; Andresen et al. 1991). Because of this collisional propensity, the collisional pumping for OH main line masers seems not to work (but for $\pi_{1/2}$, see Kylafis and Norman 1990). Because of this difficulty to create the main line inversion, line overlap mechanisms have been proposed (Bujarrabal et al., 1980; Doel et al. 1990; Cesaroni and Walmsley 1991). We reinvestigated the collisional process using the OH-H_2 collisional rates calculated by Dewangan et al. (1987) and found that the most important collisional rates responsible for the inversion in $^2\pi_{3/2}$ Λ-doublet are the rate which mixes the populations of upper and lower Λ-doublet state. In figure 1, we show the collisional rates $(\times 10^{-11}$ cm^{-3} s$^{-1})$ between J=3/2 and J=5/2 Λ-doublet in $^2\pi_{3/2}$ and the radiative rates (Einstein A coefficients [s^{-1}]). The collisional rates concerned here are C(1-4) and C(2-3). They have a factor 3 difference. If these compete with radiative transitions, the upper Λ-doublet in the ground state can be overpopulated and make main line masers when the radiation temperature is higher than the kinetic temperature as shown in figure 2. More elaborate calculations including 72 magnetic sublevels below 400 cm^{-1} and 554 radiative transitions between them confirms the main line masers can occur in the molecular cloud conditions, which is shown in figure 3.

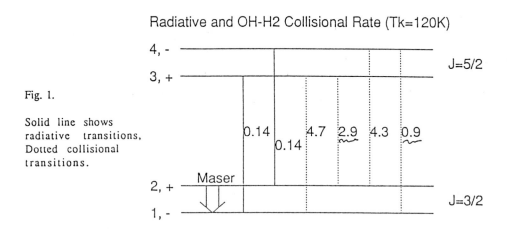

Radiative and OH-H2 Collisional Rate (Tk=120K)

Fig. 1.

Solid line shows radiative transitions, Dotted collisional transitions.

Fig. 2. The Excitation temperature of OH ground state Λ-doublet by 4-level calculation.
[opposite to Dewangan] means that the C(1-4) and C(2-3) are reversed.

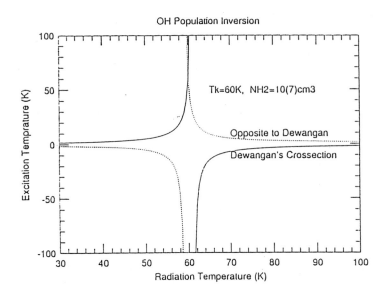

Fig. 3. Excitation temperature of OH 4 maser transitions by full calculations.
1612 MHz transition behaves differently from the other three.

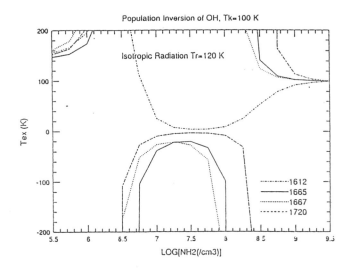

III. Effect of the Faraday Rotation

If free electrons are present in the OH maser cloud, they produce the Faraday rotation. The rotation of the plane of linear polarization along the ray suppresses the growth of the linear component, resulting dominant circular polarization. The electron number density in the OH cloud is deduced to .be in the range of 1-1000 cm^{-3}. The rotation angle due to this effect is 0.085 ($Ne/1$ cm^{-3})(H cos θ/1 milli guass) radian per 10^{16} cm at the wavelength 18 cm. Using the radiative transfer code similar to Deguchi and Watson (1985), we have calculated the influences on the polarization of overlapped Zeeman triplet (shown in figure 4). The saturation effect is not taken into account in this calculation. We can see that the linear component is well suppressed when the Ne > $10^2 cm^{-3}$.

Fig. 4. a. Polarization in Ne=100 cm^{-3}. 50% circular b. in Ne=1 cm^{-3}. Q/I still remains.

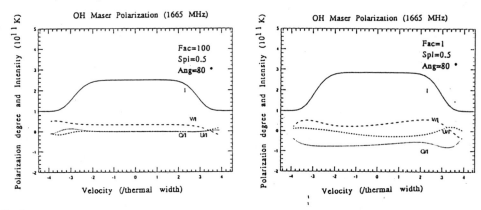

References

Andresen, P., Hausler, D., & Lulf, H. W., 1984, J. Chem. Phys. 81, 571
Andresen, P., Aristov, N., Beushausen, V., Hausler D., & Lulf, H. W., 1991,
    J. Chem. Phys. 95, 5773
Cesaroni, & Walmsley, 1991, AAp. 241, 537
Cohen, R. J., Downes, G, Emerson, R., Grimm, M., Gulkis, S., Stevens, G.,
    & Tarter, J., 1987, MNRAS. 225, 491
Bujarrabal, V., Guilbert, J., Nguyen-Q-Rieu, & Omont, A., 1980, AAp. 84, 311
Cook, A. H. 1966, Nature 221, 503
Deguchi, S., & Watson, W. D., 1985, ApJ. 289, 621
Deguchi, S., & Watson, W. D., 1986, ApJ. 300, L15
Doel, R. C., Gray, M. D., & Field, D., 1990, MNRAS. 244, 504
Dewangan, D. P., Flower, D. R., & Alexander, M. H., 1987, MNRAS. 226, 505
Kylafis, N. D., & Norman, C. A., 1990, ApJ. 350, 209
Nedoluha, G., and Watson, W. D., 1989, ApJ. 335, L20
Reid. M. J., & Moran, J. M., 1981, Ann. Rev. AAp. 19, 231

Coherence and Incoherence in Astronomical Masers

Moshe Elitzur

Department of Physics and Astronomy

University of Kentucky, Lexington, KY 40506

One of the hallmarks of laboratory laser radiation is the phase coherence across its wave front, which is maintained by resonant cavities with high Q-values ($\sim 10^8 - 10^9$), even though this coherence is not a prerequisite for the amplification process per se. For such a coherence to exist, the phase difference $\Delta\phi$ between different waves must be less than unity. The phase difference accumulated along a distance l for waves with wavelength λ propagating along two directions that differ by a small angle θ is $\Delta\phi \approx \pi\theta^2 l/\lambda$. Although this is very small in a good laboratory system, in an astronomical maser l/λ always exceeds $\sim 10^{14}$, so for $\Delta\phi$ to be small θ must be less than $\sim 10^{-7}$, much smaller than any reasonable estimate for beaming angles. Consider next waves propagating in the same direction whose frequencies are separated by $\Delta\omega$. During a time interval Δt such waves accumulate a phase difference $\Delta\phi = \Delta\omega\Delta t$. If $R = 1/\Delta t$ denotes the rate associated with the time interval Δt, coherence is possible only when $R \gg \Delta\omega$ for every relevant rate R in the statistical rate equations. Otherwise, phase differences generated across the radiation bandwidth between successive particle interactions are too large to allow buildup of coherent pulses. Line widths of astronomical masers always obey $\Delta\omega \geq 10^4$ s^{-1}, much larger than the loss rates Γ, which include both collisional and radiative losses. For any reasonable beaming angle, $\Delta\omega$ also exceeds the stimulated emission rate BJ_ν even for the brightest masers, and coherence is impossible.

Since the radiation is phase incoherent across the wave front, many properties of laboratory laser radiation that are widely perceived to be fundamental characteristics of this phenomenon are missing in astronomical masers. At any point in the source, maser photons are generated by the interaction of particles whose velocities are distributed at random (and unaffected by the radiation as long as the maser is unsaturated) with radiation that arrives randomly at that point. Still, at a given frequency and along a given ray, the different particles involved in successive interactions are all in perfect tune with the photon wave vector, resulting in remarkable coherence, as evidenced by the extreme brightness temperatures. The intensity at an arbitrary point l obeys

$$I_\nu(l) = I_\nu(l_1) \exp\tau_\nu(l_1,l), \tag{1}$$

where $\tau_\nu(l_1,l)$ is the gain from a fiducial point l_1. Since every point in the maser can be considered the input source for every other point, the entire maser is effectively coupled. How does this coupling occur and what are its consequences?

Consider first a linear maser (Elitzur 1990a). Positions are labeled by the coordinate z, which varies in the interval $[-\ell, \ell]$. The intensity $I_{\nu+}(z)$ of the rightward moving stream will be denoted by $I_\nu(z)$ and from symmetry, the intensity of the opposite stream, $I_{\nu-}(z)$, is $I_\nu(-z)$. The angle averaged intensity is $J_\nu(z) = [I_\nu(z) + I_\nu(-z)]/2$, thus the absorption coefficient $\kappa_\nu = \kappa_{0\nu}/[1 + J_\nu/J_s]$ involves non-local effects. The maser structure cannot be solved as an ordinary, initial value problem because although $I_\nu(z) = 0$ at the boundary $z = -\ell$, $I_\nu(-\ell)$ is not known. To solve for the structure, consider first a maser sufficiently small that $J_\nu \ll J_s$ everywhere (the maser in unsaturated). Since $\kappa_\nu = \kappa_{0\nu}$ there are no non-local effects and radiative transfer can be solved at once. If $J_\nu(0)$ ($= I_\nu(0)$) is the central intensity, the intensities of the two streams obey $I_{\nu\pm}(z) = J_\nu(0)\exp(\pm\kappa_{0\nu}z)$ for $z > 0$. At the exit point, the ratio of the amplified intensities is $I_{\nu+}(\ell)/I_{\nu-}(\ell) = \exp(2\kappa_{0\nu}\ell) \gg 1$. Why does one stream dominate so decisively? After all, both the source function S and the absorption coefficient are independent of direction. The reason is that *the amplified quantity is not the source function but the maser's own radiation* (as is also evident from eq. 1). The equation of radiative transfer is

$$\frac{dI_\nu}{dz} = \frac{\kappa_{0\nu}I_\nu}{1 + J_\nu/J_s} + \kappa_{0\nu}S = (h\nu\Phi_\nu/4\pi)\frac{I_\nu}{J_\nu + J_s} + \kappa_{0\nu}S, \tag{2}$$

where $\Phi_\nu = \Delta P_\nu/2$. Maser photon generation is dominated by $\kappa_\nu I_\nu$, the first term on the left in each equality, whose properties are entirely different from those of the emission coefficient for spontaneous decays, $\kappa_{0\nu}S$. Photons are generated by interaction of the existing radiation with the inverted population, which is continuously produced at the rate Φ_ν. Because the intensity of one stream is higher, photons are preferentially produced into its direction and its dominance grows.

When its length is sufficiently increased, the maser saturates and the absorption coefficient includes non-local effects. Fortunately, at that point the ratio of the amplified intensities is

$$I_{\nu-}(\ell)/I_{\nu+}(\ell) = 1/(2\gamma), \tag{3}$$

where $\gamma = J_s/S$. Typically, $\gamma \sim 10^5 - 10^7$ in astronomical masers so the subordinate stream can be neglected and the non-local effects ignored when we solve for the dominant stream in the saturated regime. In particular, the boundary condition at the right edge of the unsaturated core is $I_\nu = 2J_s$. It is an essential element of astronomical maser theory that γ is large. If γ were small, an analytic solution in the saturated regime would be impossible because of non-local effects.

The solution is displayed in the accompanying figure for masers of equal length $\kappa_{0\nu}\ell = 30$ and representative values of γ. The intensity in the $z > 0$ region and the angle-averaged intensity are essentially independent of the value of the source function S. While S cannot be completely neglected (the solution of eq. 2 is then $I_\nu = 0$), it is largely irrelevant. If the maser amplified the source function, the ratio of the two streams would vary only in proportion to length instead of the

34

exponential variation that leads to eq. 3. The maser amplifies its own radiation and the steady-state solution is the configuration determined by the self-consistency dictated by this requirement.

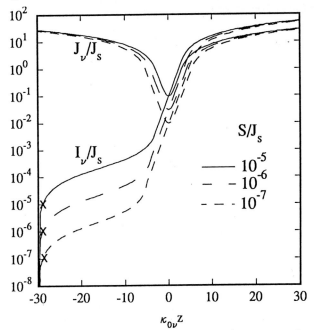

Intensities in a linear maser with length $\kappa_{0\nu}\ell = 30$ for various source functions S. The point where the intensity equals the source function is marked with an x on each plot.

Moving to three-dimensional geometries, the same properties are reflected in the high degree of beaming displayed by maser radiation. For example, on the surface of an unsaturated sphere with radius R the radiation angular distribution is Gaussian centered on the radial ray and the beam solid angle obeys $\Omega_\nu/4\pi = 1/(4\kappa_{0\nu}R)$ (Goldreich and Keeley 1972; Elitzur 1990b). As both the source function and the absorption coefficient are isotropic, the amplified source function would only display linear beaming, in proportion to the different path lengths — so how is exponential beaming produced? Again, from eq. 2, photon generation is controlled by the interaction with the amplified radiation itself. Radiation arriving along the radial ray is the most intense since it has travelled the longest distance through source. This ray therefore stimulates the background of inverted particles to produce more photons into its own direction than any other one. As a result, maser radiation is always highly beamed into a small solid angle centered on the direction determined by the peak of the radiation angular distribution at the point.

The radiation beam pattern of a sphere becomes tighter with increasing radius. By the time the sphere saturates, the beaming angle obeys

$$\Omega_\nu/4\pi \approx 1/(2\ln2\gamma) \sim (3-4)\times10^{-2}. \tag{4}$$

This remarkably tight beaming is the property that enables a solution in the saturated regime, similar to the situation for a linear maser. Again, the fact that the maser amplifies its own radiation and the large value of γ are the key properties.

Another important consequence of the self-amplifying nature of maser radiation involves its polarization. In steady state, polarization modes in tune with the population distribution among the magnetic sub-levels dictated by the pump rates get amplified the most. These are the modes of the Goldreich, Keely and Kwan (1973) polarization solution (Elitzur 1991). Again, they are selected because the maser amplifies its own radiation, not the source function. Numerical solutions by Deguchi, Watson and Western (1986) had great difficulties in producing this polarization solution. These calculations employed source terms appropriate for thermal radiation, and it is easy to verify that source terms consistent with steady-state level populations would have produced the proper polarization solution. This is a most troubling situation since the maser solution can be expected to be independent of the source terms, so the numerical approach should be thoroughly examined. Typical source terms are characterized by excitation temperature $\sim 10^2 - 10^3$ K, while maser brightness temperature is $\sim 10^{10} - 10^{14}$ K. If the polarization state of the self-amplified radiation does depend on the source terms, which are so much smaller than the quantities of interest, there is little hope we can ever understand maser polarization.

In summary, many of the salient features of astronomical maser radiation are determined from the self-consistency imposed by the self-amplified nature of the radiation; the coherence of the amplification process ensures that information about the full source structure is carried by the radiation to every point. In that sense, the solution is fully determined once the radiative transfer equation is formulated. This equation is obtained from the more fundamental equations that describe the coupled behavior of the density matrix and the field amplitudes, and this difficult procedure has not been investigated since the pioneering studies of Litvak (1970) and Goldreich, Keeley and Kwan (1973). Missing is a detailed, first-principle solution of the fundamental equations. This is one of the most important challenges for maser theory.

REFERENCES

Deguchi, S. Watson, W. D. and Western, L. R. 1986, *Ap. J.* **302**, 108.

Elitzur, M. 1990a, *Ap. J.* **363**, 628.

Elitzur, M. 1990b, *Ap. J.* **363**, 638.

Elitzur, M. 1991, *Ap. J.* **370**, 407.

Goldreich, P. and Keeley, D. A. 1972, *Ap. J.* **174**, 517.

Goldreich, P., Keeley, D. A. and Kwan, J. Y. 1973, *Ap. J.* **179**, 111.

Litvak, M. M. 1970, *Phys. Rev.* **A2**, 2107.

A MODEL FOR POLARIZED OH MASERS

Malcolm D. Gray
School of Chemistry
University of Bristol
Cantock's Close
Bristol BS8 1TS
U.K.

ABSTRACT

A semi-classical theory of the propagation of polarized maser radiation through a medium containing an ambient magnetic field is described, including processes involving the overlap of maser lines. The treatment of the partially coherent maser radiation and the problems associated with a computational implementation of the theory are considered.

1. INTRODUCTION

The majority of observations of OH masers exhibit polarization. Therefore any theoretical and computational models of OH masers which attempt accurately to reproduce the observations must include the physical processes necessary to produce polarized radiation. This work describes the development of such a model.

2. ABOUT THE NEW MODEL

Where possible, the new model has been generalized from the existing semi-classical theory of Field and Gray, 1988 which does not include magnetic effects. In certain cases, notably the overlap of two or more simultaneously masing transitions, new theory has had to be developed. Computer code for the model is to run on Bristol Chemisty Department's Meiko parallel supercomputer.

Maser line overlap intimately couples polarization properties, lineshapes and saturation effects of the masers, while Zeeman splitting of energy levels vastly increases the computational complexity. See Table 1 for a comparison.

Table 1.

Parameter	Old	New
Levels	48	384
Radiative Transitions	207	4368
Collisional Transitions	1042	14917
Order of Overlap	6	131

The model is semi-classical: radiation is treated by the classical Maxwell's equations, molecular response by quantum-mechanical density matrix formalism. The model's magnetic field is assumed to be weak, in the sense that Zeeman splittings are assumed to vary linearly with magnetic field magnitude, B. B has arbitrary orientation with respect to the line of sight to the observer, but is assumed constant in magnitude and direction over the masing region.

3. CONVENTIONS FOR ZEEMAN SPLITTING

A basic assumption is that the magnetic field is strong enough to define a good quantization axis according to the criteria of Elitzur 1991. Formulae for the Landé g-factors in OH, Palmer and Zuckermann 1967, yield a value for the energy splitting in the ground rotational state of 1.63972 kHz mG^{-1} per unit magnetic quantum number, m_F. Zeeman splittings are smaller in all higher states and are positive for positive m_F in the F1 set of levels, negative for positive m_F in the F2 set.

4. CONVENTIONS FOR POLARIZED RADIATION

The following conventions were adopted to describe the properties of the polarized radiation which propagates through the Zeeman split molecular population:

(i) Right hand polarized radiation has a positive phase angle, δ, and the tip of the electric vector, components E_x, E_y, rotates clockwise as seen by a receiving observer.

(ii) A transition with $\Delta m_F = 0$ is a pi transition, while those with $\Delta m_F = \pm 1$ are sigma $\pm$ transitions.

(iii) The change in magnetic quantum number, Δm_F, is defined as the magnetic quantum number of the lower level minus that of the higher level.

(iv) Radiation from a sigma+ transition is right-hand circularly polarized when the magnetic field points towards the observer.

(v) The polarization properties of the radiation are fully described by the four Stokes parameters whose definitions are:- $I = E_x^2 + E_y^2$; $Q = E_x^2 - E_y^2$; $U = E_x E_y \cos\delta$; $V = E_x E_y \sin\delta$.

5. SOLUTION OF THE STEADY STATE MASTER EQUATIONS

The aim of the project is to obtain level populations from a set of steady-state master equations which describe population transfer by all processes between these levels. By separating the incoherent and maser parts of these processes, we can obtain this solution in a form representing perturbation of ambient populations by maser polarization functions.

The electric field is set up in coordinates based on the line of sight, while the molecular dipole operators are set up in coordinates based on the magnetic field. The 1st order rotation matrix, i.e. for dipole radiation, is then used to couple the two systems and hence calculate the interaction Hamiltonians forming the density matrix. Equations for rates of change of its elements, combined with expressions for collisional and incoherent radiation processes, are then developed to form the master equations.

The computational procedure carried out to achieve the solution discussed above is as follows:

(i) Parameters are read in as for the model of Field and Gray 1988, with the addition of magnetic field information.

(ii) The magnetic field parameters are used to calculate Zeeman splitting of the energy levels and all transition dependent quantities (e.g. Einstein A-values) are broken up into sublevel to sublevel components using appropriate 3-j symbols.

(iii) The master equations, described above, are formed into a matrix with coefficients of the form $k_{incoherent} + J_{maser}$ where J_{maser} is an infinitessimal maser function, a combination of Stokes parameters from a single coherent radiation packet, acting at a precise frequency.

(iv) The matrix is solved by elimination and back-substitution, maintaining linearity in the infinitessimal maser terms and including the off-diagonal elements which represent coherence in transitions.

(v) Transport of the incoherent radiation is handled by the Sobolev (LVG) approximation, parameterized by single-flight escape probabilities. These escape probabilities are both part of the matrix coefficients and dependent on the solution (the populations) therefore requiring iterative solution, terminating when a self-consistent set of populations and escape probabilities has been obtained.

6. PROPERTIES OF THE SOLUTION

The solution to the set of steady-state master equations is a set of populations of the form:

$$\rho_p = \rho_p^0 \left[1 + \sum_{j=1}^{M} T_{pj} J_j \right] \tag{1}$$

taking the pth energy level as an example. In eq.(1), the level population is written in terms of an unperturbed population, ρ_p^0, and a perturbation, the sum, dependent on the maser functions. This sum is over all maser transitions which can affect the pth level; its elements are composed of an infinitessimal maser function, J_j and a multiplier, T_{pj}, which like ρ_p^0 is composed of predictable combinations of the non-maser coefficients of the original matrix.

Although the maser functions in eq.(1) are infinitessimal contributions from a single coherent packet of radiation, the result can be generalized to the effect of a combination of the macroscopic Stokes parameters of a partially coherent, broad-band maser. This is achieved by taking products of the effects of all infinitessimal contributions at a single frequency and over all frequencies which contribute to the perturbation of a given molecular subgroup (see Field and Gray 1988).

7. MASER LINE OVERLAP

In the non-magnetic model, far infra-red (FIR) line overlap is commonplace and dealt with by the theory of Doel, Gray and Field 1990. In this process a single degree of overlap can be assigned to each pair of overlapping transitions. In the Zeeman split case, however, the splittings for magnetic fields of a few milligauss are typically a few kHz, similar to the thermal widths of the maser lines themselves. This implies that overlap between masing lines, all the components of 1720 MHz for example, occurs in the Zeeman split model, in contrast to the unsplit case, where the undivided maser lines are always too widely separated in frequency to overlap.

Maser line overlap, i.e. both involved transitions inverted, needs to be treated on the level of the individual coherent packets, used in sections 5 and 6. The result is that there is a different degree of overlap for each precise frequency accross the overlapping lineshapes. The degree of overlap at a given frequency is given by a suitably normalized product of the lineshapes at the same frequency.

The effects of maser line overlap are felt through the coherent parts of the steady-state master equations and each interaction Hamiltonian in the density matrix becomes a sum of contributions to the given transition from all those with which it overlaps, each contribution depending on the frequency in the line via the overlap coefficient. Eq.(2) shows the interaction Hamiltonian for the transition pq split into its overlap contributions, where A,B and C contain the overlap coefficients.

$$(\hat{\mu}.E)_{pq} = \hat{\mu}_{pq}.AE_{ab} + \hat{\mu}_{pq}.BE_{cd} + \hat{\mu}_{pq}.CE_{ef} + \ldots\ldots \tag{2}$$

Maser gain coefficients may be similarly expanded.

8. PROPAGATION OF POLARIZED MASER RADIATION

Growth of all masers in the medium is handled by the standard propagation equation for the four Stokes parameters, which is, Goldreich, Keeley and Kwan, 1973; Elitzur 1991:

$$\frac{d(I,Q,U,V)}{dz} = \gamma_0 \begin{pmatrix} \gamma_{II} & \gamma_{IQ} & \gamma_{IU} & \gamma_{IV} \\ \gamma_{QI} & \gamma_{QQ} & \gamma_{QU} & \gamma_{QV} \\ \gamma_{UI} & \gamma_{UQ} & \gamma_{UU} & \gamma_{UV} \\ \gamma_{VI} & \gamma_{VQ} & \gamma_{VU} & \gamma_{VV} \end{pmatrix} \begin{pmatrix} I \\ Q \\ U \\ V \end{pmatrix} \tag{3}$$

where the matrix coefficients are angular dependent terms derived from the first order rotation matrix (e.g. Shore 1990) and the multiplier γ_0 represents the strength of the transition and is a function of the initial and final values of the quantum numbers J,F and m_F. Eq.(3) has three possible forms for the matrix, corresponding to pi, sigma- or sigma+ transitions.

The Stokes parameters describing the maser radiation in all transitions are 'grown' through the masing medium using eq.(3), perturbing it as they progress, according to the scheme set out in Section 6.

REFERENCES

Doel, R. C., Gray M. D., & Field, D. 1990, MNRAS, **244**, 504
Elitzur, M. 1991, ApJ, **370**, 407
Field, D., & Gray M. D. 1988, MNRAS, **234**, 353
Goldreich, P., Keeley, D. A., & Kwan, J. Y. 1973, ApJ, **179**, 111
Palmer, P., & Zuckermann, B. 1967, ApJ, **148**, 727
Shore B. W. 1990, 'The Theory of Coherent Atomic Excitation.',Wiley,New York.

SUBMILLIMETER WATER MASERS

Gary J. Melnick

Harvard-Smithsonian Center for Astrophysics
60 Garden Street, Cambridge, MA 02138, USA

ABSTRACT

The excitation of maser emission in millimeter and submillimeter transitions of interstellar and circumstellar water is considered. Using an escape probability method, the equilibrium populations in 349 rotational states of both ortho- and para-water have been determined under varying conditions of gas temperature, density, water abundance and radiation field. Results indicate that under suitable conditions, collisional excitation of warm (400 K) interstellar gas may result in significant maser action in a total of 7 centimeter, millimeter, and submillimeter transitions of H_2O, while at somewhat higher gas temperatures (1000 K) a total of 22 centimeter, millimeter, and submillimeter H_2O transitions are found to mase. Under conditions of low atmospheric humidity, several of these transitions might be observed from mountaintop altitude, and all of them are potentially observable from airborne observatories. Under the assumption that water masers from different transitions originate in the same gas and that the emission is equally beamed, the ratio of maser luminosities from different transitions can be used to set limits on the temperature as well as the bundled quantity of hydrogen density, water abundance, and velocity gradient within the masing gas.

1. INTRODUCTION

Since the first detection of the 22 GHz $6_{16} - 5_{23}$ transition in 1968 (Cheung et al. 1969), water masers have been viewed primarily as bright, readily-detectable point sources which possess a far greater potential for revealing dynamical information than for illuminating anything about the nature of the gas from which they originate. During the next 20 years, the absence of unambiguous maser action in other observed water transitions left little choice but to assess the physical conditions in the masing gas based solely upon the conditions required to get the 22 GHz transition to mase. This situation has changed markedly since 1989 - maser action has now been clearly observed in five additional millimeter and submillimeter water transitions toward both star-forming regions and late type stars (Menten et al. 1990a; Chernicharo et al. 1990; Menten et al. 1990b; Melnick et al. 1992). In this contribution we emphasize that the ratio of the strengths of these different maser lines can be used much as the ratio of non-masing lines to bracket the range of physical conditions within the emitting region. The theory underlying this assertion is briefly reviewed in Section 2 and an example is considered in Section 3.

2. THEORY

We have determined the equilibrium level populations of interstellar water in a collisionally-pumped, optically-thick, neutral medium under varying conditions of kinetic temperature, density, water abundance, and background radiation field. Full details of our method have been described in a previous paper (Neufeld and Melnick 1991, hereafter NM91). We use an escape probability method to include the effects of radiative trapping in infrared non-masing water transitions, and we obtain an initial solution for the case in which saturated maser action is absent. Our calculations apply to the case where H_2 molecules are the dominant collision partner. We consider both the criterion for saturated maser action in any given transition, and the maser luminosity that results when that criterion is met.

If the background radiation field is negligible, the solution of the equations of statistical equilibrium depends solely upon the temperature and upon the ratios of collisional rates to the net radiative rates for each transition. Under conditions where every non-masing transition is optically thick, the escape probability for each transition is inversely proportional to the optical depth, and the level populations may be written as a function of two variables: the kinetic temperature T and a combination of parameters which we represent with the variable ξ', defined by

$$\xi' \equiv n_9^2 x_{-4}(H_2O)/(dv_z/dz)_{-8} , \tag{1}$$

where $n = 10^9 n_9$ cm^{-3} is the density of hydrogen nuclei, $10^{-4}x_{-4}(H_2O)$ is the water abundance relative to hydrogen nuclei, and $10^{-8}(dv_z/dz)_{-8}$ cm s^{-1} cm^{-1} is the magnitude of the velocity gradient.

Once the level populations have been computed, possible maser transitions can be identified. Each such transition may be characterized by a maser rate coefficient, $Q(\xi',T) = 10^{-14}Q_{-14}$ cm^3 s^{-1}, defined such that for saturated maser lines the photon production rate per unit volume is given by $\Phi_{sat} = Qn(H_2O)n$, where $n(H_2O)$ is the density of water molecules. If the maser radiation is concentrated into a beam of solid angle Ω_m, the "apparent isotropic photon luminosity" seen by an observer within the maser beam is related to Q by the expression

$$L_p = \Phi_{sat}V(4\pi/\Omega_m) = \xi'Q_{-14}(dv_z/dz)_{-8}V(4\pi/\Omega_m) \text{ photons s}^{-1}, \tag{2}$$

where V is the volume of the emitting region in cm^3. Unfortunately the beam solid angle Ω_m cannot be determined observationally and is critically dependent upon the geometry of the maser. Nevertheless, it can be argued that the observed *ratio* of intensities in more than one maser line may be used as a probe of the physical conditions in the masing region *without our needing to know the exact geometry*. (We need only make the plausible assumption that the maser beaming angles are roughly the same, as expected for the case of saturated masers, in which the beam solid angle varies at most by a logarithmic factor from one transition to another.) Note also that when the integrated luminosity from a spatially unresolved *spherically*

TABLE 1. Predicted Centimeter, Millimeter, and Submillimeter Water Masers

Frequency (GHz)	Transition	Temperature[†] (K)	Accessible from: Ground (G) or Air./Space (A/S)	Observed to Date?
22.235	$6_{16} \rightarrow 5_{23}$	400/1000	G	Yes
183.310	$3_{13} \rightarrow 2_{20}$	400/1000	G	Yes
321.226	$10_{29} \rightarrow 9_{36}$	400/1000	G	Yes
325.153	$5_{15} \rightarrow 4_{22}$	400/1000	G	Yes
354.885	$17_{413} \rightarrow 16_{710}$	1000	G	No
380.197	$4_{14} \rightarrow 3_{21}$	400/1000	A/S	Maybe[‡]
439.151	$6_{43} \rightarrow 5_{50}$	1000	G	Yes
448.001	$4_{23} \rightarrow 3_{30}$	400/1000	A/S	No
470.889	$6_{42} \rightarrow 5_{51}$	1000	G	Yes
474.689	$5_{33} \rightarrow 4_{40}$	1000	A/S	No
530.429	$14_{312} \rightarrow 13_{49}$	1000	A/S	No
620.701	$5_{32} \rightarrow 4_{41}$	1000	A/S	No
826.552	$18_{415} \rightarrow 17_{512}$	1000	G	No
906.207	$9_{28} \rightarrow 8_{35}$	1000	G	No
916.172	$4_{22} \rightarrow 3_{31}$	1000	A/S	No
970.315	$5_{24} \rightarrow 4_{31}$	1000	A/S	No
1158.323	$6_{34} \rightarrow 5_{41}$	1000	A/S	No
1278.265	$7_{43} \rightarrow 6_{52}$	1000	A/S	No
1296.412	$8_{27} \rightarrow 7_{34}$	400/1000	A/S	No
1322.065	$6_{25} \rightarrow 5_{32}$	1000	A/S	No
1440.782	$7_{26} \rightarrow 6_{33}$	1000	A/S	No
1541.966	$6_{33} \rightarrow 5_{42}$	1000	A/S	No

[†] Entries listed as 400/1000 imply that the given transition exhibits significant maser emission at both 400 K and 1000 K; entries listed as 1000 imply that only gas of the order of 1000 K can yield measurable maser emission.

[‡] This transition was detected toward Orion-KL from the KAO; however, it is not clear what fraction of the measured line is due to maser emission.

symmetric source is observed, the beaming angle drops out since the observation effectively takes an average over all viewing angles.

Table 1 lists the water transitions that we predict to mase at 400 K and 1000 K, their accessibility from ground-based, airborne, or space-based telescopes, and whether each transition has been observed to date. Figures 1 and 2 display the maser emissivities at 400 and 1000 K, respectively, as a function of the variable ξ'. These results show that for every masing transition the volume emissivity increases steadily with ξ', reaches a maximum value, and then drops to zero as the level populations near thermal equilibrium and the population inversion disappears.

3. APPLICATION: THE 22/321 GHz MASER LINE RATIOS

As a example of the type of information that can be obtained by studying more than one water maser transition, we briefly consider observations of the 321 GHz ($10_{29}-9_{36}$) and the well-studied 22 GHz water masers. The range of conditions

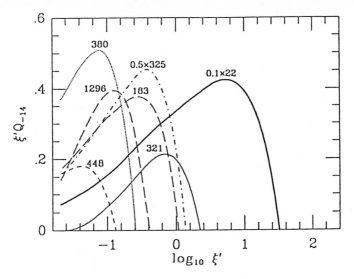

Fig. 1. Maser emissivities at 400 K, as a function of the variable ξ' (see text). Each curve is labeled with the transition frequency in GHz (after NM91).

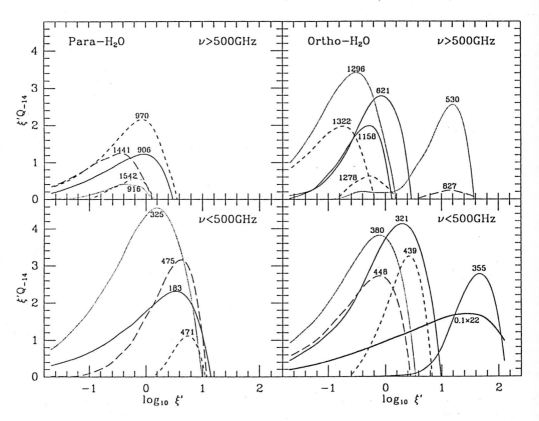

Fig. 2. Maser emissivities at 1000 K, as a function of the variable ξ' (see text). Each curve is labeled with the transition frequency in GHz (after NM91).

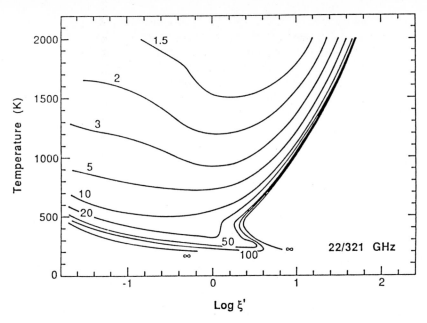

Fig. 3. Emissivity ratio: contours of the quantity $R \equiv Q(22\ GHz)/Q(321\ GHz)$ (after Neufeld and Melnick, 1990).

needed to excite these transitions is similar, but not identical. As can be seen from Figures 1 and 2, the 22 GHz maser is more easily saturated than the 321 GHz maser at low temperatures and the 321 GHz maser is "quenched" at smaller ξ' values than the 22 GHz maser, particular when the kinetic temperature is small. Thus, our calculations allow for the possibility that luminous 22 GHz maser emission will be unaccompanied by 321 GHz emission. Under conditions where both masers are saturated, the ratio of the photon emissivities in the two masing lines is given simply by the ratio of the rate coefficients. The ratio $R \equiv Q(22\ GHz)/Q(321\ GHz)$ is plotted in Figure 3. The quantity R decreases with increasing temperature, reflecting the difference in the energies of the upper states ($E_u/k = 643$ K for the 22 GHz transition and 1861 K for the 321 GHz transition).

If the maser emission in each line is identically beamed, as in the case of the cylindrical maser where the geometry of the source determines the beaming angle (Goldreich and Keeley 1972), the ratio R can be compared directly with the observed ratio of apparent isotropic photon luminosities. Table 2 lists results obtained for four galactic star-forming regions and one late-type star (after Menten et al. 1990a). This range of luminosity ratios implies the presence of masing gas of at least 200 K for $R = 150$ and at least 700 - 1000 K for $R = 3 - 5$.

Elitzur, Hollenbach, and McKee (1989) have suggested that 22 GHz water masers in interstellar star-forming regions are excited by collisional pumping of dense, neutral gas which has been compressed and heated by a dissociative shock. In this model, the maser emission originates in the molecule reformation zone behind

TABLE 2. 22/321 GHz Photon Emissivity Ratios

Source	L_{321} — 321 GHz Isotropic Photon Luminosity (s^{-1})	L_{22}/L_{321}
W3(OH)	8.6×10^{44}	70
W49N	2.9×10^{46}	150
W51 IRS2	1.7×10^{46}	3-5
W51 Main	5.7×10^{45}	39
VY CMa	1.2×10^{45}	12

the shock front which is characterized by a temperature of approximately 400 K. Our results for W3(OH), W49N, and W51 Main are consistent with this prediction; however, the observed 22/321 GHz ratio for W51 IRS2 of 3-5 may rule out such a model for this source. Instead the 22/321 GHz ratio may imply excitation behind a slower nondissociative shock in which the temperature of the molecules is $\sim$ 1000 K. Observations of other water maser transitions from this source will be highly useful in resolving this uncertainty.

4. SUMMARY

In addition to the well-studied 22 GHz water maser, there are a number of millimeter and submillimeter water transitions that we predict will mase under reasonable interstellar conditions. With the advent of good submillimeter telescopes and receivers many of these transitions are accessible for study. We show that observations of two or more of these transitions from the same region (and in the same epoch) can be used to constrain the physical conditions in the masing region.

REFERENCES

Chernicharo, J., Thum, C., Hein, H., John, D., Garcia, P., and Mattioco, F. 1990, *Astr. Ap.*, **231**, L15.

Cheung, A.C., Rank, D.M., Townes, C.H., Thornton, D.D., and Welch, W.J. 1969, *Nature*, **221**, 626.

Elitzur, M., Hollenbach, D.J., and McKee, C.F. 1989, *Ap. J.*, **246**, 983.

Goldreich, P. and Keeley, D.A. 1972, *Ap. J.*, **174**, 517.

Melnick, G.J., Menten, K.M., and Phillips, T.G. 1992, in preparation.

Menten, K.M., Melnick, G.J., and Phillips, T.G. 1990a, *Ap. J. (Letters)*, **350**, L41.

Menten, K.M., Melnick, G.J., Phillips, T.G., and Neufeld, D.A. 1990b, *Ap. J. (Letters)*, **363**, L27.

Neufeld, D.A. and Melnick, G.J. 1990, *Ap. J. (Letters)*, **352**, L9.

Neufeld, D.A. and Melnick, G.J. 1991, *Ap. J.*, **368**, 215 (NM91).

LINEAR AND CIRCULAR POLARIZATION FOR MASERS WITH WEAK ZEEMAN SPLITTINGS

Gerald E. Nedoluha
NRC Cooperative Research Associate
Naval Research Lab
Washington DC 20375-5000

William D. Watson
Loomis Lab of Physics
University of Illinois
Urbana IL 61801

ABSTRACT

Theory is reviewed for the linear and circular polarization of radiation from astrophysical masers in which the Zeeman splitting $g\Omega$ is much less than the spectral linewidth $\Delta\omega$. When $g\Omega \gtrsim \Gamma$ (the decay rate for the molecular states), polarization — mainly linear — can be generated. New results are presented that describe the generation of circular polarization by intensity dependent (or "false") Zeeman effects. As a result, the magnetic field strengths inferred from the circular polarization of circumstellar SiO masers may be smaller than previous estimates by a factor of 100 or more.

1. LINEAR POLARIZATION

The general features of linear polarization have been delineated by Goldreich, Keeley, & Kwan (1973; hereafter GKK) by considering an angular momentum $J = 1 - 0$ masing transition and obtaining the polarization for the limiting (or asymptotic) solutions to the equations of radiative transfer that are appropriate when certain strong inequalities are satisfied that involve $g\Omega$, Γ, and the rate R for stimulated emission. This description has been extended in several significant ways. In particular, these limiting solutions do not describe the polarization when the strong inequalities are not satisfied. It also is not evident how well satisfied they must be in order for the limiting solutions to be a satisfactory approximation. To answer these questions, equations similar to those of GKK have been numerically integrated in a series of investigations (Western & Watson 1984; Nedoluha & Watson 1990a). Representative results are displayed in Figure 1 of the latter citation where the general behavior as a function of saturation evidently is what is expected from GKK for a $J = 1 - 0$ transition.

There are qualitatively important new features, however. The presence of the large Stokes U-component in the numerical solutions indicates that the polarization is neither parallel nor perpendicular to the magnetic field at these intensities. Here, the direction of the linear polarization rotates from parallel to perpendicular relative to the magnetic field. This demonstrates how the change in direction occurs for propagation angles $< 55°$ as predicted by GKK when $R > g\Omega$. Unlike for the $J = 1 - 0$ transition, the polarization disappears rapidly for transitions with higher angular momenta (e.g., for $J = 2-1$) when $R > g\Omega$. This was expected from previous studies in which the limiting fractional polarization for these transitions in this regime ($g\Omega << R << (g\Omega)^2/\Gamma$) was found to be only a few percent (Deguchi & Watson 1990) instead of the one-third as found by GKK for the $J = 1 - 0$ transition. In the regime ($g\Omega >> R >> \Gamma$), the limiting polarizations for the states of higher angular momenta are identical with those obtained for $J = 1 - 0$ by GKK (Deguchi, Watson, & Western 1986). However, for realistic strengths of the magnetic field, the peak value of the linear polarization for states of higher angular momenta is significantly smaller. Specifically, we find that for $g\Omega/\Gamma \lesssim 2000$, the fractional linear polarization does not exceed 30 percent even though the limiting polarization in this regime is 100 percent for certain propagation angles.

Claims have been made that the linear polarization is independent of saturation (Elitzur 1991; 1992), a completely different behavior from that given by GKK and by our calculations. These claims are invalid as they are based upon erroneous simplifications in the calculations such as assuming that the populations of the magnetic substates are always equal and ignoring off-diagonal elements in the density matrix which are important when $R \gtrsim g\Omega$. The additional assertion that leads to polarization in the unsaturated *limit* ($g\Omega >> \Gamma >> R$) – that the fractional polarization here is required to be the same as in the saturated *limit* ($g\Omega >> R >> \Gamma$) to allow the populations to be independent of substate in the as in the saturated limit (between these two limits populations generally depend on substate) – is clearly false. The radiation does not influence the populations in the unsaturated limit and any incident polarization (usually none) simply is amplified in this regime (GKK). The results of Elitzur are appropriate only in the saturated *limit*, and then only when $R << g\Omega$ (case 2a of GKK) where they duplicate the solutions of GKK.

Circumstellar $J = 3-2$ and $2-1$ SiO masers, and flares of interstellar 22 GHz H_2O masers, have fractional linear polarization that exceeds 30%. In some cases, 60-70% and higher is reported. Such high fractional polarization is most likely a result of anisotropic excitation by infrared radiation due either to an incident "beam" or to anisotropic escape. Explicit calculations for the circumstellar SiO masers have demonstrated how such anisotropic excitation can occur (Western & Watson 1983b). In recent observations of the $J = 1-0, 2-1$, and $3-2$ SiO masers at the same location, the fractional linear polarization has been reported to increase with the angular momentum of the molecular states (McIntosh & Predmore 1991). We have performed computations in which this variation is reproduced for plausible anisotropic pumping. All of the calculations of the foregoing discussion treat the maser radiation as a single or bidirectional beam. When the finite cross section of the maser is treated, geometrical effects can cause linear polarization of the emergent radiation (Western & Watson 1983a). However, it seems unlikely that such geometrical effects can be the cause for the observed polarizations that are > 30%.

2. CIRCULAR POLARIZATION

The calculations discussed above describe the polarization only at the center of the spectral line. The circular polarization (e.g., the Stokes V-parameter) commonly has a strong, antisymmetric variation about the center of the spectral line, and is zero at the center. To perform calculations for the circular polarization in this regime of weak Zeeman splitting, it is thus necessary to treat explicitly enough frequencies to delineate the spectral line profile. In determining the variation across the spectral line, it also is necessary to consider deviations from Maxwellian in the velocity distribution of the masing molecules. This deviation is the cause for the rebroadening of the spectral lines after the maser becomes saturated and the rate for stimulated emission exceeds the rate for relaxation of the molecular velocities. Excitation properties (i.e., the elements of the quantum mechanical density matrix for the molecular states) must then be computed at a sufficient number of velocities to delineate the distribution at each spatial location in the numerical integration.

H_2O Masers. Calculations have previously been performed (Nedoluha & Watson 1990b, 1992a) to aid in interpreting the weak circular polarization detected by Fiebig & Güsten (1989) [also Zhao, Goss, & Diamond (1992)] in the spectra of the 22 GHz H_2O masers in regions of star formation. As long as $g\Omega \gg R$ and R does not exceed the rate for relaxation of the molecular velocities, the ordinary Zeeman effect seems to provide a satisfactory basis for inferring the strength of the magnetic field from these observations. The presence of 3 closely spaced hyperfine transition makes the spectral linewidth a useful indicator for the ratio of R to the rate for velocity relaxation, and hence for the isotropic luminosity, even in the absence of accurate thermal linewidth estimates (Nedoluha & Watson 1991). Based on the spectral linewidth, it seems likely that a number of the observed spectra do yield reliable evidence for magnetic fields near 30 mG in these regions of star formation.

SiO Masers. Fractional circular polarization has also been detected in the spectra of the circumstellar SiO masers (Barvainis, McIntosh, & Predmore 1987) and is much larger that that of the 22 GHz H_2O masers (5-10% vs. 0.1%). Based on the ordinary Zeeman effect, magnetic field strengths of ~100 G are inferred. For VY CMa, measurements of the circular polarization in the 22 GHz H_2O masers are available and yield much lower upper limits ($\lesssim$ 100 mG). These inferences about the magnetic fields are not necessarily incompatible since the SiO masers are closer to the star than are the H_2O masers. Although the profile of the Stokes V component of the SiO masers does not exhibit the characteristic antisymmetric behaviour about line center, this can be understood as a result of velocity gradients within the maser (e.g., Deguchi & Watson 1986; Nedoluha & Watson 1990c).

We have recently performed calculations to see whether intensity dependent modifications to the ordinary Zeeman effect can be significant for the SiO masers (Nedoluha & Watson 1992b). The methods are the same as those that were used to examine such effects in the spectra of H_2O masers (Nedoluha & Watson 1990b). There are key differences in the relevant parameters, however. The magnetic moments for the SiO transition are almost a factor of ten smaller than for the strongest hyperfine component of the 22 GHz transition. When the SiO masers are just saturated ($R \approx 5$ s^{-1}) as seems likely, modifications to the ordinary Zeeman effect seem plausible since $g\Omega \approx 10$ radians s^{-1} for a field of 10 mG. Another key difference is the larger linear polarization of SiO masers (tens of percent is common) which tends to be conducive to modifications in the circular polarization. Since a velocity gradient seems to be present, our calculations include this effect as well. Our conclusion is that the main properties of the observations − the fractional circular and linear polarization as well as the spectral line profiles of the intensity and polarizations − can be reproduced either by the ordinary Zeeman effect with magnetic fields of 10-100 G or by "false" Zeeman features with magnetic fields of 10-100 mG. Surprisingly, we cannot distinguish between these possibilities from the available information, though the latter seems preferable since it is clearly compatible with the data from the H_2O masers and places less severe demands upon the strength of the magnetic field. If there are circumstellar regions where the SiO and H_2O masers are in similar locations, a comparison would be valuable.

REFERENCES

Barvainis, R. E., McIntosh, G. C. & Predmore, C. R. 1987, Nature, 329, 613

Deguchi, S. & Watson, W. D. 1986, ApJ, 300, L15

Deguchi, S. & Watson, W. D. 1990, ApJ, 354, 649

Deguchi, S., Watson, W. D., & Western, L. R. 1986, ApJ, 302, 108

Elitzur, M. 1991, ApJ, 370, 407

Elitzur, M. 1992, Astronomical Masers (Dordrecht: Kluwer), 184

Fiebig, D. & Güsten, R. 1989 A&A, 214, 333

Goldreich, R., Keeley, D. A., & Kwan, J. Y. 1973, ApJ, 182, 55 [GKK]

McIntosh, G. C. & Predmore, C. R. 1991, in Proc. Third Haystack Obs. Conf. (ASP Vol. 16), ed. A. D. Haschick & P. T. Ho, 83

Nedoluha, G. E. & Watson, W. D. 1990a, ApJ, 354, 660

Nedoluha, G. E. & Watson, W. D. 1990b, ApJ, 361, L53

Nedoluha, G. E. & Watson, W. D. 1990c, ApJ, 361, 653

Nedoluha, G. E. & Watson, W. D. 1991, ApJ, 367, L63

Nedoluha, G. E. & Watson, W. D. 1992a, ApJ, 384, 185

Nedoluha, G. E. & Watson, W. D. 1992b, ApJ, in preparation

Western, L. R. & Watson, W. D. 1983a, ApJ, 274, 195

Western, L. R. & Watson, W. D. 1983b, ApJ, 275, 195

Western, L. R. & Watson, W. D. 1984, ApJ, 285, 158

Zhao, J.-H., Goss, W. M., & Diamond, P. 1992, preprint

RADIATIVE TRANSFER IN THREE-DIMENSIONAL MASERS

David A. Neufeld
Department of Physics and Astronomy
The Johns Hopkins University, Baltimore MD 21218

ABSTRACT

The standard theory of three-dimensional masers is based upon inadequate approximations. In particular, the standard treatment adopts equations for the steady-state molecular level populations and for the transfer of maser radiation which apply strictly only to one-dimensional masers. I consider here the specific case of a partially saturated spherical maser in which the rate of cross-relaxation between molecules of different velocities is small. For that case the standard approach leads to a relationship between the maser intensity and the molecular level populations which differs significantly from that obtained using the exact equations for the transfer of maser radiation. The correct solution to the problem of the spherical maser can be obtained numerically by an iterative method and will be considered in future work.

1. BASIC EQUATIONS

The transfer of radiation within a sphere of masing material is a classic problem in the theoretical study of astrophysical masers. The standard treatment of this problem (e.g. Elitzur 1990), makes use of equations for the steady-state level populations and for the transfer of radiation which apply strictly only to one-dimensional masers, with the claimed justification that the maser radiation at any point is strongly beamed. I argue here that this standard approach may lead to an erroneous result.

Let us consider a three-dimensional maser in which (i) there is no velocity gradient; (ii) neither elastic collisions nor the radiative trapping of infrared radiation result in significant cross-relaxation between molecules of different velocities; and (iii) any hyperfine splitting of the maser transition may be neglected.

In general the matter in the system is completely described by the fractional level populations at every point in phase-space. Thus we may say that the number of molecules in state i, at position $\bar{\mathbf{r}}$, and with velocity $\bar{\mathbf{v}}$ is given by $N_i = \rho_i(\bar{\mathbf{r}}, \bar{\mathbf{v}}) d^3 \bar{\mathbf{r}} d^3 \bar{\mathbf{v}}$, where $\rho_i(\bar{\mathbf{r}}, \bar{\mathbf{v}})$ is the "phase space density". The quantity of interest is always the difference in the level populations, $\Delta\rho(\bar{\mathbf{r}}, \bar{\mathbf{v}}) \equiv \rho_u(\bar{\mathbf{r}}, \bar{\mathbf{v}}) - \rho_l(\bar{\mathbf{r}}, \bar{\mathbf{v}})$. The equations of statistical equilibrium imply that

$$\frac{\partial}{\partial t}\Delta\rho(\bar{\mathbf{r}}, \bar{\mathbf{v}}) = [P_u(\bar{\mathbf{v}}) - P_l(\bar{\mathbf{v}})] - [\Gamma_u \rho_u(\bar{\mathbf{r}}, \bar{\mathbf{v}}) - \Gamma_l \rho_l(\bar{\mathbf{r}}, \bar{\mathbf{v}})] - 2BJ_*(\bar{\mathbf{r}}, \bar{\mathbf{v}})\Delta\rho(\bar{\mathbf{r}}, \bar{\mathbf{v}}), \quad (1)$$

where P_u and P_l are the pump rates and Γ_u and Γ_l are the decay rates for the upper and lower states, respectively. The quantity J_* in this equation is the mean intensity

51

as measured in the frame of a molecule at position $\bar{r}$ and with velocity $\bar{v}$. That mea intensity is given by

$$J_*(\bar{r}, \bar{v}) = \frac{1}{4\pi} \int I(\bar{r}, \hat{k}, \nu_0[1 - \bar{v}.\hat{k}/c])d\Omega,$$

where ν_0 is the rest frequency of the maser transition and where $I(\bar{r}, \hat{k}, \nu)$ is the speci intensity *in the laboratory frame* at position $\bar{r}$, at frequency ν, and for a ray in directic $\hat{k}$. *Note in particular, that there is no way of computing the value of this integral if u are given only the quantity*

$$J_\nu(\bar{r}) \equiv \frac{1}{4\pi} \int I(\bar{r}, \hat{k}, \nu)d\Omega.$$

The steady-state solution to equation (1) is

$$\Delta\rho(\bar{r}, \bar{v}) = \frac{J_s}{J_* + J_s}\Delta\rho_0(\bar{v}),$$

where $\Delta\rho_0(\bar{v}) \equiv [P_u(\bar{v})/\Gamma_u - P_l(\bar{v})/\Gamma_l]$ is value of $\Delta\rho(\bar{r}, \bar{v})$ in the limit of small intensit and where $J_s \equiv \Gamma_u\Gamma_l/B(\Gamma_u + \Gamma_l)$ is the "saturation intensity".

Equation (4) describes how the matter responds to a given radiation field $I(\bar{r}, \hat{k}, \nu$ The radiation field is related to the properties of the matter by the transfer equation

$$\frac{d}{ds}I(\bar{r}, \hat{k}, \nu) = j(\bar{r}, \nu) + \kappa(\bar{r}, \hat{k}, \nu)I(\bar{r}, \hat{k}, \nu),$$

where $j(\bar{r}, \nu)$ is the true emissivity, where

$$\kappa(\bar{r}, \hat{k}, \nu) = \frac{hcB}{4\pi} \int \Delta\rho(\bar{r}, \bar{v})\delta(\bar{v}.\hat{k} - c\Delta\nu/\nu_0)d^3\bar{v},$$

and where $\Delta\nu = \nu - \nu_0$ is the frequency shift from line center. *Equations (2), (4) a (6) imply that the opacity of the masing material to radiation at frequency ν may affected by radiation at frequencies other than ν, even though the stimulated emissic process is coherent in the rest frame of the stimulated molecule.* An iterative methc may prove useful in obtaining a simultaneous solution of equations (4) and (5).

In the standard approach to the problem of three-dimensional masers, two sir plifying approximations have been made. First, the opacity $\kappa(\bar{r}, \hat{k}, \nu)$ in the transf equation (5) is replaced by $\kappa(\bar{r}, \hat{e}_d(\bar{r}), \nu)$, where $\hat{e}_d(\bar{r})$ is the direction in which the r diation is predominantly beamed at position $\bar{r}$. For the case of the sphere, $\hat{e}_d(\bar{r})$ everywhere directed radially outwards. Second, $J_*(\bar{r}, \bar{v})$ in equation (4) is replaced I $J_\nu(\bar{r})$ evaluated at the frequency $\nu = \nu_0[1 - \hat{e}_d.\bar{v}/c]$. For a general three-dimension maser with no special symmetry, this simplification reduces by two the number of c ordinates upon which $\Delta\rho$ and κ depend; thus instead of being functions of $(\bar{r}, \bar{v})$ a $(\bar{r}, \hat{k}, \nu)$ respectively, they become functions of $(\bar{r}, \hat{e}_d.\bar{v})$ and $(\bar{r}, \nu)$. In the standard trea ment of three-dimensional masers, we effectively assume that radiation at frequency may interact with distinct sample of molecules which can interact with each other b with no other molecules. That assumption is strictly true *only* for a one-dimension maser. In justifying these approximations, it was argued (Goldreich and Kwan 197

Elitzur 1990) that the maser radiation is strongly beamed, so that only rays for which $\hat{e}_d$ and $\hat{k}$ are nearly parallel need be considered.

2. THE SPHERICAL MASER

In spherical geometry - because of the symmetry - the radiation field, opacity and level populations are each a function of 3 rather than 6 variables. Thus $I = I(r, \alpha, \nu)$ and $\kappa = \kappa(r, \alpha, \nu)$, where α is the angle between the ray and the radial vector. Similarly, $\Delta\rho = \Delta\rho(r, v_r, v_t)$, where v_r is the radial component of the velocity and v_t is the tangential component. These are the most general expressions consistent with the spherical symmetry. It is convenient to divide phase-space into an "unsaturated" region, for which $J_*(\bar{r}, \bar{v}) \leq J_s$, and a "saturated" region, for which $J_*(\bar{r}, \bar{v}) \geq J_s$. The boundary between these two regions occurs at $r = r_s(v_r, v_t)$, where $r_s(v_r, v_t)$ is the solution to $J_*(r_s, v_r, v_t) = J_s$. In general the "saturation radius " $r_s(v_r, v_t)$ must be expressed as a function of two variables.

In considering the adequacy of the standard approximations, I have used the exact equations of radiative transfer to estimate the maser spot profile which results from the level populations derived by Elitzur (1990) for the spherical maser using the standard approach. I have also made a *dichotomous* approximation in which $\Delta\rho(\bar{r}, \bar{v})$ is taken as $\Delta\rho_0(\bar{v})$ in the unsaturated region of phase-space $r \leq r_s(v_r, v_t)$ and is taken as zero elsewhere. An analogously dichotomous approximation was used to obtain the beaming angle (or equivalently, the maser spot profile) in the treatment of Elitzur (1990).

Making use of equations (5) and (6), I have computed the emergent intensity as a function of impact parameter. The results apply to a partially-saturated sphere of radius $R = 30/\kappa_0(\nu_0)$ and in which $\kappa_0(\nu_0)J_s/j(\nu_0) = 10^6$, where $\kappa_0(\nu)$ is the opacity in the limit of zero maser intensity. This computation allows the solid angle of the emergent beam, $\Omega_\nu \equiv [4\pi I(0, \nu)]^{-1} \int I(\theta, \nu)d\Omega$, to be obtained using both the exact equations and the standard approximations. Figure 1 (overleaf) compares the results yielded by the two methods, as a function of the dimensionless frequency shift, $x_\nu \equiv \Delta\nu/\Delta\nu_D$, where $\Delta\nu_D$ is the Doppler width. The beam solid angle, Ω_ν, is of course proportional to the projected area of the observed maser spot.

3. DISCUSSION

The results plotted in Figure 1 demonstrate clearly that the standard treatment of spherical masers makes an inadequate approximation in neglecting the difference between $\hat{k}$ and $\hat{e}_d$. For the given level populations I assumed, the standard approach *under*estimates Ω_ν by a factor ~ 6 when $x_\nu = 0$ (i.e. at line center) but *over*estimates it by a factor ~ 9 when $x_\nu = 1$. Furthermore, the standard equations yield a beam solid angle which is an increasing function of frequency shift, whereas the exact equations imply an Ω_ν which decreases with x_ν.

It must be emphasized that the objections to the standard approach raised here apply only when the rate of cross-relaxation between molecules of different velocities is negligible (cf. assumption (ii) in §1.) In the opposite limit where cross-relaxation is very rapid, the effective radiation field is independent of the molecular velocity, and the opacity is isotropic. In that limit - which may indeed apply in many cases of astrophysical interest (Goldreich and Kwan 1974) - the standard approach adopted by Goldreich and Keeley (1972) and by Alcock and Ross (1985) is correct.

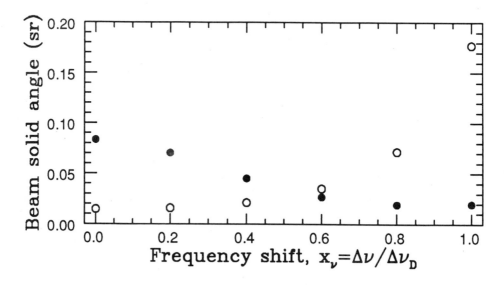

Figure 1:   Beam solid angle resulting from the level populations of Elitzur (1990). Results are computed as a function of the frequency shift $x_\nu \equiv \Delta\nu/\Delta\nu_D$, both with (open circles) and without (filled circles) the use of the standard approximations.

Furthermore, the results presented in Figure 1 are not to be regarded as the correct solution for the case of a spherical maser, even when the cross-relaxation rate is small. These results, after all, were computed assuming the dichotomous approximation and the saturation radius derived by Elitzur (1990) using the standard approximations which I have shown to be inadequate. Thus the discussion presented here is merely to be regarded as a demonstration that the standard approximations lead to an erroneous relationship between the level populations and the apparent maser spot size. The correct solution must obtained by the iterative numerical solution of the exact equations given in §1; this will be considered in future work.

It is a pleasure to acknowledge helpful discussions with M. Elitzur and C. F. McKee.

REFERENCES

Alcock, C., and Ross, R. R. 1985, *Ap. J.*, **290**, 433.
Elitzur, M. 1990, *Ap. J.*, **363**, 638.
Goldreich, P., and Keeley, D. A. 1972, *Ap. J.*, **174**, 517.
Goldreich, P., and Kwan, J. 1974, *Ap. J.*, **190**, 27.

A SPECIAL DATABASE FOR GAS-PHASE COMPOUNDS STUDIED BY MICROWAVE SPECTROSCOPY AND RADIO ASTRONOMY

Jürgen Vogt
Sektion für Spektren- und Strukturdokumentation
Universität Ulm, Postfach 4066, W-7900 Ulm, Germany

ABSTRACT

A machine-readable literature compilation on free molecules has been established for microwave spectroscopy. The new bibliographic numerical resource also comprises molecular radio astronomy. The inhouse database, which can be run on personal computers with the well known Messenger retrieval language, enables the users to trace back literature by means of bibliographic, chemical and physical search terms.

1. INTRODUCTION

Molecular physical and spectroscopic investigations on gas-phase molecules are often published in special, not well known sources and consequently quite often overseen. Many of these papers cannot even be located by means of databases, which generally cover chemistry and physics or which are specialised in particular fields. Thus, in order to facilitate the access to structural and related properties of free molecules, the Sektion für Spektren- und Strukturdokumentation at the University of Ulm has compiled and critically evaluated for more than two decades literature in the field of high resolution molecular spectroscopy (especially in the long wavelength region) and gas-phase electron diffraction as well [1-7].

This complete and worldwide unique compilation has been the basis for the computerized database MOGADOC [8]. The acronym stands for Molecular Gas-Phase Documentation. The project has been carried out in cooperation with Fachinformationszentrum in Karlsruhe, the European host for STN International online databases.

2. SCOPE OF DATABASE

The information has been systematically compiled from scientific journals, periodicals and books. Moreover a lot of grey literature such as dissertations, theses, reports and conference proceedings is also included. In particular, relevant abstracts from the Austin Symposia on Molecular Structure, Ohio Symposia on Molecular Spectroscopy as well as from the European High Resoltution Molecular Spectroscopy Conferences are taken into account. The literature is recorded back to:

-    1945 for microwave spectroscopy, and

-    1960 for molecular radio astronomy.

Thus the MOGADOC database covers the literature from the very beginning for each method. In total about 17300 references (until 1991) are presently included for about 5300 inorganic, organic and organometallic compounds (neutral molecules, ions and radicals) as well.

For the quoted references the bibliographic entries do not contain abstracts as given in the literature. However, the content of the publications is characterized by means of keywords, which form an hierarchical controlled vocabulary of about 1900 items. Experts who review the documents select appropiate keywords. The keywords cover the following topics:

-    spectral ranges,

-    rotational constants and rotation vibration interaction,

-    structure of free molecules,

-    nuclear quadrupole and electronic hyperfine structure,

-    Stark and Zeeman effect,

-    dipole moments and magnetic susceptibilities,

-    collisional effects (i.e. pressure broadening, line shifts etc.),

-    large amplitude motion (internal rotation, ring puckering, inversion, quasi-linear and quasisymmetric top molecules, loosely bound molecules), and

-    specific isotopic labels,

-    astrophysical and environmental observations.

3. FEATURES OF DATABASE

The MOGADOC database is not accessible online, that means by international computer network. However, it was developed as an inhouse database, which can be run on IBM compatible personal computers under MS-DOS with the Messenger retrieval language, well known from STN online databases. The inhouse database enables the users to trace back literature by means of bibliographic, chemical and physical search terms. Nested Boolean expressions, proximity and numeric operators as well as thesaurus functions may be applied. The database structure, computational details and retrieval features have been developed over many years and are the result of extensive and very valuable interactions with the community of prospective users [9-10].

As a special feature the MOGADOC database contains structural and stereo formulas, conformational descriptions, and numerical data sets for structural parameters such as internuclear distances and angles. Presently about 700 numerical data sets are available in MOGADOC. It is emphasized that no other parameters are given numerically. However, keywords, which form an hierarchical, controlled vocabulary, give hints to appropiate references for rotational and centrifugal distortion constants, hyperfine structure parameters, dipole moments, internal rotation potential, mean amplitudes of vibration, etc.

4. OUTLOOK

Now the MOGADOC database is available and can be distributed by floppy diskettes (3.5" or 5.25") [8]. It is planned to update the database once every year. Hereby the implementation of further recent and even older structural numerical data will be continued. Furthermore high-resolution infrared spectroscopic publications are being added to MOGADOC. Further information may be obtained from the author.

REFERENCES

1.   B. Starck: Molecular Constants from Microwave Spectroscopy. Landolt Börnstein New Series II, Vol. 4, Springer, Berlin 1967.

2.   J. Demaison, W. Hüttner, B. Starck, I. Buck, R. Tischer, and M. Winnewisser: Molecular Constants from Microwave, Molecular Beam, and Electron Spin Resonance Spectroscopy. Landolt Börnstein New Series II, Vol. 6, Springer, Berlin 1974.

3.   J. Demaison, A. Dubrulle, W. Hüttner, and E. Tiemann: Molecular Constants Mostly from Microwave, Molecular Beam, and Electron Resonance Spectroscopy. Landolt Börnstein New Series II, Vol 14a, Springer, Berlin 1982.

4. J. M. Brown, J. Demaison, A. Dubrulle, W. Hüttner, and E. Tiemann: Molecular Constants Mostly from Microwave, Molecular Beam, and Electron Resonance Spectroscopy. Landolt Börnstein New Series II, Vol. 14 b, Springer, Berlin 1983.

5. J. M. Brown, J. Demaison, W. Hüttner, E. Tiemann, J. Vogt, and G. Wlodarczak: Molecular Constants Mostly from Microwave, Molecular Beam, and Electron Resonance Spectroscopy. Landolt Börnstein New Series II, Vol. 19, Springer, Berlin, in press.

6. J. H. Callomon, E. Hirota, K.Kuchitsu, W. J. Lafferty, A. G. Maki, C. S. Pote, I. Buck, and B. Starck: Structure Data of Free Polyatomic Molecules. Landolt Börnstein New Series II, Vol. 7, Springer, Berlin 1976.

7. J. H. Callomon, E. Hirota, T. Iijima, K. Kuchitsu; W. J. Lafferty: Structure Data of Free Polyatomic Molecules. Landolt Börnstein New Series II, Vol. 15, Springer, Berlin 1987.

8. STN International, c/o FIZ Karlsruhe, P. O. Box 2465, D-7500 Karlsruhe, Federal Republic of Germany.

9. J. Vogt: Rev. Mex. Astron. Astrof., in press

10. J. Vogt: Acta Chim. Hung. Mod. Chem., accepted

3. MASER SURVEYS

CLASSIFICATION AND STATISTICAL PROPERTIES OF H_2O MASERS

R. Cesaroni[1], G. Comoretto[1], M. Felli[1], V. Natale[2], F. Palagi[3]
[1]Osservatorio Astrofisico di Arcetri, [2]C.N.R., C.A.I.S.M.I.
[3]C.N.R., Gruppo Nazionale di Astronomia, U.d.R. Arcetri
Largo E. Fermi 5, I-50125 Firenze, Italy

ABSTRACT

The revised version of the Arcetri H_2O maser Atlas (AA2) and the IRAS Point Source Catalog were used to associate each maser to one IRAS source. For 442 masers out of a total of 505 (87.5%) an association was established. However, only ~3% occur in regions where no FIR emission can be found.

Based on this association an unbiased classification scheme was developed that uses all IRAS colours. A colour combination was derived which best characterizes the sources already identified as HII or STAR. This combination was then applied to the whole sample resulting in 262 HII and 229 STAR.

The following properties of each class were then studied: 1) the distribution of maser and far infrared luminosities, 2) the variability of the H_2O maser emission, 3) the correlation between maser and far infrared luminosities, 4) the detectability of high velocity features in the spectrum and 5) the galactic distribution and luminosity function of the H_2O masers of HII class.

1. THE H_2O MASER ARCETRI ATLAS.

The AA2 contains an homogeneus set of observations of H_2O maser obtained using the Medicina 32-m antenna of the Istituto di Radioastronomia C.N.R., Bologna, Italy (see Comoretto et al. 1990, A&AS, 84, 179). The HPBW is 1.9 arcmin and the pointing accuracy 25 arcsec. With a system temperature of 160° K, the sensitivity is 11 Jy (3σ). All the known H_2O masers north of δ > -30° (505 sources) were observed and 203 were detected.

2. COMPARISON WITH THE IRAS POINT SOURCE CATALOG

In order to establish a correspondence between a maser and an IRAS Point Source the two list were cross correlated using a box 120 arcsec wide around the H_2O position. A one to one correspondence was found for 442 sources. The remaning 63 are divided in the following cases: 1) 12 are in regions not covered by IRAS, 2) 15 have an IRAS source within the box already associated with a closer maser, 3) 36 fall in confused regions of the IRAS survey or represent true lack of correspondence. It is estimated that H_2O masers truly without an IRAS source should not exceed 3%.

3. CLASSIFICATION OF MASER IN HII AND STAR CLASSES

IRAS colour-colour plots have often been used to discriminate between HII and STAR. We have used a multivariate statistical analysis which finds the best combination of all four IRAS fluxes to perform this separation. With a principle component analysis, based on the intrinsic sample statistics, and with a linear discriminant analysis, which uses the masers with known classification as a training sample, 229 sources were classified as STAR. 262 as HII , while for 14 no classification was possible. The distribution of sources with the variable u_i, linear combinations of the IRAS fluxes, is shown in Fig. 1.

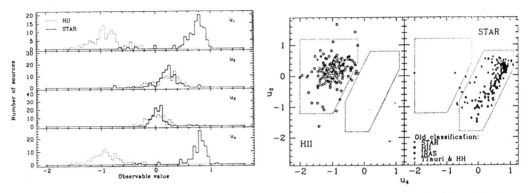

Fig. 1 Distribution of sources with u_1 - u_4 for sources with known *a priori* classification *(left)*, distribution in the u_2 - u_4 plane of sources classified as STAR and HII *(right)*.

4. DISTRIBUTION OF MASER AND FIR LUMINOSITIES

The distributions of the H_2O maser luninosities and FIR luminosities for HII and STAR, shown in Fig. 2. indicates a marked difference in the two classes of objects.

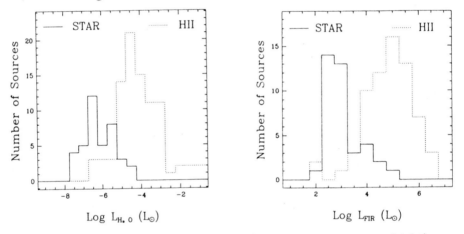

Fig. 2 Distribution of H_2O luminosities *(left)* and of FIR luminosities *(right)*

5. VARIABILITY OF THE H_2O MASER EMISSION

The comparison between peak flux densities at two epochs (S_r from the literature and S_a from AA2) offers the possibility to study of the variability in the two classes. The large spread around the full line (Fig. 3 left, upper part) and the distributions of the S_a/S_r (Fig. 3 right) indicates that large variability are a common aspect of the H_2O

maser emission. For 55 % of the sources (independent of class) the emission varies by more than a factor 2. while for 10% the variation is greater than a factor 10.

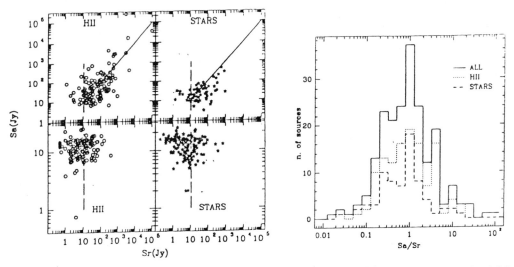

Fig.3 Comparison between S_r and S_a for detections *(upper left)* and upper limits in AA2 *(lower left)*; distribution of S_a/S_r *(right)*

6. CORRELATION BETWEEN MASER AND FIR LUMINOSITIES

Fig. 4 shows the correlation between maser and FIR luminosities for STAR and HII . Upper limits in AA2 are also shown. The correlation coefficient between the luminosities is rather low, however one should consider that FIR emission, which comes from dust over an area much larger than the maser, is at least 9 orders of magnitude greater than the maser emission. The best fitting lines are $L_{H_2O} = 10^{-10.5} L_{FIR}^{1.4}$ for STAR and $L_{H_2O} = 10^{-9.9} L_{FIR}^{1.2}$ for HII .

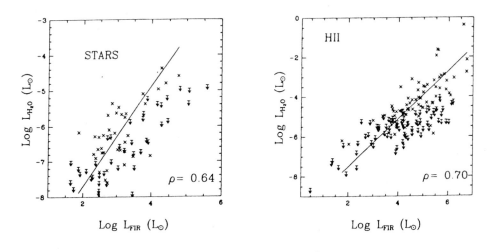

Fig. 4 Correlation between H_2O maser and FIR luminosities

7. SPECTRAL SHAPE OF H_2O MASERS OF HII CLASS

Many H_2O spectra of the HII class show only one component (i.e $\Delta v \leq 2$ km s^{-1}), while other show also High Velocity Features (HVF). To examine if this is the result of the sensitivity of the observations and of the intrinsic ratio of the HVF to the main peak we have plotted in Fig. 5 Δv as a function of $F_{peak}/3\sigma$, as well as the distributions of sources with $\Delta v \geq$ or ≤ 2 km s^{-1} as a function of $F_{peak}/3\sigma$. HVF can be detected only if greater than 0.1 F_{peak}.

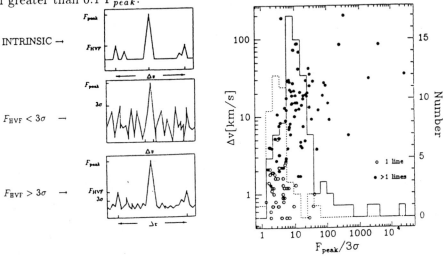

Fig. 5 Detectability of HVF as a function of $F_{peak}/3\sigma$

8. GALACTIC DISTRIBUTION AND LUMINOSITY FUNCTION OF HII

In Fig. 6 we show the galactic distribution of H_2O masers of class HII with known distance and the observed luminosity function (dashed histogram). To obtain the intrinsic luminosity function this last was corrected taking into account that sources of a given luminosity L_{H_2O} are observed only within a distance $d_{max} = \sqrt{L_{H_2O}/4\pi F_{min}}$.

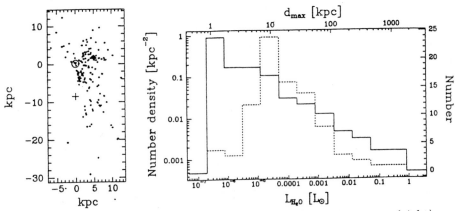

Fig. 6 Galactic distribution *(left)* and luminosity function of H_2O masers *(right)*

OBSERVATIONS OF NEW SUBMILLIMETER MASER LINES OF WATER AND METHANOL

P.A. Feldman, H.E. Matthews, T. Amano, F. Scappini
Herzberg Institute of Astrophysics, NRC of Canada
Ottawa, Canada K1A 0R6

and

R.M. Lees
Physics Department, Univ. of New Brunswick
Fredericton, NB, Canada E3B 5A3

We have used the 15-m James Clerk Maxwell Telescope (JCMT)[1] in Hawaii to search for highly excited maser lines of water towards late-type stars and star-forming regions. In February 1991 we detected quasi-maser emission from the $17_{4,13}$ - $16_{7,10}$ transition of H_2O at 354,809 MHz towards the oxygen-rich supergiant VY CMa. This line was subsequently confirmed in November 1991. It is at the $\approx$2-Jy level in flux density and arises from an upper level 5781 K above the ground state. A second quasi-maser line of H_2O was also detected in November 1991 from the circumstellar envelope of VY CMa. It is the $5_{2,3}$ - $6_{1,6}$ rotational line within the ν_2 excited vibrational state and occurs at a frequency of 336,228 MHz. This transition arises from a level 2955 K above the ground state. Both its flux density and its line shape were found to be very similar to what was observed for the higher excitation 354,809 MHz line. Both lines peak at LSR velocities of approximately +22 km s[-1], the LSR velocity characteristic of the stronger H_2O maser lines observed toward this source. In the case of the very high excitation 354,809 MHz line, there is also a narrow emission feature at an LSR velocity of approximately +60 km s[-1]; this maser component presumably arises from the most redshifted part of the circumstellar envelope of VY CMa. Our observations of these two highly excited (quasi-)maser lines of H_2O can be regarded as confirming the essential validity of the collisional excitation model for water masers (Neufeld and Melnick, 1991).

We have also attempted to detect methanol maser emission from a number of submillimeter lines of both the E- and A- symmetry species toward some of the stronger Class II 6.6-GHz and 12.2-GHz methanol maser sources detected by Menten (1991) and Batryla *et al.* (1987), respectively. In November 1991 we used the JCMT to detect four new CH_3OH maser transitions toward star-forming regions; these are the first methanol maser lines in the submillimeter band reported to date. They are as follows: the $7_{-4} - 6_{-4}$ E_2 transition at 338,504 MHz toward S231 (≈ 130 Jy line at an LSR velocity of approximately -9.2 km s^{-1}); the $7_3 - 6_3$ A^+ and A^- transitions at 338,541 and 338,543 MHz, respectively, toward S252 (≈ 130 Jy lines at LSR velocities of approximately +6.0 and +6.7 km s^{-1}, respectively); and either the $7_3 - 6_3$ E_1 transition at 338,583 MHz with an LSR velocity of approximately +26.5 km s^{-1}, or the $7_{-3} - 6_{-3}$ E_2 transition at 338,560 MHz with an LSR velocity of approximately +6.0 km s^{-1} toward S269 (flux density ≈ 210 Jy). The ambiguity of the line identification in the latter case of S269 is due to the fact that previously observed methanol masers in this source have LSR velocities in the range +13 to +20 km s^{-1}.

[1] The James Clerk Maxwell Telescope is operated by the R.O.E. on behalf of the S.E.R.C. of the United Kingdom, the Netherlands Organisation for Scientific Research, and the N.R.C. of Canada.

REFERENCES

Batryla, W., Matthews, H.E., Menten, K.M., Walmsley, C.M., 1987, Nature **326**, 49
Menten, K.M., 1991, Ap. J. Letts. **380**, L75
Neufeld, D.A., Melnick, G.J., 1991, Ap. J. **368**, 215

THE CRL 34-M RADIO TELESCOPE AT KASHIMA
– A NEW STRONG TOOL IN MASER RESEARCH –
AND THE FIRST RESULTS OF A 22 GHz H_2O MASER SURVEY

Hiroshi Takaba, Takahiro Iwata
Communications Research Laboratory
Kashima Ibaraki, Japan

Takeshi Miyaji, Noriyuki Kawaguchi, and Masaki Morimoto
Nobeyama Radio Observatory
Minamisaku Nagano, Japan

ABSTRACT

The CRL 34-m radio telescope at Kashima was completed in 1988. The telescope is equipped with low noise receivers from 1.4 GHz to 44 GHz and is able to observe most of the important astronomical masers in this frequency range. We made an automatic maser survey software and started an H_2O maser survey at November 1991 for IRAS color selected objects. Until the end of January 1992, about 930 sources were observed with the rms noise level of 0.1 Jy and about 300 sources were detected.

1. Introduction

The CRL have been conducting VLBI observations for geodesy and earth rotation study since 1986 by using the 26-m antenna at the Kashima Space Research Center (KSRC). As the 26-m antenna became old, we had constructed a new 34-m radio telescope at the KSRC. This new telescope has wide frequency range receivers to make many scientific researches. Besides VLBI observations for geodesy and earth rotation study, we started a domestic radio astronomical VLBI observation between Kashima 34-m and Nobeyama 45-m telescope to investigate SiO masers at 43GHz (Morimoto et al. 1992).

The 34-m radio telescope has a great potential in maser research also as a single dish because of its flexible system. To make a statistical study of maser sources and to find some interesting sources for VLBI observations, we had started a 22 GHz H_2O maser survey based on the IRAS point source catalog.

2. The CRL 34-m radio telescope

The mechanical structure of the CRL 34-m telescope is very similar to those of the 34-m DSS15(Goldstone, USA) and DSS45(Tidbinbilla, Australia) antennas, both recently built by TIW Systems, Inc. However, the CRL 34-m telescope at Kashima uses very accurate main reflector panels and has a very stiff main-reflector backstructure for use in mm-wave regions. The accuracy of the surface panels is better than 0.1 mm (rms), and at a 45° elevation the paraboloidal shape has been adjusted to 0.17 mm (rms). The pointing was calibrated at the 22 GHz band by observing H_2O maser sources with the rms residuals from the 13-term pointing model being 7" (Takaba 1991).

Receiving frequencies and its performances are listed in Table 1. An acousto-optical spectrometer (AOS) using an optical diode laser was developed in the Nobeyama Radio Observatory and was installed in the KSRC. The spectrometer has the frequency resolution of 37 kHz and the total band width of 40 MHz.

Besides an automatic VLBI observation software which control the telescope and the VLBI backend system according to the observation schedule, we made an automatic maser observation software. The maser observation software selects an object from the observation list in observable elevation, less observed, and shortest time required to turn the telescope. We only need the object's list and after set the integration time, observations are done automatically with a nearly maximum time efficiency.

Table 1. Receivers and efficiencies of CRL 34 m radio telescope

Band	Frequency	Receiver noise	Tsys.	HPBW	efficiency	maser
1.5 GHz	1.35 – 1.75 GHz	10K(cooled FET)	38K	24'	68%	OH
2.2 GHz	2.12 – 2.32 GHz	11K(cooled FET)	71K	16'	65%	
5.0 GHz	4.60 – 5.10 GHz	25K(cooled FET)	55K	7'.5	70%	OH
8.2 GHz	7.86 – 8.68 GHz	13K(cooled FET)	52K	4'.4	68%	
10 GHz	10.2 – 10.7 GHz	43K(cooled FET)	70K	3'.6	65%	
12 GHz	12.0 – 12.3 GHz	300K(room temp HEMT)	350K	3'.0	63%	CH_3OH
15 GHz	14.4 – 15.5 GHz	45K(cooled FET)	100K	2'.4	60%	
22 GHz	21.9 – 24.0 GHz	55K(cooled HEMT)	145–200K	1'.6	57%	H_2O, NH_3
43 GHz	42.0 – 43.5 GHz	500K(cooled HEMT)	800K	48"	40%	SiO

3. A 22GHz H$_2$O maser survey

The sources which exhibit masers are well studied with the color-color diagram of the IRAS 12, 25, and 60 micron intensities (e.g. Lewis and Engels 1988, Deguchi et al. 1989, and Haikala 1990).

We had selected maser candidates from the IRAS point source catalog as shown in Figure 1. The sky distribution of the selected sources is plotted in Figure 2, the sources of the declination lower than $-40°$ are omitted.

We had started a 22GHz H$_2$O maser survey at November 1991. At first, we observed most of the all known stellar sources, then we started an IRAS color selected sources survey in order of the infrared intensity. The integration time is 30 minutes in position switching mode and the rms noise levels is 0.1 Jy with the 1 km/s velocity resolution, typically.

Figure 3 shows the sky distribution of the 22GHz H$_2$O maser detected sources. Because the sources have strongest infrared intensity, they are the nearest and the detectability is rather high (about 30%).

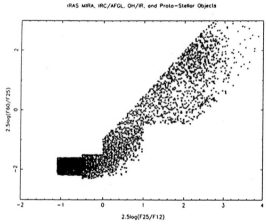

Figure 1. IRAS color-color diagram of the 22GHz H$_2$O maser candidates

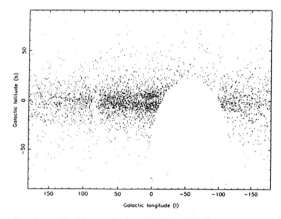

Figure 2. The sky distribution of the 22GHz H$_2$O maser candidates

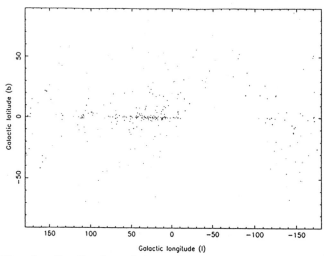

Figure 3. The sky distribution of the 22GHz H_2O maser detected sources

4. Future plan

To provide an unbiased sample data, we will extend the survey up to 2000 sources in 1992. Also surveys of CH_3OH masers at 6.7 GHz and 12 GHz band, SiO (J=1-0) masers at 43 GHz band are planning to compare masers' characteristics in stellar and interstellar maser sources, respectively.

References

Deguchi, S., Nakada, Y., and Forster, J.R., 1989, M.N.R.A.S. 239, 825

Haikala, L.K., 1990, Astr. Ap. Suppl. 85, 875

Takaba, H., 1991 J. of the Comm. Res. Lab. Vol.38, No.3

Lewis, B.M., and Engels, D., 1988, Nature, 332, 49

Morimoto et al., 1992 submitted to P.A.S.J.

4. EXTRAGALACTIC MASERS

MOLECULAR MEGAMASERS AFTER TEN YEARS

Willem A. Baan
Arecibo Observatory
Arecibo, PR 00613

ABSTRACT

Megamaser activity for the molecules OH, CH, H_2CO, and H_2O is reviewed in terms of a consistent picture of exponential amplification of radio continuum by foreground molecular gas. The different characteristics of the various types of megamasers can all be related to the density and pumping conditions of the molecular gas. These conditions strongly depend on the evolutionary stage of the nuclear activity.

1. CLASSES OF MEGAMASERS

Powerful extragalactic masers are known for four molecules: OH, H_2O, H_2CO, and CH. Unusual physical circumstances in nuclear regions of galaxies at various stages of activity conspire to produce a beamed emission pattern of spectral line emission. A working model of the extragalactic (mega-) maser action is the *exponential amplification of background radio continuum by inverted foreground molecular material* (Baan 1985,1989; Haschick and Baan 1985; Henkel and Wilson 1990). The implications for this model are rather simple as compared with those for the alternative picture of individual saturated maser sources. Maser action for OH, CH, and H_2O is quite common in the Galaxy but formaldehyde masers are uncommon. There are more extragalactic formaldehyde masers than Galactic ones. The physical conditions for the extragalactic amplifying regions are thus quite different from molecular maser regions within the Galaxy.

This paper gives a brief overview of what is known about the different molecular maser sources and discusses their common aspects and their differences. The physics of the OH and H_2O megamasers are best understood because of the relatively large number of sources and because their strength allows for interferometric studies. Much of the discussion will thus center on megamasers for these two molecules. The representative numbers for the four groups of sources have been given in Table 1. The apparent gains are largest for H_2O and OH with H_2CO and CH following far behind. The pumping process for H_2O and OH must be very effective. A stringent requirement for megamasers is the availability of sufficient column densities of molecular gas with a density range susceptible for pumping. The beaming angles for the H_2O is likely to be very small, while for the other molecules amplification occurs throughout a large fraction of a molecular disk. The weakness and apparent lower occurence rate for formaldehyde and CH sources points at a less efficient pump and the non-availability of molecular gas at the right density and radial distance.

Reviews of the properties of extragalactic OH, H_2O, H_2CO and CH maser sources have been published in Baan (1985, 1989, 1991), Henkel and Wilson (1990) and Henkel, Baan, and Mauersberger (1991).

Table 1. Characteristics of Extragalactic Megamasers

	Number Sources	Apparent Gain	Pumping Source	References
OH	53+	0.05 - 3.4	IR	1-5
H_2O	9	7 or more	coll./shock	1,5,6
H_2CO	9	0.006-0.02	radio	5,7
CH	9	0.001	coll./IR + radio	5,8

References: 1. Baan 1985, 2. Baan 1989, 3. Henkel and Wilson 1990, 4. Baan 1991, 5. Henkel, Baan, and Mauersberger 1991, 6. Haschick *et al.* 1990, 7. Baan, Güsten and Haschick 1986, 8. Bottinelli *et al.* 1991a.

2. PROPERTIES OF OH MEGAMASERS

2.1 The L_{OH} - L_{IR} relation

The understanding of the physical characteristics of OH megamasers is primarily based on the tight correlation of OH and FIR properties of the galaxies (Baan 1989; Henkel and Wilson 1991). A correlation between OH line luminosity and FIR luminosity was noticed quickly after the discovery of the first few megamasers (Baan 1985) but later a quadratic L_{OH} - L_{FIR} relation was recognized (Martin *et al.* 1989; Baan 1989). A quadratic relation follows simply from the radiation transfer solution for low-gain amplification, where both the amplifying optical depth and the amplified radio continuum flux vary with the FIR flux density. Since the main infrared pumping lines for OH are at 35 and 53 μm, the IRAS 60 μm luminosity is representative of the pumping mechanism. A comparison of an up-to-date L_{OH} - $L_{60\mu m}$ diagram (Figure 1; Baan *et al.* 1992) with earlier published diagrams (Baan 1989; Martin *et al.* 1989) shows that the correlation between the two luminosities becomes much tighter with the use of the 60 μm luminosity. An additional curve in the diagram represents the non-linear (exponential) L_{OH} - L_{IR} relation expected without the low-gain assumption; there is (yet) no evidence for a non-linear L_{OH} -L_{IR} behavior.

2.2 OH Correlations

A number of simple parameter relations follows from the model of exponential amplification for OH megamasers. Two optical depth variables may be devised for the OH lines (Henkel and Wilson 1990): *an apparent optical depth* $\tau(app)$ based on the ratio of line and continuum intensities and *an unsaturated optical depth* $\tau(uns)$ obtained from the line ratio of the 1667 and 1665 MHz OH lines. These two optical depths also may define *the covering factor* F_C for the fraction of the amplified continuum flux: $F_C = (e^{\tau(app)}-1) / (e^{\tau(uns)}-1)$. The two optical depths are strongly correlated (Henkel and Wilson 1990), which is to be expected without any specific dependence of the covering factor on the optical depth. The covering factor of the present sample of sources with known radio fluxes and clear line ratios varies from 0.3 to 3 x 10^4. No clear correlation has been found for this covering factor with any of the other variables. An unsaturated amplification model involving a clumpy molecular foreground structure would not predict a correlation of F_C with anything.

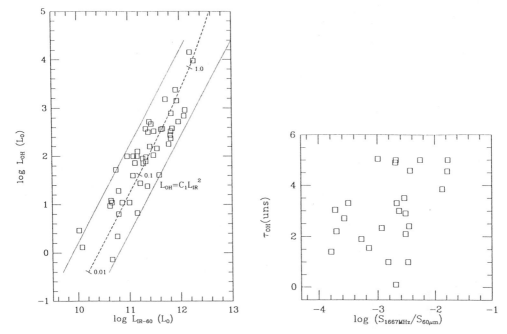

Figure 1. The relation of OH and 60 micron luminosities for all known OH megamasers. The dotted lines represent the quadratic relation for low-gain amplification between L_{OH} and L_{FIR}. The dashed line represents an exponential relation expected for amplification without the low-gain simplification. Values for the optical depth are indicated with the exponential curve.

Figure 2. The unsaturated optical depth obtained from the OH line ratio and the flux density ratio in the 1667 MHz OH line and at 60 micron. The flux ratio is another measure of the FIR-to-OH photon conversion efficiency.

The efficiency of conversion of 60 μm photons into 1667 MHz line photons may be expressed as $S_{1667}/S_{60\mu m}$. This photon conversion efficiency should be directly related to the OH hyperfine ratio (Baan 1989). Higher values of the conversion efficiency are indeed found in sources with higher hyperfine ratios (for a recent assessment see Henkel and Wilson 1990). An increase of τ(uns) with $S_{1667}/S_{60\mu m}$ can be observed, but a tight correlation is not present (Figure 2).

The observational data also suggest that the halfwidth ΔV of the maser lines varies with the apparent optical depth (Martin *et al.* 1989). The sample of megamasers indeed shows a weak correlation of τ(app) with ΔV (Figure 3); the narrowest lines occur in sources with high apparent gain. Assuming a comparable (doppler) linewidth for all sources, one would expect that exponential amplification results in line narrowing for the sample as a whole. The data is thus consistent with the unsaturated (exponential) amplification picture. he optical depth for the amplifying molecular gas is expected to depend on the availability of the pumping photons. Bottinelli *et al.* (1991b) suggest the following dependence for the gain of the amplification process: $(e^{\tau(app)} - 1) \propto (L_{FIR}/\Delta V^3)^{0.7}$. However, to first order the amplifying optical depth simply depends on the number of photons per unit frequency (or velocity) along the line-of-sight in the foreground medium or $F_{FIR}/\Delta V$. Figure 4 shows that $\tau(app)$ varies strongly with $L_{60\mu m}/\Delta V$. Large optical depths occur in sources with large IR fluxes and small velocity gradients in the amplifying medium. This correlation supports the initial assumption of a dependence of τ_{OH} on the FIR flux leading to the quadratic L_{OH} - L_{FIR} relation.

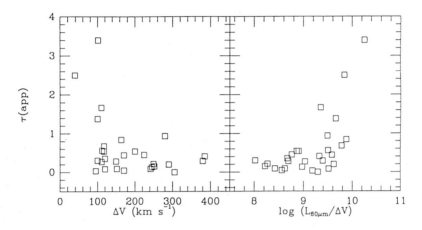

Figure 3. The relation between the apparent optical depth and the half power line width of the OH emission line. The line width increases with decreasing apparent optical depth.

Figure 4. The relation between the apparent optical depth and the pumping photon density expressed as $L_{60\mu m}/\Delta V$. The optical depth increases with higher values of the pumping photo density.

2.3 OH Pumping Models

Various infrared pumping models for OH masers and megamasers have appeared in the literature. The densities required for inversion in the ground state of OH vary among the studies: from 10^{2-4} cm^{-3} to $10^{6.5-8}$ cm^{-3} (Burdyuzha and Vikulov 1990; Henkel, Güsten, and Baan 1987; Elitzur, Baan, and Lockett 1991; Deguchi 1992). A detailed discussion of the various results is not apppropriate at this time, but it is clear that with the use of updated collision rates OH inversions are obtained at relatively high densities. Densities of $\geq 10^6$ cm^{-3} for the amplifying clouds are consistent with the growing body of molecular characteristics for luminous FIR/megamaser galaxies.

2.4 Optical Characteristics of OH Megamasers

The OH megamaser sample is a subgroup of the luminous FIR galaxies. Spectroscopic results show that Seyfert nuclei are very dominant among the OH megamasers as compared with the whole IRAS galaxy sample (Baan and Salzer 1992; Meadows *et al.* 1990). The percentages of Seyfert and starburst nuclei are almost reversed for the two samples (see Table 2). The predominance of Seyfert nuclei among the OH megamasers invokes several correlations.

Table 2. Optical Classification of OH Megamasers

Spectral Designation	OH Megamasers[a]	General Sample[c]
	35 sources[b]	500 IRAS sources
Starburst Nucleus	37 %	61 %
Seyfert 2 or Liner	57 %	34 %
Seyfert 1	5.7 %	0.6 %

Comments: a) Baan and Salzer (1992); b) 35 out of the sample of 42 could be classified; c) Meadows *et al.* (1990).

a) The radio nuclei of Seyferts tend to be more compact that those of SBN's (Ulvestad and Wilson 1984: Norris *et al.* 1990) and have a higher brightness temperature. The nuclear molecular structure in the Seyferts could thus be more compact than in starburst nuclei due to a possibly deeper potential well. The conditions in Seyferts may thus be more conducive for amplification than in SBN's.

b) The low-gain amplification model requires the presence of relatively large molecular column densities in front of the radio nucleus. Although interacting systems could have molecular gas in many directions, the preferential orientation of all OH megamasers would thus be edge-on. The predominance of Sy2's among the Seyferts in the OH megamaser sample confirms an earlier Seyfert model with Sy2's being more edge-on than Sy1's (de Zotti and Gaskell 1985; Barthel 1989).

c) The optical results also indicate that some (or all) of the megamasers classified as SBN might contain an almost completely hidden AGN and that more Seyferts possess a completely hidden broad-line region (Baan and Salzer 1992).

d) The optical data are consistent with the suggestion that OH megamasers and the very luminous FIR galaxies represent the very early stage in the evolution of an AGN (Baan 1988; Sanders *et al.* 1988). The luminous FIR stage would occur at an evolutionary time when a pre-existing AGN is re-activated or a new one is being formed, but before the intense UV radiation can dissociate the molecular gas surrounding the nucleus.

2.5 Molecular Characteristics of OH Megamasers

The molecular properties of FIR galaxies have been found to be exceptional (for reviews see Henkel, Baan and Mauersberger 1991; Young and Scoville 1991). The OH megamaser activity occurs in the most extreme of the prominent FIR/molecular galaxies.

a) The CO(1-0) properties of 32 megamasers show that the star formation efficiency (SFE) ratio $F_{FIR}/M(H_2)$ is a factor of 2.5 larger than for the Bright Galaxy Sample in the same luminosity range (Baan, Freund, and Haschick 1992; Sanders *et al.* 1991). These large values for the molecular content of the nuclei and their SFE could indicate the presence of large amounts of dense molecular gas in the nuclear regions. The initial mass function for the star formation process in these extreme nuclear regions could be modified significantly in order to produce higher mass stars.

b) Interferometric data for a limited number of prominent molecular sources indicate extremely high surface densities in the nuclear regions (Scoville *et al.* 1991). Among these sample sources the four OH megamaser sources are found to have the highest surface densities ($\Sigma_{H_2} > 10^4$ M$_\odot$ pc^{-2}).

c) The line luminosities of a low- and high-density tracers indicate the presence of abundant high-density gas. For a limited sample of prominent FIR sources the L_{HCN}/L_{CO} ratio, which signifies a mass ratio for high- and low-density gas, is found to be highest for those galaxies with OH megamaser activity (Solomon, Downes and Radford 1992).

The nuclear molecular gas in luminous FIR galaxies and particularly in megamaser galaxies show strong evidence for very unusual conditions. Not only are large quantities of molecular gas present in their nuclear regions, but also the excitation conditions of the gas as a result of the presence of the nuclear activity are also extreme. On the basis of pumping considerations (above discussion and Baan 1991), the amplifying OH molecules are thought to trace the higher density component of the gas. The occurence of OH megamaser activity specifically among the most extreme molecular sources confirms the presence of some (surface-) density threshold for the radiative inversion of the OH molecules.

3. H$_2$CO MEGAMASERS

After the initial detection of formaldehyde megamaser emission in IC 4553 (Arp 220) (Baan, Güsten and Haschick 1986), several recent detections have been made with the Mini-Gregorian at Arecibo (Baan, Haschick, and Uglesich 1992). The formaldehyde emission lines are not as spectacular as the OH megamasers; the line strength of the detected sources is typically only a few millliJansky. The requirements for the occurence of formaldehyde megamasers appears to be less stringent than those for OH megamasers. Formaldehyde emission has among others been seen in several OH absorbers and a partial OH emitter/absorber. The inversion of the lowest K-doublet of formaldehyde may be achieved by pumping the J=2 K-doublet with a radio continuum (Baan *et al.* 1986). Amplifying formaldehyde clouds are likely to trace a density of 10^4 cm^{-3}, while the OH amplifying gas traces densities close to 10^6 cm^3.

A quadratic L_{H_2CO} - L_{6cm} relation may also be expected for formaldehyde megamasers. If the amplifying optical depth depends on the pumping radio continuum, the formaldehyde line luminosity would vary as the square of the radio luminosity. Since the 60 μm FIR flux and the 6-cm flux of FIR galaxies are also correlated, this relation also translates into a quadratic L_{H_2CO} -L_{FIR} relation. Such a relation can already be seen for the limited sample of nine formaldehyde (mega-) masers.

4. CH MEGAMASERS

CH emission has been detected in a total of nine galaxies (Whiteoak *et al.* 1980; Bottinelli *et al.* 1991b). CH is a known low-density tracer and CH emission could be expected in galaxies with OH and H_2CO emission. Instead an anti-correlation has been found between the occurrence of CH and OH amplification. All CH emission has been found in strong OH absorbers; two of the sources are also H_2O emitters and two others are H_2CO emitters.

5. H_2O MEGAMASERS

The powerful H_2O megamasers observed in a relatively small number of nearby galaxies have allowed some detailed studies of the maser structure. The masers occur in the nuclear regions of the galaxies although they need not coincide with the central component of the radio source. At low resolution (VLA-A) the maser and the continuum are superposed but with VLBI observations much of the extended continuum structure has been resolved. VLBI studies of the NGC 3079 and M33 / IC 133 show multiple maser features making up the maser lines (Haschick *et al.* 1991; Greenhill *et al.* 1991, 1992). These multiple features resemble Galactic HII-regions and have similar sizes. The presentation by Lincoln Greenhill will present more details about the VLBI results (these proceedings).

The excitation requirements for H_2O megamasers has not been explored specifically. Evidence exists from a number of megamaser sources that collisional pumping in shocked molecular regions provide the necessary inversion of the water molecules (Elitzur, Hollenbach, and McKee 1989). Several of the maser features are offset from the centroid velocity of the galaxy but there is no clear correlation between the direction of the offset and the occurence of the masers. Monitoring studies show time variability of the velocity, the linewidth, and the strength of individual extragalactic maser features (Haschick and Baan 1990; these proceedings; Baan and Haschick 1992). An extreme case of such variation has been presented for the strongest feature in NGC 4258 (Haschick and Baan 1990). The variability of this feature has been explained in terms of a foreground variable stellar source periodically pumping the surrounding molecular material, which in turn amplifies the background radio continuum. The variation of line width and the velocity characteristics are consistent with collisional shock pumping of foreground material and unsaturated maser amplification.

Various interpretations besides background amplification have been put forward for the extragalactic water masers (see Greenhill, these proceedings). Although superposition of line emission and continuum has not been shown in detail for individual maser components in extragalactic sources, the notion of amplification within narrow cones by foreground pockets of gas is very viable (Haschick *et al.* 1990). A weaker background radio continuum only requires a slightly higher amplifying optical depth for the feature to be observable.

An optical depth of 8 would amplify a 0.1 mJy continuum to 0.3 Jy, which is a typical strength for an individual feature in NGC 3079 (Greenhill *et al.* 1992). Eight gain lengths do not anymore represent low-gain amplification as found for the other megamasers, but it is still much less than the 25 gain lengths required for the standard saturated maser picture. Even clusters of maser spots would be expected in extragalactic sources because only the densest clumps could be inverted. Extragalactic H_2O megamasers are not easily explained in terms of Galactic maser phenomenology; equivalent masers of 10^6 Jy are not observed in the Galaxy. The extraordinary strength of extragalactic masers must indeed be connected with the radio continuum of the galaxy.

6. MEGAMASER COSMOLOGY

The prominence of OH and CO in luminous IRAS galaxies provides important tools for study of their nuclear regions, even at great distances. If the OH and CO line luminosities continue to rise with the second and first power of L_{FIR}, the observable universe will be very large for both molecular lines. How many such sources are to be found depends strongly on the space density of IRAS sources at large luminosities. When the space density at high FIR luminosities behaves as $\phi_{IRAS} \propto L_{FIR}^2$, the quadratic L_{OH} -L_{FIR} relation would make OH observable to larger redshifts than CO, which shows a linear dependence (Henkel et al. 1991). An exceptional CO source has already been detected at redshift 2.286 (Brown and vandenBout 1991). OH sources with $L_{FIR} = 10^{13}$ $L_\odot$ would also be visible to redshifts of 2-3 (Burdyuzha and Komberg 1990; Baan 1991; Henkel et al. 1991). The OH gigamaser IR 14070+0525 is presently the maser source with the highest redshift of $\zeta = 0.265$ (Baan et al. 1992).

The redshift distribution of OH sources still shows large gaps (Baan 1991) indicating that large numbers of OH sources are still to be found. An obvious handicap for searches at high redshift has always been the absence of redshifts for distant IRAS galaxies. The existing optical redshift surveys did not favor highly obscured galaxies with most of their luminosity in the FIR. Recent redshift surveys of IRAS galaxies provide new selections of luminous high redshift IRAS galaxies to be searched for molecules. The second gigamaser IR 14070+0525 has been found among these new samples and has a redshift of 0.265 (79600 km s^{-1}) and an isotropic OH line luminosity of 1.4 x 10^4 $L_\odot$ (Baan et al. 1992). OH and CO molecular studies are very complimentary because the two molecules trace quite different physical aspects of the same luminous IRAS galaxies.

7. MEGAMASER FAMILY TIES

The different observational characteristics of the various molecular megamasers result from satisfying the different physical requirements for pumping and amplification. A number of partially correlated factors dictate the occurence of such megamaser activity in galaxies.

a) There must be *a radio continuum* to be amplified. The integrated radio flux of the source is correlated with the FIR fluxes but its compactness is related to the nature of the nuclear activity. The tendency of OH megamasers to associate with Seyfert nuclei rather than starburst nuclei could be related not only to the compactness of the radio source but also to the depth of the potential well experienced by the molecular gas.

b) *A powerful pump* must be available to provide sufficient inversion in the molecular gas. The radio FIR radiation fields pumping the OH and H_2CO are centrally located in the galaxy and cause an inversion in a significant part of the central molecular structure. The collisional pump required for water vapor may require strong shocks passing through the molecular disk but also in foreground starforming regions. H_2O megamaser regions could have local pump sources and thus be uncorrelated with nuclear activity, although the nuclear radio source is needed for amplification.

c) Sufficient *molecular gas of the right density* must be present along the line of sight to the continuum source. This criterium could be correlated with the nature of the central source and its nuclear potential well. Mergers and interactions will facilitate the buildup of the large molecular column densities in the nuclear region through accretion and direct deposits.

d) The occurence of megamaser activity during *a particular stage in the galactic evolution* is critically dictated by a *combination of density and pumping requirements*. Radial density gradients and varying central radiation fields strongly affect the population inversions. The nuclear activity cannot be very advanced such that the nuclear molecular gas has been destroyed by the ionizing radiation fields. The OH and H_2CO megamasers are confined to luminous FIR/radio galaxies during early stages of nuclear evolution. On the other hand, the CH and H_2O galaxies are in galaxies with more evolved nuclear activity and may exhibit population inversions and amplification in non-nuclear regions.

The occurence of formaldehyde and hydroxyl megamasers appears correlated. H_2CO emission occurs in OH emitters but also in some OH absorbers, where the FIR radiation field does not invert the foreground gas. The CH and H_2O megamaser activity appears anti-correlated with the OH and formaldehyde megamaser activity; both molecules are mostly found in galaxies with strong OH absorption. The water masers represent localized pockets of shocked high-density gas, while the CH emission represents large scale lower-density gas at possibly large radii. The density and pumping criteria for the different masing molecules provide new diagnostic probes for the molecular gas in the various stages of nuclear activity.

The line strength of OH and CO in luminous IRAS galaxies holds great potential for detailed studies of the extreme circumstances in nuclear regions. OH and CO emissions provide complementary data as they sample different density ranges and different excitation conditions. High redshift OH and CO detections confirm the presence of previously unrecognized populations of sources and provide precious information about nuclear and galaxy evolution in the early universe.

Arecibo Observatory is part of the National Astronomy and Ionosphere Center and is operated by Cornell University under contract with the National Science Foundation.

REFERENCES

Baan, W.A. 1985, *Nature*, 315, 26.
Baan, W.A. 1988, *Ap.J.*, 330, 743.
Baan, W.A. 1989, *Ap.J.*, 338, 804.
Baan, W.A. 1991. in *Skylines, the Third Haystack Conf., P.A.S.P. Conf. Proc.*, eds. A.D. Haschick, P.T. Ho, p. 45.

Baan, W.A., Freund, R., Haschick, A.D. 1992, in preparation.

Baan, W.A., Güsten, R., Haschick, A.D. 1986, *Ap.J.*, 305, 830.

Baan, W.A., Haschick, A.D. 1992, *Ap.J.*, submitted.

Baan, W.A., Haschick, A.D., Uglesich, R. 1992, in preparation.

Baan, W.A., Salzer, J. 1992, in preparation.

Baan, W.A., *et al.* 1992, *Ap.J. (Letters)*, in the press.

Barthel, P.D. 1989, *Ap.J.*, 336, 606.

Bottinelli, L. *et al.* 1991a, in *Dynamics of Galaxies and their Cloud Distributions*, IAU Symp. 146, eds. F. Combes and F. Casoli, (Kluwer: Dordrecht), p. 201.

Bottinelli, L. *et al.* 1991a, in *Dynamics of Galaxies and their Cloud Distributions*, IAU Symp. 146, eds. F. Combes and F. Casoli, (Kluwer: Dordrecht), p. 442.

Brown, R.L., vandenBout, P.A. 1991, *Astron. J.*, 102, 1956.

Burdyuzha, V.V., Komberg, B.V. 1990, *Astron. Ap.*, 234, 40.

Burdyuzha, V.V., Vikulov, K.A. 1990, *M.N.R.A.S.*, 244, 86.

de Zotti, G., Gaskell, C.M. 1985 *Astron. Ap.*, 147, 1

Deguchi, S. 1992, these proceedings.

Elitzur, M., Baan, W.A., Lockett, P. 1991, unpublished.

Elitzur, M., Hollenbach, D., McKee, C. 1989, *Ap.J.*, 367, 333.

Greenhill, L., *et al.* 1990, *Ap.J.*, 364, 513.

Greenhill, L., *et al.* 1992, *Ap.J.*, submitted.

Haschick, A.D., Baan, W.A. 1990, *Ap.J. (Letters)*, 355, L23.

Haschick, A.D., *et al.* 1990, *Ap.J.*, 356, 149.

Henkel, C., Baan, W.A., Mauersberger, R. 1991, *Astron. Ap. Rev.*, 3, 47.

Henkel, C., Wilson, T. 1991, *Astron. Ap.*, 229, 431.

Martin, J.-M. *et al.* 1989 *C.R. Acad. Sci. Paris*, 308(II), 287.

Meadows, V.I., *et al.* 1990, *Proc. ASA*, 8 (3), 246.

Norris, R.P., *et al.* 1990, *Ap.J.*, 359, 291.

Sanders, D.B., *et al.* 1988, *Ap.J.*, 325, 74.

Sanders, D.B., Scoville, N.Z., Soifer, B.T. 1991, *Ap.J.*, 370, 158.

Scoville, N.Z., *et al.* 1991, *Ap.J. (Letters)*, 366, L5.

Solomon, P., Downes, D., Radford, D. 1992, *Ap.J. (Letters)*, 387, L55

Whiteoak, J., Gardner, F.F., Höglund, B. 1980, *M.N.R.A.S.*, 190, 17p.

Young, J.S., Scoville, N.Z. 1991, *Ann. Rev. Astron. Astroph.*, 29, 581.

VLBI OBSERVATIONS OF THE WATER VAPOR MEGAMASER IN IC10

A.L. Argon, M.J. Reid, J.M. Moran, and K.M. Menten
Harvard-Smithsonian Center for Astrophysics

C. Henkel
Max-Planck-Institut für Radioastronomie

L.J. Greenhill
U. of California, Berkeley

C. Gwinn
U. of California, Santa Barbara

H. Hirabayashi and M. Inoue
Nobeyama Radio Observatory

ABSTRACT

We present VLBI synthesis observations of the water vapor megamaser in the south-eastern region of IC10. We compare our spectrum with earlier single dish spectra and conclude that the maser is highly variable.

1. INTRODUCTION

IC10 is an IBm-type dwarf galaxy at a distance of 1−3 Mpc. Optical and radio continuum maps show two centers of activity. In 1985 a very luminous 22.2 GHz water vapor maser was discovered in the more prominent south-eastern center, coincident *with* or close *to* the continuum peak (see Henkel *et al.*, 1986 and references therein for further details). In this paper we present VLBI observations of the south-eastern water vapor megamaser. We compare our spectrum with Henkel *et al.*'s spectra and conclude that the maser is highly variable. We offer several explanations for its variability, small size, and high luminosity.

2. OBSERVATIONS AND RESULTS

The Mark III VLBI observations of the water vapor megamaser in IC10 were made on 12 November 1990 from 04:07 UT to 05:51 UT and from 20:25 UT to 22:09 UT. The quasars 0224+671 and 3C345 were also observed to provide delay calibration information. Six stations participated in this experiment: the Max-Planck-Institut für Radioastronomie (MPIfR) at Effelsberg, Germany (100m); Haystack Observatory in Westford, MA (37 m); the National Radio Astronomy Observatory (NRAO) in Green Bank, WV (43 m); the Very Large Array (VLA) in Socorro, NM (27 25−m antennas in the C configuration); the Owens Valley Radio Observatory (OVRO) in Big Pine, CA (40 m); and the Nobeyama Radio Observatory (NRO) in Nobeyama, Japan (45 m).

In this paper we present observations from a single 2 MHz wide band. The band was centered at 22259.99 MHz, which corresponds to an LSR velocity (velocity with respect to the Local Standard of Rest, V_{LSR}) of -332.3 km s^{-1}. Data were correlated with the Mark III processor at Haystack Observatory and reduced with the spectral–line VLBI software package at the Center for Astrophysics in Cambridge, MA. Post–correlation processing included a delay calibration, which corrected for inaccurate *a priori* station clocks; an amplitude calibration, whereby a single calibrated (in units of flux density) MPIfR scan, a "template", was used to calibrate the scans at all other times and stations; and a phase referencing, which corrected for atmospheric variations by subtracting the phase of one spectral channel, called the reference feature, from the phases of all other spectral channels (see Greenhill, 1990 and references therein for further details).

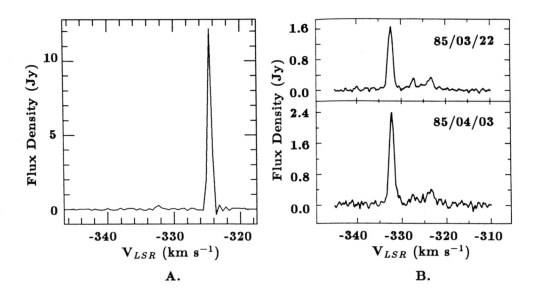

Fig. 1.– A. Synthesized VLBI spectrum of water vapor megamaser in south-eastern region of IC10 – observations made in November 1990. B. Single dish (MPIfR) spectra of same water vapor megamaser – observations made in March and April, 1985. See Tables 1A and 1B for detailed information about velocity components. Figure 1B and Table 1B taken from Henkel *et al.*, 1986.

After this, data were variance weighted to maximize sensitivity and mapped and CLEANed with the NRAO AIPS task MX. The FWHM of the interferometer beam was 0.28x0.20 milliarcseconds (mas), the search area was 46x46 mas, and the image size was 512x512 pixels. We found two features with flux densities above a detection limit of 50 mJy (5σ). Both features persisted over at least two adjacent channels at the 10σ level or above. Table 1A shows the peak flux densities, position offsets, and number of channels over which the feature persisted. Each channel has a width of 35.7 kHz or 0.48 km s^{-1}. An earlier VLA "snapshot" observation gives the position of the reference feature to be $\alpha = 0^h20^m27.00^s$, $\delta = 59°17'29.1''(\pm0.2'', J2000)$. A synthesized VLBI spectrum, produced by summing up the flux in the innermost 64x64 pixels of the image, is shown in Figure 1A.

Figure 1B and Table 1B were taken from Henkel *et al.*'s 1986 IC10 discovery paper. Their single dish (MPIfR) observations were made in March and April of 1985 and show three velocity components. The two weak components remain constant in flux density between the March and April observations, while the flux density of the strongest component increases by 40% within the same time period. The authors claim that the difference in flux density is considerably larger than their amplitude calibration uncertainty. They, therefore, conclude that IC10 is variable.

Tables 1A and 1B summarize the data on IC10. Our main velocity component, with a peak flux density of 12.14 Jy and a V_{LSR} of -324.6 ± 0.1 km s^{-1}, does not seem to correspond to any of their velocity components. Our secondary velocity component with a peak flux density of 0.26 Jy and a V_{LSR} of -332.3 ± 0.1 km s^{-1} probably corresponds to their primary feature. The errors in our amplitude calibration are due mainly to uncertainties in calibrating the MPIfR "template" and are estimated to be no more than 25%. Clearly, IC10 is highly variable.

The correlated flux densities of our two velocity components are about the same for all baselines, indicating that the maser "spots" are unresolved. Assuming a distance of 2 Mpc, we estimate an upper limit for the spot size of $\sim 10^{15}$ cm and a spot separation of $\sim 10^{16}$ cm. If we have not missed any far-flung emission, this maser seems to be associated with a single object. The brightness temperature of the maser is $\sim 10^{12}$ K and the integrated luminosity of the main velocity component (assuming isotropic emission of radiation) is $\sim 0.7L_\odot$.

The IC10 water vapor maser is comparable in luminosity to the brightest water vapor maser in our Galaxy, i.e., the one in W49N. Such luminous masers are difficult to explain by conventional pumping schemes, but models that invoke special conditions (such as collisional pumping in a region where free electrons and neutral particles have different temperatures) do work (see Reid and Moran, 1988 and references therein).

Another possibility is that the maser is unsaturated and highly beamed. The variability could be due to the maser being unsaturated. The output of an unsaturated maser is an exponential function of the optical depth of the masing region, so small changes in the latter could lead to dramatic changes in the former. Saturated masers are far less sensitive to changes in optical depth. Since the brightness temperature of the IC10 maser is $\sim 10^{12}$ K, it would also have to be highly beamed. Haschick *et al.* (1990) have shown that if a water vapor masing region with an inverted population lies in front of a nuclear continuum source and amplifies its radiation, the maser emission could indeed be highly beamed. The degree of beaming depends on the size of the continuum source as seen from the maser. In their simplest model, the radiation will be highly beamed for all reasonable distances of the continuum source from the maser. The true luminosity of such a maser could then be many orders of magnitude less than the isotropic luminosity. Until simultaneous high resolution observations of the IC10 maser and nuclear continuum source are obtained, we cannot comment on the likelihood of this model.

TABLE 1

A. VLBI SYNTHESIS RESULTS

Peak V_{LSR} (km s^{-1})	Peak Flux Density (Jy)	Feature Width (channels)	$\Delta\alpha$ (mas)	$\Delta\delta$ (mas)
-324.6 ± 0.1	$12.14 \pm 25\%$	3	0	0
-332.3 ± 0.1	$0.26 \pm 25\%$	2	0.178	-0.717

B. SINGLE DISH RESULTS
(from Henkel *et al.*, 1986)

Peak V_{LSR} (km s^{-1})	Peak Flux Density (Jy)	$\Delta V_{1/2}$ (km s^{-1})	Observation Date
-332.33 ± 0.01	1.62 ± 0.03	1.26 ± 0.03	85/03/22
-327.27 ± 0.07	0.27 ± 0.03	1.21 ± 0.16	
-323.64 ± 0.11	0.25 ± 0.02	2.63 ± 0.26	
-332.22 ± 0.01	2.29 ± 0.06	1.22 ± 0.04	85/04/03
-327.40 ± 0.20	0.21 ± 0.05	1.83 ± 0.50	
-323.66 ± 0.16	0.31 ± 0.04	2.50 ± 0.40	

NOTE—In part A the relative position offset of the secondary feature is accurate to 0.010 mas. The position and position error of the reference feature is given in the text.

REFERENCES

Greenhill, L.J.: 1990, Ph.D. Dissertation, Harvard University.
Haschick, A.D., Baan, W.A., Schneps, M.H., Reid, M.J., Moran, J.M., and Güsten, R.: 1990, *Astrophys. J.* **356**, 149.
Henkel, C., Wouterloot, J.G.A., and Bally, J.: 1986, *Astron. Astrophys.* **155**, 193.
Reid, M.J., and Moran, J.M.: 1988, in *Galactic and Extragalactic Radio Astronomy*, ed. G.L. Verschuur and K.I. Kellermann (2d ed.; New York: Springer), Chap. 6.

EXTRAGALACTIC H₂O MASERS

Lincoln J. Greenhill
Deptartment of Astronomy
University of California
Berkeley, CA 94720

ABSTRACT

Twenty-five extragalactic H_2O masers are known, classified as *normal* masers and *megamasers*. The first VLBI synthesis images of a normal maser, the one associated with the IC 133 region of M33, as well as observations of its environs, show it to be comparable physically to the most powerful galactic masers. Inference from past observations of megamasers has been that megamasers are intrinsically different than normal masers, arising from uncommon or extreme physical conditions. However, the first ever synthesis images of a megamaser, the one toward the nucleus of NGC 3079, show that this source too is structurally similar to galactic maser sources.

1. INTRODUCTION

Twenty-five extragalactic H_2O maser are known (see table). These may be divided into two categories, *normal* masers and *megamasers*, based largely on apparent luminosity, computed assuming isotropic emission of radiation. Normal extragalactic masers are comparable to galactic masers, where the strongest one, associated with W49N, routinely has a total isotropic luminosity of $\sim 1\ L_\odot$, of which up to 0.1 $L_\odot$ arises from a single 2 km s⁻¹ wide spectral line (Liljeström *et al.* 1989). These masers lie away from galactic nuclei or in the arms of spiral galaxies. The megamasers lie toward the nuclear regions of their host galaxies.

2. THE IC 133 MASER

The closest extragalactic H_2O maser outside the Magellanic Clouds lies toward the IC 133 H II complex in the galaxy M33. Properties of the H II complex are typical of massive star forming regions in our Galaxy. The estimated total luminosity of the region is at least $10^7\ L_\odot$ (Israel *et al.* 1990), comparable to that of the W49 and W51 regions in our galaxy. The radio continuum emission in the immediate vicinity of the maser arises from a compact H II region, of diameter ~ 1 pc, presumably powered by several O4-type stars. The maser is offset ~ 0.5 pc from the continuum peak (Greenhill *et al.* 1990).

The maser is composed of two centers of activity, IC 133 Main and West, which are separated by $\sim 0.''3$ (1 pc). It is likely that these two centers of activity have different energetic and dynamic centers (see Figure 1). Such multiple structure is similar to that found toward numerous large galactic star forming regions. Examples include W49N and W49S, and W51 Main, North, and South. The individual angular extents of IC 133

Galaxy	No. of Masers[a]	Location[b]	Distance (Mpc)	Luminosity[c] ($L_\odot$)
LMC	5	irregular	0.05	0.0008–0.003
SMC	2	irregular	0.06	0.001, 0.002
M33	5	disk	0.72	0.005–0.2
IC 342	2	disk	2.5	0.07, 0.1
NGC 253	1	nucleus	2.5	0.3
IC 10	1	irregular	3.0	1.0
M51	1	nucleus	6.5	0.8
NGC 6946	1	nucleus	4.5	1.0
M82	1	nucleus	3.5	3.0
Circinus	1	nucleus	4.0	37
NGC 4945	1	nucleus	4.0	85
NGC 4258	1	nucleus	6.9	120
NGC 6240	1	nucleus	100	200
NGC 1068	1	nucleus	16.5	350
NGC 3079	1	nucleus	16.5	520

[a] Greenhill *et al.* (1990) list the discovery references in their Table 1.
[b] Positions toward which masers lie. Most coincidences are accurate to only $\sim 15''$.
[c] Total luminosity assuming isotropic emission of maser radiation.

Main and West are comparable to the extent of the low radial velocity H$_2$O masers in W49N, ($\sim 2''$ at a distance of 10 kpc), scaled to the distance of M33 (720 kpc). The total radial velocity spread of masers in W49N is greater than that for IC 133, due largely to a population of high radial velocity masers, which in IC 133 are probably too weak to observe (Greenhill *et al.* 1990). The peak flux density of IC 133 Main, (historically ~ 2 Jy; Huchtmeier, Eckart, and Zensus 1988), is also comparable to the strength of persistent strong features in W49N (10^4 Jy), scaled to the distance of M33. The proper motions of four maser features, measured over 479 days, are consistent with a distance of 600 ± 300 kpc (Greenhill *et al.* 1992).

3. THE NGC 3079 MEGAMASER

The most luminous known megamaser lies toward the weakly active nucleus of the edge-on spiral NGC 3079 (Filippenko and Sargent 1992 and references therein). Early estimates of the maser region size ($\lesssim 0.02$ pc) and the relative position of the compact nuclear continuum source suggested that the megamaser arises from amplification of high brightness temperature background emission (Haschick *et al.* 1990).

Recent VLBI observations of the maser show two centers of activity, NGC 3079 Main and North, that are separated by ~ 0.7 pc. Each center is ~ 0.1 pc in size. These scales are similar to those of IC 133 and galactic sources (see Figure 2). The continuum source was not detected with VLBI (~ 1 mas beamwidth), from which we infer that emission is extended and of low brightness temperature. The maser may still amplify

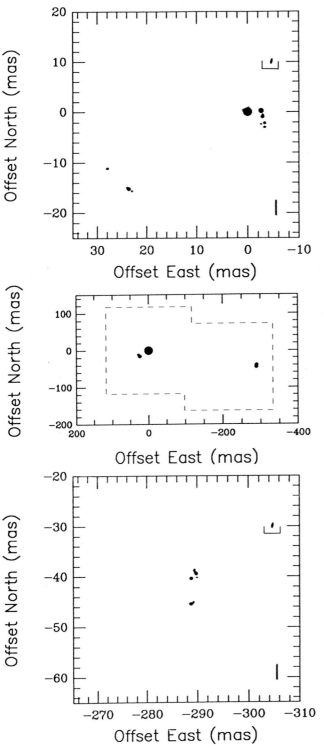

Fig. 1.—VLBI maps of the IC 133 maser. (*center*)–The relative positions of IC 133 Main and IC 133 West. The dashed box indicates the region searched for emission. (*top*)– Expanded view of IC 133 Main. (*bottom*)– Expanded view of IC 133 West. Each maser feature is denoted by a circle the area of which is proportional to its peak flux density. The FWHM of the synthesized beam is shown in the northwest corner of the latter two maps. Position uncertainties of the features are $\lesssim 10$ μas in right ascension. The bars denote 0.01 pc at an assumed distance of 720 kpc. The celestial coordinates of the origin are

$$\alpha_{2000} = 01^h 33^m 16\overset{s}{.}55 \pm 0\overset{s}{.}01$$
$$\delta_{2000} = 30°52'50\overset{''}{.}0 \pm 0\overset{''}{.}1.$$

The radial velocity extents of IC 133 Main and West individually are about 20 km s^{-1}. The maser features in IC 133 West are blueshifted with respect to those in IC 133 Main, by 20–30 km s^{-1}, with only scant overlap. Based on their angular separation and velocities it is likely that these two centers of activity have separate energetic and dynamic centers. (For purposes of this discussion, we have absorbed into IC 133 Main the center of maser activity that was designated IC 133 Southeast by Greenhill *et al.* (1990), since both are probably associated).

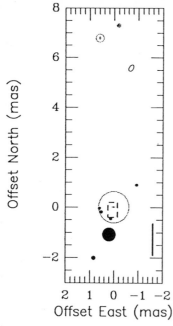

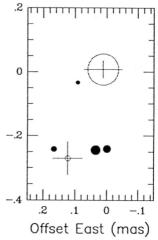

Fig. 2.—VLBI maps of the H_2O megamaser in NGC 3079. (*left*) A view of the maser source, showing both NGC 3079 Main and North. The dashed box indicates the field of view expanded and shown to the right. (*right*) An expanded view of the area in the immediate vicinity of the origin. The legend is the same as for Figure 1, except that unfilled circles represent unresolved knots of features and the scale bar represents 0.1 pc at a distance of 16.5 Mpc. Error bars associated with knots indicate the angular spread. The velocity range of the emission shown is $941 < V_{LSR} < 1039$ km s^{-1}. The most red and blueshifted emission arises from NGC 3079 Main. The weakest features have isotropic luminosities of $\sim 0.5\ L_\odot$.

background emission that was too weak to detect. Other megamaser mechanisms rely on the presence of enhanced ambient magnetic fields or foreground amplification of normal maser emission by shock excited H_2O (Greenhill *et al.* 1992).

I would like to emphasize the important contributions of my collaborators in this work, J. M. Moran, M. J. Reid, C. R. Gwinn, K. M. Menten, H. Hirabayashi, W. A. Baan, A. D. Haschick, and A. Eckart.

REFERENCES

Filippenko, A. V. and Sargent, W. L. W. 1992, *A. J.*, 103, 28

Greenhill, L. J., Moran, J. M., Reid, M. J., Gwinn, C. R., Menten, K. M., Eckart, A., and Hirabayashi, H. 1990, *Ap. J.*, 364, 513

Greenhill, L. J., Moran, J. M., Reid, and Hirabayashi, H. 1992, in preparation

Haschick, A. D., Baan, W., Schneps, M. H., Reid, M. J., Moran, J. M., and Güsten, R. 1990, *Ap. J.*, 356, 149

Huchtmeier, W. K., Eckart, A., and Zensus, A. J. 1988, *Astr. Ap.*, 200, 26

Israel, F. P., Hawarden, T. G., Geballe, T. R., Wade, R. 1990, *MNRAS*, 242, 471

Liljeström, T., Mattila, K., Toriseva, M., and Anttila, R. 1989, *Astr. Ap. Suppl.*, 79, 19

H$_2$O MEGAMASER VARIABILITY

Aubrey D. Haschick
Haystack Observatory

Willem A. Baan
Arecibo Observatory

ABSTRACT

A long term monitoring program of extragalactic water vapor sources has been conducted at Haystack Observatory. The sources monitored are NGC 1068, NGC 3079, NGC 4258, IC 10, and M 33. Variability of line strength, line width, and line velocity of individual line features has been found for all sources. The variability data of extragalactic H$_2$O megamasers is consistent with models of exponential amplification and shock-induced pumping of the molecules.

1. MEGAMASER VARIABILITY

Systematic observations of a small number of Galactic water maser sources have revealed strong variability of the sources. Similarly the inspection of published spectra of extragalactic H$_2$O masers reveals distinct intensity variations over time intervals of even months (Whiteoak and Gardner 1986; Claussen and Lo 1986). Systematic variability data for the extragalactic source NGC 4258 has revealed evidence for rapid periodic variability of one of the megamaser features (Haschick and Baan 1990). The period of 85 days suggests a scenario, where a Mira-like variable star periodically pumps surrounding molecular material and causes an inversion. The inverted H$_2$O molecules amplify the background radio continuum of NGC 4258.

The powerful maser source NGC 4258 has been part of a systematic (monthly) observing program at Haystack Observatory of a small number of extragalactic sources. Monitoring data have been obtained for the sources NGC 1068 (1984-1992), NGC 3079 (1984-1992), NGC 4258 (1982-1992), IC 10 (1986-1992), and M 33 (1987-1992). A full discussion of the data will be published elsewhere (Baan and Haschick 1992).

2. MONITORING MEGAMASERS

Data has been obtained for the program sources at nearly monthly intervals and the total number of epochs for the sources varies between 40 and 90. The parameters determined for each identifiable feature in the spectra are: 1) the peak line intensity, 2) the line width at the half power points, and 3) the line center velocity at half power. No gaussian fitting has been done because the line shapes of individual features were generally not gaussian.

The data obtained for the extragalactic H_2O maser sources has been searched for *temporal variations of amplitude and line width* and for *systematic changes in the center velocity of individual features. All of these effects have been detected for each of the five sample sources.*

2.1 Intensity Variations

The individual maser features in the various sources show line intensity variations on both short and long time scales. Calibration inaccuracies may have contributed to flux variations on timescales equal or less than the sampling time, but would not have affected the systematic long term variations. Due to appearing and disappearing features, the spectra at different epochs can be very different.

2.2 Velocity Variations

Amplitude variations reflect variations in the gain due to variable pumping conditions. For NGC 4258, a variable stellar source has been postulated to explain the observed periodic variations (Haschick and Baan 1990). The velocity of the periodic feature in NGC 4258 varies systematically and may be explained in the context of a shock pumping mechanism and resulting systematic change in the velocity of the amplifying gas. The velocities of the other two main velocity components in NGC 4258 have been found to show the same systematic trend as found for the periodic feature.

All sample sources show systematic variations of the center velocity for at least one of their features. The changes in velocity can be very small and can be both positive and negative. When systematic velocity shifts have been observed, they are significantly larger than the instantaneous line width of the feature. The observed shifts cannot easily be explained with the randomly popping up of individual features making up a blended broad line. Some velocity variations due to contributing features may have been observed for the prominent feature of NGC 3079, which is known to be a blended line (Greenhill *et al.* 1992). The velocity jitter in NGC 4258 is much smaller than the systemic variations observed for all three prominent features.

2.3 Line Width Variations

Variations of the gain may also be accompanied by variations of the line width. In the exponential amplification stage of a masing region, the center of a thermal line profile grows significantly faster than the line wings. As a result, the line width decreases as the gain increases. In the standard theory, once the line center begins to saturate, the wings will get relatively stronger and the line width increases to the thermal width (Sullivan 1973; Goldreich and Kwan 1974; Elitzur 1990). In the low gain region the line width varies as $\Delta V \propto \tau^{-0.5}$. Exponential (unsaturated) amplification with $I = I_o(e^\tau - 1)$ will thus give a $\Delta V \propto (\log I)^{-0.5}$ dependence. Systematic variations of the line width may also result from a decrease of the thermal width due to cooling of the masing clouds (Strelnitskii 1992).

During saturation and after full saturation, the standard theory predicts that the line width increases rapidly until it reaches the thermal width. A line width vs. line intensity dependence of $\Delta V \propto (\log I)^n$ may be expected, where n can be large (Sullivan 1973). For OH and H_2O masers some environmental influences may delay the saturated (re)broadening of the lines (Goldreich and Kwan 1974). In particular, photon scattering processes may cause the line to narrow a little further before the broadening starts (Elitzur 1990).

The observed variations of the line width for all individual features show a similar trend: the line width increases with decreasing line intensity. This result is consistent with the unsaturated amplification picture for extragalactic H_2O masers. The most extreme and conclusive case for such variation is found for a flaring maser feature in IC 10 (Baan and Haschick 1992). Considering the fact that the broader features are likely to be blends of many uncorrelated weaker features, the anticorrelation could disappear. However, if all contributing features behave the same way and are pumped by the same source, the ensemble could again behave as a single line.

3. DISCUSSION

The variability database for the five powerful extragalactic H_2O megamasers is of great help for understanding the physics of these sources. The extraordinary strength of the extragalactic maser features distinguishes them from observed Galactic sources and their physical conditions are likely to be distinctly different. Variations of strength, velocity width, and center velocity of extragalactic maser features are certain clues about the physical conditions of the masing regions.

The important conclusions to be drawn from these monitoring data is that no spectral parameter appears to remain constant for the sources. The physical conditions for the extragalactic maser regions are varying rather rapidly. The observed characteristics are all consistent with a picture of exponential amplification of background continuum by foreground pockets of pumped molecular gas (Baan 1985; Haschick and Baan 1985; Haschick *et al.* 1990). Further conclusions to be drawn from the observed variability in extragalactic megamaser sources are:
1) Variability can be found in all extragalactic maser sources. The line strength, the half power line width, and the center velocity of individual features show both systematic and erratic variations.
2) Broader line features composed of many weaker contributing features appear to behave like single features. Clusters of contributing features vary coherently in strength and show systematic velocity drifts. These broad features appear not to be affected by randomly varying contributing features.
3) The line width of individual features becomes smaller with increasing line strength. This behavior is consistent with unsaturated exponential amplification and has been observed for all features showing significant variability in their line strength.
4) Systematic changes in the line velocities and the observed velocity gradients have been

93

detected for at least one feature in each of the sources. The variations are consistent with shock-driven pumping schemes (Elitzur, Hollenbach, and McKee 1989). Shocked pockets of gas contributing to the amplification process are changing velocity steadily; the masing pockets appear and disappear as dense clouds in the molecular structure are being sequentially pumped by the passing shocks. Clusters of features are to be expected for a shock pumping scheme in an inhomogeneous medium.

Radio astronomy at Haystack Observatory of the Northeast Radio Observatory Corporation is supported by the National Science Foundation.

The Arecibo Observatory is part of the National Astronomy and Ionosphere Center and is operated by Cornell University under a cooperative agreement with the National Science Foundation.

REFERENCES

Baan, W.A. 1985, *Nature*, 315, 26.
Baan, W.A., and Haschick, A.D. 1992, *Ap.J.*, sumbitted.
Claussen, M. and Lo, K.Y. 1986, *Ap.J.*, 308, 592.
Elitzur, M. 1990, *Ap.J. (Letters)*, 350, L17.
Elitzur, M., Hollenbach, D., and McKee, C. 1989, *Ap.J.*, 346, 983.
Goldreich, P., and Kwan, 1974, *Ap.J.*, 190, 27.
Greenhill, L., *et al.* 1991, *Ap.J.*, 364, 513.
Greenhill, L., *et al.* 1992, *Ap.J.*, submitted.
Haschick, A.D., and Baan, W.A. 1985, *Nature*, 314, 144.
Haschick, A.D., and Baan, W.A. 1990, *Ap.J. (Letters)*, 355, L23.
Haschick, A.D., *et al.* 1990, *Ap.J.*, 356, 149.
Strelnitskii, V. 1992, personal communication.
Sullivan, W. 1973, *Ap.J. Suppl.*, 25, 393.
Whiteoak, J.B., and Gardner, F.F. 1986, *M.N.R.A.S*, 222, 513.

5. MASERS IN STAR FORMING REGIONS (GENERAL)

INTERSTELLAR MASERS: THEN AND NOW

Nels Anderson and Reinhard Genzel
Max-Planck-Institut für extraterrestrische Physik
W-8046 Garching bei München, Germany

1. INTRODUCTION

The large attendance at this conference, given its specialized theme, attests to the resurgence of interest in masers. Some of the reasons will be pinpointed herein. We begin with a brief history of the field and then highlight major topics of current research—masers as tracers of star-formation, outflows in star formation, magnetic fields, recently discovered maser transitions, and extragalactic masers. Finally, we touch on the emerging field of thermal emission from gas in masers.

2. MASERS THEN

Twenty-five years ago masers were accidentally discovered in interstellar space. The first paper in the field is that by Weaver et al. (1965) announcing the discovery of "mysterium" in emission towards W3. Mysterium was eventually recognized as OH. The second key paper (Cheung et al. 1969) announced the discovery of water vapor by Charles Townes' group at Berkeley. Even this early detection showed the very broad velocity range (10–100 km s^{-1}) that is characteristic of water masers. Interferometry was crucial in showing that the observed emission came in fact from masers. The first interferometry experiments found source sizes significantly less than the resolution of 10–20″ (Cudaback, Read, & Rougoor 1966; Rogers et al. 1966). This in turn implied brightness temperatures of $\sim 10^6$ K or more and demonstrated that the emission could not be thermal. In 1967 very long baseline interferometry (VLBI) established that the OH emission originated in many spots with sizes of just a few milliarcseconds and brightness temperatures of $\sim 10^{12}$ K (Moran et al. 1968). When VLBI became possible at the relatively high frequency of 22 GHz a few years later, water masers were also found to be smaller than a few milliarcseconds in size (Burke et al. 1970).

The observations stimulated a flurry of theoretical activity. The classic series by Peter Goldreich and collaborators (Goldreich & Keeley 1972; Goldreich, Keeley, & Kwan 1973a,b; Goldreich & Kwan 1974) is still a must for the student of maser research. Another early contribution which has proven very durable is de Jong's (1973) pumping scheme for water masers, which forms the basis for current models. As mentioned above, even the early observations of H_2O showed lines with widths of hundreds of kilometers per second. Many papers were written about these high-velocity features, which had never before been seen in star-forming regions. Strel'nitskij & Sunyaev (1972) were the first to point out that the high-velocity features indicated fast mass outflow. Although widely recognized in the maser field, this paper has not been justly appreciated by the molecular line community as a whole. (Those in the maser community knew by 1975 or so that many star-forming regions had high-velocity outflows; by the time these outflows were seen and studied in the millimeter transitions of CO, the maser results had been virtually forgotten.)

3. DISTRIBUTION OF MASERS AND RELATION TO STAR FORMATION

Surveys of water masers conducted by many researchers have been assembled into a "survey of surveys" by Cesaroni et al. (1988). The few hundred circumstellar and interstellar sources included in this compendium are closely correlated with the galactic plane and with regions of star formation (Fig. 1). In fact, in the entire region covered by the Orion and Monoceros molecular clouds, there happens to be a maser in every region of star-forming activity. There is a good spatial correlation between maser sources and outflows, at least the more compact ones, that have been mapped in thermal lines by Fukui et al. (1986) and others.

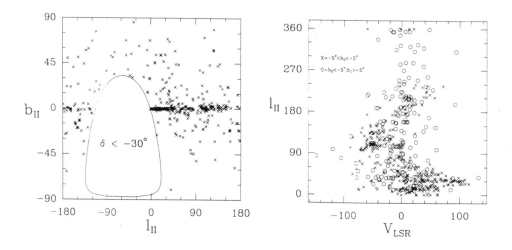

FIG. 1.—Galactic distribution of known H_2O masers (Cesaroni et al. 1988).

In addition to the spatial correlation, there is a close temporal correlation between masers and outflows from young stars. By the end of the 1970's, it was realized that typical OH and H_2O masers had lifetimes of a few times 10^4 to perhaps 10^5 years, which happens to be approximately the time a massive star and its surrounding H II region require to form and affect their parental molecular cloud, as estimated from the occurrence of thermal outflows (Lada 1985). Every star more massive than 5 to $10 M_\odot$ is likely to go through a maser phase. The maser community might have used these correlations to find sites of star formation by performing maser surveys. In fact, however, this has never been done and the current style is to search for masers in objects drawn from some other kind of survey, e.g., the IRAS catalogue. In this sense, the community may have missed an opportunity to point the way in the study of star formation.

4. KINEMATICS

Masers are a powerful tool for determining the dynamics of very dense gas in outflow regions. This is perhaps best demonstrated in the very well studied case of the Orion-KL star-forming core (Fig. 2).

Proper motions of the masers can be determined with VLBI and show a pattern of expansion in the vicinity of the most luminous energy source, IRc2. On a smaller scale, IRc2 is seen to contain a concentration of water masers and a clump of SiO masers. Recent work by Plambeck, Wright, & Carlstrom (1990) shows that the redshifted and blueshifted SiO masers are clearly separated and show a clear velocity gradient. The velocity field is plausibly explained in terms of an inclined, expanding, and rotating circumstellar disk a few tens of astronomical units in size.

SiO masers are rarely seen in star-forming regions, but other masers may also be used as kinematical tracers. OH masers are correlated with compact H II regions, the best example being W3(OH). Not only does it feature 1665-MHz masers, but also excited-state OH masers and many class 2 methanol masers (covered elsewhere in these proceedings by Menten). Elsewhere in this volume Moran reviews proper motion of H_2O masers and Bloemhof et al. present recent results on the motion of OH masers. Bloemhof et al. find that OH masers, like H_2O masers, show a pattern of expansion, but at rather lower velocity than in the case of H_2O. This seems to settle a long-standing dispute as to whether OH masers reside in an expanding compressed shell. There may be some complications, however, arising from the fact that, as in all VLBI work, one can add an arbitrary velocity along the line of sight without changing the proper motions. Bloemhof et al. show that this uncertainty can transform the simple expansion pattern into a streaming motion combined with expansion, i.e., a cometary H II region. In recent years, compact H II regions have often been supposed to be streaming into dense environments, making them last longer than they otherwise would.

5. MAGNETIC FIELDS

Masers are ideal for probing magnetic fields. One example is the recent detection of Zeeman splitting in H_2O by Fiebig & Güsten (1989), showing that the high densities typical of water masers (about 10^9 cm^{-3}) are associated with fields of a few tens of milligauss to 100 mG. These measurements are consistent with the common wisdom that the field scales as the square root of the density, but it is still not certain that the trend is smooth and continuous, principally because recent measurements in the vicinities of dense cloud cores have had difficulty establishing magnetic fields on the order of 100 μG. It could still be that field strength is nearly independent of density in this region, perhaps because of ambipolar diffusion, and then increases rapidly at higher densities. Alternatively, OH and H_2O masers may reside in regions possessing unusually strong fields. In any event, there is obviously a trend toward stronger fields in maser regions.

Masers can also be used to trace the magnetic field on the galactic scale. Reid has found that fields in OH masers around the galaxy show magnetic field vectors oriented in the direction of galactic rotation. This leaves us with a something of a puzzle, as some

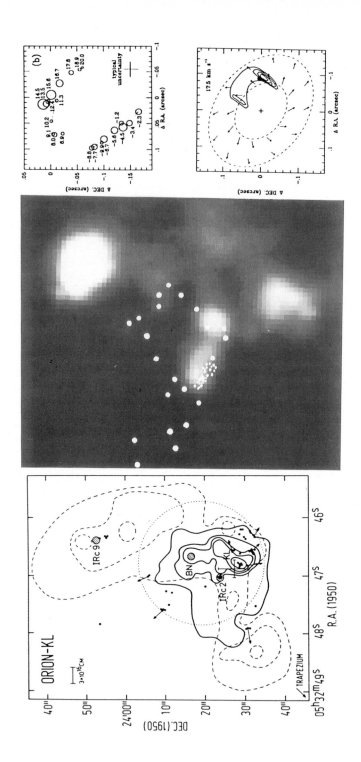

FIG. 2.—(*Left*) H$_2$O masers in Orion-KL with contours of 20-μm continuum (*heavy lines*) and of H$_2$ (*dashed lines*) (from Downes et al. 1981 and Genzel et al. 1981). (*Center*) Locations of OH maser clumps (*large filled circles*) from Johnston, Migenes, & Norris (1989), and H$_2$O masers (*small filled circles*) from Wright et al. (1990), superimposed upon 8.7-μm continuum map (Gezari 1992) featuring IRc2 (*leftmost spot*) and BN (*brightest, upper right*). (*Top right*) SiO masers from Plambeck et al. (1990) and (*bottom right*) rotating, expanding disk model.

measurements indicate that while magnetic fields are perhaps dynamically important for cloud support, it is their random, turbulent nature that counts, not their large-scale, homogeneous component. Then it is unclear why one should see such a tight correlation between large-scale fields and local magnetic fields. This is a very interesting problem for future research.

6. NEW MASER TRANSITIONS

Perhaps the key development which has led to the revival of interest in the field is the discovery of new masing molecules and new masing transitions in molecules previously known to mase. Observation of multiple masing transitions in a single species offers an opportunity to use masers themselves to obtain the physical conditions of the medium, as Melnick points out elsewhere in this volume. The long-standing inability to obtain column densities from maser observations is eliminated, at least in some cases. In this sense the discovery of millimeter and submillimeter water and methanol masers is very important. It is also clear, however, that more accurate cross sections are needed for proper modeling, especially in the case of methanol.

7. MASERS IN THE STAR-FORMATION CONTEXT

The picture of masers in the context of star formation that has emerged from 25 years of maser work is summarized in Figure 3. At the center one has a newly-formed massive star, perhaps surrounded by an accretion disk. A clumpy outflow emanates from the star. The outflow sweeps clean a cavity around the star. Clumps striking the boundary of the cavity either create shocks in the surrounding cloud or are themselves shock compressed. Masers form within the compressed gas, with preferential beaming perhaps perpendicular to the the clumps' motions. Turbulence destroys the clumps on a dynamical time scale, which is to say on a time scale of a few years. This is then a rudimentary model for H_2O masers. The maser energy is derived from the outflow and then converted by some mechanism, call it a shock, into maser radiation. From an observational point of view, this picture is very natural in that it produces small maser spots and it also explains the lifetimes of individual spots.

OH masers may be an entirely different story. They arise in the expanding dense environment surrounding an H II region. If the star causing the H II region is moving through the ambient medium, then the H II region will be cometary in nature, and it is probably the resulting high-density medium in which the chemistry has been altered by shocks, high temperature, or evaporation of dust grains that gives rise to OH and methanol masers.

8. PHYSICAL PROCESSES IN MASERS

The idea of masers in shocks has been in the literature for a very long time (Litvak 1969). Elitzur, Hollenbach, & McKee (1989) have used a detailed shock code to analyze masers in a dissociative shock (J-shock). The key point in the J-shock picture is the

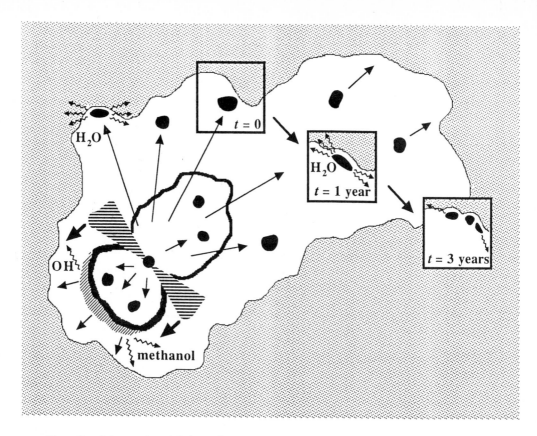

Fig. 3.—Masers in vicinity of young high-mass star. Methanol and OH masers reside near boundary of cometary H II region (*heavy lines*), while H_2O masers arise as outflowing clumps strike ambient medium.

reformation plateau behind the shock front, in which, after all molecules have been destroyed by the shock, H_2 begins to reform on dust grains. The heat of formation produces a constant-temperature plateau in which, they argue, H_2O masers form. Since shocks naturally produce long filaments or disks, the resulting masers can be highly beamed, and this helps to generate large brightness temperatures. There are several caveats, however. One is that most of the energy in J-shocks emerges in lines of species other than H_2O. Therefore, J-shocks are an energetically inefficient means of producing masers; C-shocks would have a much higher conversion efficiency. Another problem is the high degree of beaming required to explain the brightest masers. If beaming is purely geometric, as Elitzur et al. assume, then brightness temperatures of 10^{13} K can be explained with aspect ratios of 40. On the other hand, as Kylafis & Norman (1991) point out, a ratio of velocity transverse and longitudinal velocity gradients 20 times larger than this is needed if velocity coherence causes the beaming. Still, this mechanism seems to be the most promising.

An intriguing scenario invented by Strel'nitskij (1980, 1984), namely two-temperature pumping, appeared attractive for a while in a particular form (Kylafis & Norman 1987), but it fails to produce significant power with current cross sections for rotational excitation of H_2O by H_2 (Elitzur & Fuqua 1990; Anderson & Watson 1991).

For the very brightest masers, especially the Orion flares having temperatures exceeding 10^{14} K (Abraham et al. 1979; Matveenko, Graham, & Diamond 1988; Garay et al. 1989), it has been proposed that two intrinsically fainter masers some distance apart may happen to illuminate one another (Deguchi & Watson 1989). The beaming angle of the combined maser then becomes very small, and a high brightness results. An independent argument based on the relatively narrow line widths observed in the 22-GHz transition of H_2O despite its substantial hyperfine splitting also suggests that at least water masers are tightly beamed (Nedoluha & Watson 1991).

There has also been substantial progress in OH masers. Collisions, because of reassignment of parities, do not seem to be a plausible pumping agent. Cesaroni & Walmsley (1991), among others, show that line overlap and pumping by dust radiation can explain the 1665-MHz masers and numerous other lines of OH seen in absorption and emission, both thermal and maser, in W3(OH).

9. EXTRAGALACTIC MASERS

The very luminous kilo- and mega- OH, H_2O, and H_2CO masers in external galaxies are a particularly intriguing phenomenon (see Baan's piece elsewhere in this volume). As an example, Claussen & Lo (1986) have found a strong, high-velocity megamaser in the nucleus of NGC 1068. Figure 5 displays H_2 emission superimposed upon a 6-cm VLA map which shows a double-sided jet and three sources in the nucleus. It is widely believed that the nucleus is at the tip of the cone-like distribution of narrow-line clouds. The water maser appears coincident with the nucleus, although whether the two are actually close to one another is another question. There is, however, a peak in the H_2 very near the maser, suggesting that there is molecular gas close to the maser, and perhaps this material is in front of the nucleus, explaining why the narrow-line region is heavily obscured, and the maser amplifies radiation emitted by the nucleus.

10. THERMAL EMISSION

Most information acquired thus far about masers has come from observations of masing transitions. Given a model of maser emission, the characteristics of the environment may then to some extent be inferred from observations. Full confrontation between observation and theory is only possible, however, with direct measurements of the maser environment through observations of thermal transitions. Initial steps in this direction have been taken by Melnick, Genzel, & Lugten (1987), Betz & Boreiko (1989), and Melnick et al. (1990). They have studied a number of far-IR rotational transitions of OH in Orion-KL, both within and across the $^2\Pi_{1/2}$ and $^2\Pi_{3/2}$ ladders. Complex velocity structures (P-Cygni profiles) are seen (Fig. 5), which demonstrate the need for careful treatment of radiative transfer. Analysis of these observations reveals the importance of IR pumping. Thermal $H_2{}^{18}O$ has been detected in water maser sources (Jacq

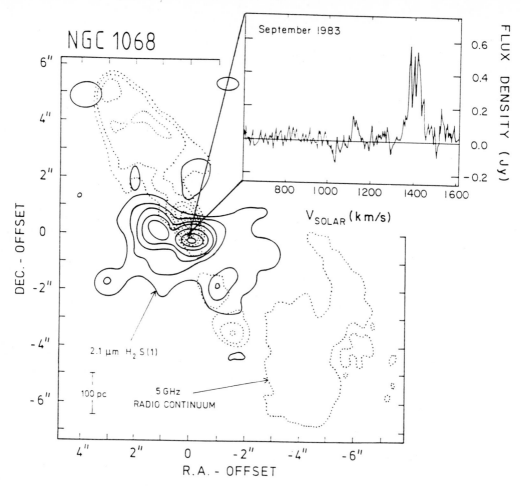

FIG. 4.—H_2O megamaser in NGC 1068 (Claussen & Lo 1986) with H_2 emission and contours of 6-cm continuum.

et al. 1988; Menten et al. 1990), although contamination by other lines poses significant obstacles to interpretation.

Maser environments may also be studied in transitions of non-masing molecules. Particularly promising in this respect are the mid-J rotational transitions of HCN at several hundred gigahertz which, when optically thin, trace densities of $\sim 10^9$ cm^{-3} and temperatures of ~ 200 K or higher—the regime expected for water masers. In addition, resolutions of $10''$ or so may be attained with the large submillimeter telescopes now in operation, as opposed to resolutions of 30–$60''$ generally achieved in lower-frequency observations from the ground or in far-IR work from the Kuiper Airborne Observatory. Although this size is still much larger than the few AU's characteristic of individual maser spots, it allows one to single out individual maser clumps even at distances of

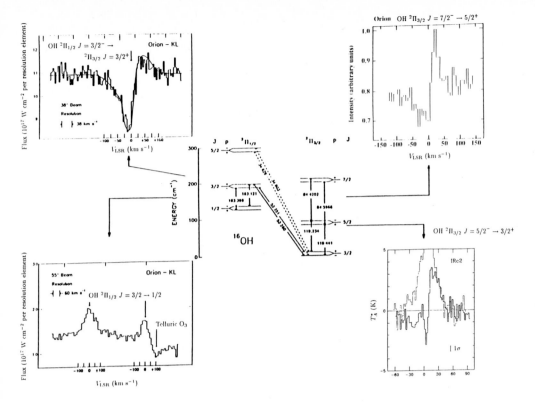

FIG. 5.—Far-IR observations of thermal emission from OH in Orion-KL region. Left-hand panels from Melnick et al. (1990), top right from Melnick et al. (in preparation), bottom right (also showing CO $J = 22 \rightarrow 21$ with light line) from Betz & Boreiko (1989).

several kiloparsecs. Future interferometric observations of submillimeter transitions will clearly be most desirable.

Several specific issues can be addressed with such multiline observations. First of all, temperatures and densities may be established and compared with those required by the models. As velocity coherence is necessary in a maser, line profiles are also of great interest. Given VLBI estimates of maser sizes and assuming that the maser power we see is representative of that beamed in other directions, masing clumps may also be compared with neighboring non-masing areas; similarity of emission in the two types of regions might suggest that some factor aside from density, temperature, and velocity gradient was crucial in producing masers.

Thermal emission, unlike maser emission, is not beamed. In interpreting thermal emission, one must therefore keep in mind that the region under observation may be criss-crossed by many masers in addition to those few that happened to be pointed toward the observer. If approximately equal maser power is emitted in all directions, tighter beaming of individual spots implies the existence of more spots and more masing gas. Determination of the amount of gas capable of masing may therefore allow the setting of limits on beaming.

Stutzki et al. (1988) have already found strong HCN $J = 9 \to 8$ emission in Orion. Much of the emission appears to originate in a thin shell of $10''$ radius which is centered on IRc2, the approximate center of the outflowing masers. Because the optical depths here are so large, radiative trapping reduces density of the gas traced to $\sim 10^8$ cm^{-3}. Isotopic observations are needed to push to higher densities.

REFERENCES

Abraham, Z., Cohen, N. L., Opher, R., Rafaelli, J. C., & Zisk, S. H. 1979, A&A, 100, L10

Anderson, N., & Watson, W. D. 1990, ApJ, 348, L69

Betz, A. L. & Boreiko, R. T. 1989, ApJ, 346, L101

Burke, B. F., Papa, D. C., Papadopoulos, G. D., Schwartz, P. R., Knowles, S. H., Sullivan, W. T., Meeks, M. L., & Moran, J. M. 1970, ApJ, 160, L63

Cesaroni, R., Palagi, F., Felli, M., Catarzi, M., Comoretto, G., Di Franco, S., Giovanardi, C., Palla, F. 1988, A&AS, 76, 445

Cesaroni, R., & Walmsley, C. M. 1991, A&A, 241, 537

Cheung, A. C., Rank, D. M., Townes, C. H., Thornton, D. D., & Welch, W. J. 1969, Nat, 221, 626

Claussen, M. J., & Lo, K.-Y. 1986, ApJ, 308, 592

Cudaback, D. D., Read, R. B., & Rougoor, G. W. 1966, Phys. Rev. Lett., 17, 452

Deguchi, S., & Watson, W. D. 1989, ApJ, 340, L17

de Jong, T. 1973, A&A, 26, 297

Downes, D., Genzel, R., Becklin, E. E., Wynn-Williams, C. G. 1981, ApJ, 244, 869

Elitzur, M., & Fuqua, J. B. 1989, ApJ, 347, L35

Elitzur, M., Hollenbach, D. J., & McKee, C. F. 1989, ApJ, 346, 983

Fiebig, D. & Güsten, R. 1989, A&A, 214, 333

Fukui, Y., Sugitani, K., Takaba, H., Iwata, T., Mizuno, A., Ogawa, H., & Kawabata, K. 1986, ApJ, 311, L85

Garay, G., Moran, J. M., & Haschick, A. D. 1989, ApJ, 338, 244

Genzel, R., Reid, M. J., Moran, J. M., & Downes, D. 1981, ApJ, 244, 884

Gezari, D. Y. 1992, ApJ (Letters), in press for 1 September 1992

Goldreich, P., & Keeley, D. A. 1972, ApJ, 174, 517

Goldreich, P., Keeley, D. A., & Kwan, J. 1973a, ApJ, 179, 111

————. 1973b, ApJ, 182, 55

Goldreich, P., & Kwan, J. 1974, ApJ, 190, 27

Jacq, T., Jewell, P. R., Henkel, C., Walmsley, C. M., & Baudry, A. 1988, A&A, 199, L5

Johnston, K. J., Gaume, R., Stolovy, S., Wilson, T. L., Walmsley, C. M., & Menten, K. 1992, ApJ, 385, 232

Johnston, K. J., Migenes, V., & Norris, R. P. 1989, ApJ, 341, 847

Kylafis, N., & Norman, C. 1987, ApJ, 323, 346

————. 1991, ApJ, 373, 525

Lada, C. J. 1985, ARA&A, 23, 267

Litvak, M. M. 1969, Sci, 165, 855

Matveenko, L. I., Graham, D. A., & Diamond, P. J. 1988, Pis'ma AZh, 14, 1101 (Engl. trans. SvAL, 14, 468)

Melnick, G. J., Genzel, R., & Lugten, J. B. 1987, ApJ, 321, 530

Melnick, G. J., Stacey, G. J., Genzel, R., Lugten, J. B., & Poglitsch, A. 1990, ApJ, 348, 161

Menten, K. M., Melnick, G. J., Phillips, T. G., & Neufeld, D. A. 1990, ApJ, 363, L27

Moran, J. M., Burke, B. F., Barrett, A. H., Rogers, A. E. E., Carter, J. C., Ball, J. A., & Cudaback, D. D. 1968, ApJ, 151, L99

Nedoluha, G. E., & Watson, W. D. 1991, ApJ, 367, L63

Plambeck, R. L., Wright, M. C. H., & Carlstrom, J. E. 1990, ApJ, 348, L65

Rogers, A. E. E., Moran, J. M., Crowther, P. P., Burke, B. F., Meeks, M. L., Ball, J. A., & Hyde, G. M. 1966, Phys. Rev. Lett., 17, 450

Strel'nitskij, V. S. 1980, Pis'ma AZh, 49, 704 (Engl. trans. SvAL, 6, 196)

———. 1984, MNRAS, 207, 339

Strel'nitskij, V. S., & Sunyaev, R. A. 1972, AZh, 49, 704 (Engl. trans. SvA, 16, 579)

Stutzki, J., Genzel, R., Harris, A. I., & Herman, J. 1988, ApJ, 330, L125

Weaver, H., Williams, D. R. W., Dieter, N. H., & Lum, W. T. 1965, Nat, 208, 29

Wright, M. C. H., Carlstrom, J. E., Plambeck, R. L., & Welsh, W. J. 1990, AJ, 99, 1299

SOME INTERESTING OH/H_2O MASERS

JAMES R FORSTER

Hat Creek Radio Observatory, Cassel CA 96016

Several sources from a VLA study of OH/H_2O maser associations in the galactic plane are presented. The VLA sample was taken from a Parkes 64 m telescope survey of OH and H_2O masers with angular separation less than 15″. The relative position accuracy is estimated to be better than 0.2″ for maser spots of the same species and for HII regions with respect to the water masers. The absolute position accuracy is estimated to be 0.5″ rms for the OH and H_2O reference features. Details of the observations are given in Forster & Caswell (1989).

The VLA survey showed that most associations are simple, consisting of a single group of OH and H_2O masers spatially coincident within the errors and extending over ~30 mpc (10^{17} cm). However, fourteen (20%) of the maser associations were classified as complex, containing multiple groups of OH and H_2O masers extending over ~150 mpc, and usually including a compact HII region. Several of the most interesting of these complex maser associations are discussed below. The maser spot maps are shown in the accompanying figure.

9.62+0.19 (a) The masers in this association consist of a linear string of intermixed OH and H_2O masers extending over ~15″ on the sky. No HII region was detected in this field. This is one of the most striking examples of a linear arrangement of masers in our sample. The highest velocity masers occur in the south and the lowest in the north; however, intermediate velocities occur throughout the association.

19.61-0.23 (b) These masers are located in a well known complex of HII regions at a distance of about 4 kpc. The H_2O masers have a velocity extent of ~50 km s^{-1} while the OH masers extend over only a few km s^{-1}. The spatial extent is ~30 mpc, typical of a young association. An HII region was detected 2″ SE of the maser groups, and a smaller HII region was found by Garay et al. (1985) near the NW edge of the association. No HII region is detected at the position of the OH maser group, which resides at the center of a fairly linear string of water masers. Red and blue-shifted water masers are found on both sides of the OH group. This association has the general appearance of a bipolar structure with origin at the OH maser group, although the radial velocities of the water masers do not entirely support this interpretation.

<u>5.88-0.39</u> (c) This is the best example of a shell-like HII region in our sample. It has also been recently recognized as one of the most powerful molecular outflow sources known. The H_2O masers in this source have a rather narrow velocity range while the OH masers extend over ~40 km s^{-1}. The maser spots tend to be located in a broad belt across the southern half of the HII region. G5.88 bears a remarkable similarity to the northern HII region W3(OH), both in its continuum structure and its OH maser distribution. At an assumed distance of 2.6 kpc it is five times larger, however. There is a general velocity gradient in the OH maser distribution across G5.88 with velocity increasing toward the SW. The water masers show no systematic pattern however, and the sense of the OH velocity gradient is opposite to that seen in mm-line maps made with the Hat Creek interferometer (Forster & Wilner in preparation).

<u>351.42+0.64</u> (d) This source is located in the northern part of the massive star-forming region NGC 6334. The water masers have a wide velocity spread (~50 km s^{-1}) with most of the emission blueward of the ambient (NH_3) gas. The OH masers lie near the leading edge of a cometary-shaped HII region, perhaps arising in a compressed molecular zone as suggested by Gaume and Mutel (1987). The H_2O masers lie in a straight line extending north from the HII region, probably in three or four compact groups (the apparently continuous string of maser spots is due to beam blending of masers from separate groups with overlapping velocities). The maser group nearest the HII region has the widest velocity spread, covering the central 30 km s^{-1} of the spectrum. The northern group contains mainly red velocities, and the central group is predominantly blue. The OH masers are all near the ambient velocity, and there is little spatial-velocity correlation between the two maser species. The overall size of the HII region and maser association is ~90 mpc, suggestive of a more massive or evolved system. The linear distribution of water masers gives the impression of a one-sided jet originating near the core of the HII region.

<u>351.58-0.35 and 10.62-0.38</u> (e and f) These are two more examples of the one-sided jet morphology for water masers. The H_2O velocities extend over ~20 km s^{-1}, much less than in NGC 6334, although the overall extent of ~200 mpc is much greater. The OH masers have velocities near the ambient, and do not show the high-velocity tails characteristic of the water masers. The OH masers tend to lie near the periphery of the HII region, and do not share the jet-like morphology of the water masers. It is interesting to note that the HII

regions in these sources often have cores which are elongated in the direction of the H_2O jets; this could be significant in the context of bipolar stellar winds. Neither of these sources show systematic velocity gradients in the OH or H_2O maser distributions. In 10.62 the H_2O group nearest the HII region covers the full velocity extent, while in 351.58 the group nearest the HII region comprises the reddest velocities, and the widest velocity group is further out in the jet.

CONCLUSIONS

These six sources exhibit many of the characteristics common to the "complex" class of OH/H_2O maser associations found in the VLA survey. The tendency for H_2O and OH masers to occur in linear arrangements suggests that they may be associated with shock-compressed boundary layers, dense molecular disks, or jet-like ejecta from young stellar objects. If a proper interpretation of these maser spots can be arrived at, masers should prove to be excellent probes of the dense and energetic environment within a few milliparsecs of massive young stars.

In most cases the maser radial velocities do not show systematic gradients. This is particularly true of H_2O masers, but also holds in general for OH masers. This suggests that the wide velocity spreads observed in many H_2O maser groups is caused by shocks or impact phenomena. The range of velocities observed in a compact maser group frequently contains both ambient and high-velocity components, consistent with maser formation at the interface between dense ambient gas and high-velocity shocks or ejecta from a central source.

Finally, maser spot maps of these regions are often complex and difficult to interpret. It seems likely that many of the process inferred to be going on in these regions, e.g. expanding shocks, rotating disks, energetic outflows, gravitational infall, etc., may all be occurring together. It is therefore not surprising that a complex and somewhat chaotic distribution of maser spots is often found in massive star forming regions.

Forster, J.R. and Caswell, J.L. 1989, Astron. Astrophys. 213, 339
Garay, G., Reid, M.J. and Moran, J.M. 1985, Ap. J. 289, 681
Gaume, R.A. and Mutel, R.L. 1987, Ap. J. Suppl. 65, 193

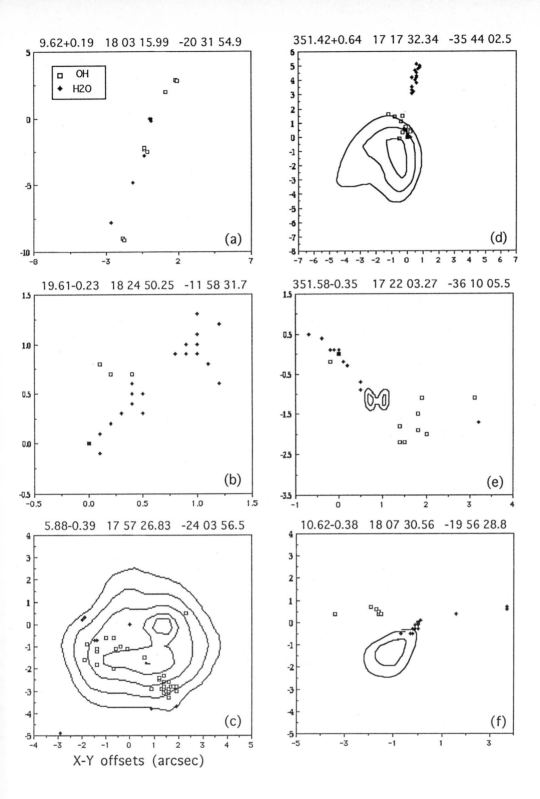

HIGH RESOLUTION CENTIMETER RADIO CONTINUUM AND AMMONIA MASER OBSERVATIONS OF THE W51 REGION OF STAR FORMATION

R. A. Gaume and K. J. Johnston
Center for Advanced Space Sensing
Code 4200, Naval Research Laboratory
Washington, D.C. 20375–5320

T. L. Wilson
Max-Planck-Institut für Radioastronomie
Auf dem Hügel 69
53 Bonn, 1, Germany

ABSTRACT

High angular resolution and high sensitivity observations at wavelengths of 1.3 and 3.6 cm have detected three new ultracompact emission regions in the core of the W51 (regions d and e). In total, five ultracompact continuum objects of diameter <300 to 3000 AU are located in this region. These sources may best be explained as photoionized stellar winds. This suggests that there may exist a quasi stable point in early stellar evolution where observable H II regions are formed by stellar winds around massive stars.

The $J,K=(9,8)$ NH_3 maser associated with W51 d has been shown to have a minimum brightness temperature of $2.7 \ 10^6$ K. This maser is most probably saturated.

1. INTRODUCTION

W51 is a region of on-going star formation in the constellation Aquila. At a distance of about 7 kpc (1″ is 0.03 pc) the W51 complex has been studied by numerous investigators. The centimeter continuum emission from the W51 region was shown by Martin (1972) to consist of eight distinct components, W51 a-h. The strongest of the eight components, W51 e, and nearby W51 d were studied by Scott (1978) who found two compact continuum components, W51 e_1 and e_2, in addition to the larger H II regions. OH and H_2O masers, which trace active regions within star forming complexes, have been found toward the W51 d and e regions (Gaume & Mutel 1987; Schneps et al. 1981; Genzel et al. 1981). Some of the clusters of OH and H_2O masers are closely associated with radio continuum components. Others are found in regions without detected continuum sources. Studies of thermal emission from molecules such as NH_3 (Mauersberger, Henkel and Wilson 1987, and Ho, Genzel and Das 1983) have traced warm, dense regions within W51 d and W51 e.

Madden et al. (1986) found the first intense NH_3 maser emission in the $(J,K)=(9,6)$ inversion line. Monitoring of the (9,6) line in W51 by Wilson and Henkel (1988) and

Wilson *et al.* (1990) showed time variability on a period of less than one year. In 1986, the most intense (9,6) maser line was located near W51 d, but after July 1987, this maser became a factor of 10 weaker, and the intensities of the (9,6) maser features in the general vicinity of W51 e_1 and e_2, approximately 40″ SW of W51-IRS2, increased by more than a factor of 10. Very long baseline interferometry observations (Pratap et al. 1991) have shown that the (9,6) line emission arises in W51 e from regions of less than 0.1 mas in size, in two areas NW of W51 e_1, and NE of W51 e_2.

The (9,6) NH_3 maser arises from orth-NH_3. Mauersberger, Henkel and Wilson (1987) detected other NH_3 maser lines from para-NH_3. Confirmation that the NH_3 emission toward W51 d in the (9,8) para transition could be attributed to the maser process was provided by Wilson, Johnston and Henkel (1990), placing a lower limit on the brightness temperature of the emission at 1600K. Wilson, Gaume and Johnston (1991) were able to set a lower limit to the brightness temperature of this (9,8) maser of $\geq$ 260,000K, and found the NH_3 maser projected onto a previously undetected ultra-compact H II region. To limit the brightness temperature of the (9,8) NH_3 maser yet further, and to investigate the continuum emission in the W51 d and e regions we have employed the VLA.

The data reported here are discussed in greater detail in Gaume, Johnston, & Wilson (1993).

2. OBSERVATIONS

The observations occurred on 20 July 1991 in both spectral line and continuum modes of the VLA. All available antennas were used in the A array, which provides baseline lengths from 0.7 to 36 km.

Observations of W51 were conducted in continuum mode using the 3.6 cm receiver system. The FWHM resolution of the 3.6 cm images is 0.″21 by 0.″20 (uniform weighting). The line mode observations were conducted with the 1.3 cm receiver system in a dual IF mode. The first IF channel was 0.781 MHz wide and was separated into 128 spectral line channels, each with a width of 6.104 KHz (0.077 km s^{-1}). Channel 64 of the 127 narrow channels was tuned to the frequency of the NH_3 (9,8) inversion transition, 23,657.471 MHz, at an LSR velocity of 55.4 km s^{-1}. The second 1.3 cm IF channel was 25 MHz wide, and was separated into 8 spectral line channels. The first channel (channel 0) of the 8 line channels was a wide band channel 18.750 MHz wide averaged over the inner 75% of the 25 MHz band. It was tuned to a central frequency 13 MHz red-shifted from the (9,8) NH_3 maser line. The FWHM resolutions of the restored 1.3 cm images are 0.″075 by 0.″073 (uniform weighting) and 0.″094 by 0.″093 (natural weighting).

3. RESULTS

The detailed results and discussion of the continuum data is found in Gaume, Johnston, & Wilson (1993).

Wilson, Gaume, and Johnston (1991) found the $(J,K)=(9,8)$ NH_3 maser emission projected NW of continuum component d_2. Our result is consistent with their lower spatial resolution data. A Gaussian fitted position of the peak NH_3 maser line channel is $\alpha(1950)= 19^h\ 21^m\ 22.1666^s\ \delta(1950)= 14°\ 25'\ 11.''980$. This is $0.''05$ NE of the Gaussian fitted position of d_2. There is a $0.''003$ scatter in the fitted positions of the strongest NH_3 maser line channels. This scatter is within the channel-to-channel measurement uncertainty of the $0.''07$ beam. There is no systematic trend in the fitted positions of the channel features as a function of velocity. A Gaussian fit to the data is consistent with a beam deconvolved size of $<0.''04$. This implies a minimum Rayleigh Jeans brightness temperature of $2.7\ 10^6$ K for the maser. Within measurement errors, a single component Gaussian fit to the velocity spectrum of the NH_3 maser emission produces a peak flux density (1.3 Jy), linewidth (0.35 km s^{-1}) and central velocity (55.3 km s^{-1}) consistent with those reported by Wilson, Gaume, and Johnston (1991).

4. THE $(J,K)=(9,8)$ NH_3 MASERS

The main result in regard to spectral lines is a better upper limit for the size of the ammonia maser region (40 mas). This size gives a lower limit of $2.7\ 10^6$ K for this maser. From our measurements of the brightness temperature and Eq(1) of Johnston et al. (1992), it appears that this maser must be saturated, unless the beaming angle is much smaller than 0.1 steradians. The saturated nature of this emission would explain the small (less than 30% over 5 years) amount of time variability observed. It appears that this maser amplifies the compact continuum source in this direction. Then, if the maser is saturated, the optical depth must be of order 60. Following the analysis of Wilson et al. (1991), we then estimate that the excitation temperature across the inversion doublet levels is -15 ± 8. This gives rise to an overpopulation in the upper doublet level of 8%. It is of interest to compare the situation for the (9,8) maser, from para-ammonia, with that for the (9,6) maser from ortho-ammonia, as measured by Pratap et al. (1991). The (9,6) maser has a peak brightness temperature of more than 10^{12} K, is highly time variable, and arises from a region of less than 0.1 mas.

With the exception of NGC 7538 (Gaume et al. 1991), ammonia masers arise only from non-metastable (J>K) levels. The non-metastable levels are populated by a combination of IR radiation or collisions, or both. Details of the excitation of non-metastable NH_3 masers is not clear. From observations, the ortho-ammonia masers show a higher degree of time variability than para-NH_3 masers in W51. From this observation it would seem that there is an intrinsic difference between masers in ortho and para-NH_3. If so, the maser excitation process must be intrinsic to the NH_3, and *not* the result of an overlap with lines from other molecules. As stated often, however, a method has not been found thus far to invert only a few selected non-metastable levels of NH_3.

REFERENCES

Gaume, R., Johnston, K., & Wilson, T., 1993, ApJ, submitted

Gaume, R., Mutel, R. 1987, ApJS, 65, 193

Gaume, R., Johnston, K., Nguyen, H., Wilson, T., Dickel, H., Goss, W., and Wright, M. 1991, ApJ, 376, 608

Genzel, R., Downes, D., Schneps, M., Reid, M., Moran, J., Kogan, L., Kostenko, V., Matveyenko, L., and Rönnäng, B. 1981, ApJ, 247, 1039

Ho, P., Genzel, R., Das, A. 1983, ApJ, 266, 596

Johnston, K., Gaume, R., Stolovy, S., Wilson, T., Walmsley, M., Menten, K. 1992, ApJ, 385, 232

Madden, S., Irvine, W., Matthews, H., Brown, R., Godfrey, P. 1986, ApJ, 300, L79

Martin, A. 1972, MNRAS, 157, 31

Mauersberger, R., Henkel, C., Wilson, T. 1987, A&A, 173, 352

Pratap, P., Menten, K., Reid, M., Moran, J., Walmsley, C. 1991, ApJ, 373, L13

Schneps, M., Lane, A., Downes, D., Moran, J., Genzel, R., Reid, M. 1981, ApJ, 249 124

Scott, P. 1978, MNRAS, 183, 435

Wilson, T., Henkel, C. 1988, A&A, 206, L26

Wilson, T., Johnston, K., Henkel, C. 1990, A&A, 229, L1

Wilson, T., Gaume, R., Johnston, K. 1991, A&A, 251, L7

Masers in star-forming regions associated with cold, bright IRAS sources

G. C. MacLeod[1], M. J. Gaylard[1], E. Scalise Jr.[2]
& V. A. Hughes[3]
[1]Hartebeesthoek Radio Astronomy Observatory,
South Africa
[2]Instituto de Pesquisas Espaciais, Brazil
[3]Queen's U. at Kingston, Canada

Abstract

Twenty-four regions of massive star-formation have been searched for OH, H_2O, 6.6-GHz and 12.2-GHz CH_3OH masers. Chosen from the IRAS Point Source Catalog by their infrared colours, these objects are the coolest objects in the catalog. Masers were detected towards all but two of these IRAS sources. Seven sources were observed in the ^{12}CO (J=1-0) line and all show evidence of molecular outflow. In six of the CO sources we have found maser emission with peaks in both the red and the blue wings of the outflow. We infer that the conditions for masing in these regions are probably produced by the mechanical energy of the outflowing material.

Introduction

The criteria used to select young regions of massive star-formation from the IRAS Point Source Catalog were : $\log_{10}[S_{25\mu}/S_{12\mu}] \geq 1.3$, $\log_{10}[S_{60\mu}/S_{25\mu}] \geq 0.8$ and $S_{100\mu} \geq 500$ Jy (Hughes & MacLeod 1989). These criteria were used because they selected sources which have IR colours similar to Cepheus A, believed to be one of the youngest regions of massive star-formation known (Hughes 1988), and they resulted in a sample of 24 objects. From radio observations of twelve of the regions, together with the IRAS data, MacLeod (1990) showed that these regions are excited by early B- or O-type stars. Evolutionary models of HII regions indicate that they are in a early epoch of star-formation (Fukui 1988; MacLeod & Hughes 1991).

Many of these sources have previously been observed in various maser transitions. Ground state hydroxyl masers are associated with 16 of the 24 sources, water masers with eight sources, 6.6-GHz A^+-methanol masers with nine, and 12.2-GHz E-methanol masers with three.

We made high spectral resolution searches for masers in the 1.6-GHz OH, 22-GHz H_2O, 6.6-GHz A^+-CH_3OH and 12.2-GHz E-CH_3OH transitions. The OH and CH_3OH observations were made with the 26-m Hartebeesthoek telescope, the 22-GHz H_2O observations were made with the 14-m Itapetinga telescope, and CO observations were made at 115-GHz with the 14-m FCRAO telescope. We discuss some implications of the results.

Results and Discussion

Many of the sources have not been previously observed at all four OH ground state frequencies, or do not have published spectra, hence the OH observations of 22 of the sample represent an almost complete set. Two new OH masers were detected in this search, towards IRAS07427-2400 and IRAS18265-1517 (MacLeod 1991). An OH cloud was found associated with IRAS16596-4012. Six new, and two possible new H_2O masers were found. Two new, and one possible 6.6-GHz A^+-methanol masers were seen, but no new 12.2-GHz methanol masers were detected. CO outflows were detected in all seven observed objects. The results are summarized in Table 1.

The detection rates for OH, H_2O, 6.6-GHz and 12.2-GHz methanol masers towards these sources are 78%, 63%, 48% and 13% respectively. No masers were found towards two of these sources, IRAS 16596-4012 and 22272+6358A. Towards the latter neither OH nor CH_3OH were observed, being too far north, while Palla *et al.* (1991) searched for H_2O and failed to detect it. If we consider that of the $\sim$600 presently known HII regions only $\sim$210 have associated OH masers (33%), then our sample has a much higher detection rate. However the total sample of known HII regions represent a very large range in age of massive star-formation. In the Palla *et al.* (1991) colour-selected sample of objects the H_2O maser detection rate increases with increasing redness. Five out of eight of their sources ($\sim$62%) with IRAS colours satisfying our criteria have associated H_2O masers. Within statistical limits our H_2O detection rate matches this result. Menten (1991) stated that the 6.6- and 12.2-GHz class II methanol masers are associated with ultracompact and hence very young HII regions. His 6.6-GHz maser detection rate towards sites of star-formation was 68%, ours being 48%. Menten (1991) also found that the detection rate of 6.6-GHz to OH masers was 76%, while here we find only 55%. These differences in the methanol detection rate are probably due to our lower sensitivity, one tenth that of Menten (1991).

Radio data have been obtained on the UC HII regions embedded in the infrared sources for twelve of the 24 sources and compared to the IRAS data (MacLeod 1990). The spectral type of the exciting star of each region was determined independently from the radio and IRAS flux densities, using kine-

117

matic distances estimated from maser and CO radial velocities (adopting R_0 = 8.5 kpc). For two objects, IRAS 17233-3606 and IRAS 17352-3153, kinematic distances cannot be used because they are within 10^0 of the Galactic centre. We place 17233-3606 at 1 ± 0.5 kpc, and 17352-3153 (V_{lsr} = -54 km s^{-1}), in the 3 kpc arm at a distance of 5.5 kpc. We assume that only one ZAMS star excites each region. In all cases the estimates of the spectral types derived from the radio and from the infrared fluxes were found to be consistent given the uncertainties. The derived spectral types are in the range B2.5 to O9.5.

We have compared the CO velocity profiles to those of the four types of masers. In six of the seven cases masers are detected within the velocity ranges of both the blue and the red CO wings. An example is IRAS 05480+2545, in which the CO peak lies at -8.9 km s^{-1} and has a weak blue wing, while the double peaked 6.6-GHz maser spectrum has components at -14.8 km s^{-1} and -5.0 km s^{-1}. The gaussian FWHM of the CO line is $\sim$7 km s^{-1} (see Fig. 1a, b). In Cep A the velocity range of the OH masers is nearly centered on the CO peak, but the 6.6-GHz masers lie only in the red CO wing.

The CO and maser velocity structures in these objects imply that the mechanical energy of the outflow may be responsible for the formation of the masing regions. This would occur by direct compression of the molecular cloud or by the creation of shock waves in the cloud, so that the required high densities are obtained, the emission being stimulated through collisional excitation. However it is also possible that the maser emission lies in a disk collimating the outflow.

References

Fukui, Y., 1988. Vistas in Astronomy, **31**, 217.

Hughes, V.A., 1988. Astrophys. J., **333**, 788.

Hughes, V.A. & MacLeod, G.C., 1989. Astr. J., **97**, 789.

MacLeod, G.C., 1990. Ph.D. thesis, Queen's University at Kingston, Canada.

MacLeod, G.C., 1991. Mon. Not. R. astr. Soc., **252**, 36P.

MacLeod, G.C. & Hughes, V.A., 1991. Astr. J., **102**, 658.

Menten, K.M., 1991. Astrophys. J. (Lett.), **380**, L75.

Palla, F., Brand, J., Cesaroni, R., Comoretto, G. & Felli, M., 1991. Astr. Astrophys., **246**, 249.

Table 1: Results of the Maser observations

IRAS Name	$S_{100\mu m}$ (Jy)	v (kms^{-1})	D_{kin} (kpc)	Sp	Amplitude(Jy) 1.6-GHz	22-GHz	6.6-GHz	12.2-GHz	CO
05480+2545	643	−10.0	2.1	B0.75[1]	≤0.3	[2]36	[2]20	≤3	wbw
05553+1631	525	6.0	1.2	B2.3[1]	≤0.3	32	≤1	≤2	outflow
07427−2400	744	63.5	5.0	B0.0[1]	[2]1.9	[2]28	≤1	≤2	
12091−6129	795		2.3	B0.75	≤0.3	[2]60	[3]9	≤3	
12326−6245	9245	−39.0	4.3	O6.0	25.0	[2]84	≤4	≤3	
14498−5856	2209	−53.5	3.9	O9.5	31.2	180	[3]5	≤3	
15360−5554	1411		1.3	B2	≤0.3	[3]24	[3]≤3	[3]≤3	
15520−5234	16390	−45.0	3.2	O6.0	98.0	≤20	363	23	
16060−5146	19170	−85.0	5.3	O5.0	121.2	[2]264	≤6	≤3	
16445−4459	1977	−51.0	4.0	O9.5	6.7	≤20	[3]≤5	≤3	
16547−4247	6471	−30.5	3.0	O7.5	77.8	92	≤4	≤3	
16594−4137	1126	−5.5	15.6	O5.5	23.3	76	3	≤5	
16596−4012	525	−10.7			[2]abs	≤20	≤5	≤5	
17016−4124	7000	−26.5	2.9	O7.5	38.1	≤20	148	7.1	
17233−3606	20410	1.5	0.5	B0.7[1]	749.6	96	190	≤4	
17352−3153	748	−54.0	5.5	B0.0[1]	3.6	≤20	[3]abs	≤5	
18265−1517	1297	17.0	2.0	B1.0[1]	[2]2.2	≤20	[2]19	≤4	outflow
19035+0641	922	33.0	2.4	B0.75[1]	91.0	≤20	14	≤3	
19088+0902	666	10.5	2.9	B0.75	105.6	92	≤5	≤4	
19095+0930	2744	42.0	3.0	B0.0[1]	117.4	[2]512	152	≤2	
20081+3122	3120	12.5	2.3	B0.0[1]	15.3	[2]120	91	≤4	outflow
20126+4104	1947	−6.5	4.5	B0.5[1]	4.6	40	12	≤3	
22272+6358A	702	−8.5	1.0	B3.0[1]		≤1.1			bw
22543+6145	20450	−12.0	0.73	B0.5[1]	10.0	4700	1420	302	outflow

Notes: wbw Weak Blue Wing; bw Blue Wing; abs absorption
[1]Spectral type determined from both the VLA radio
data and IRAS data. [2]New detections. [3]Possible detections.

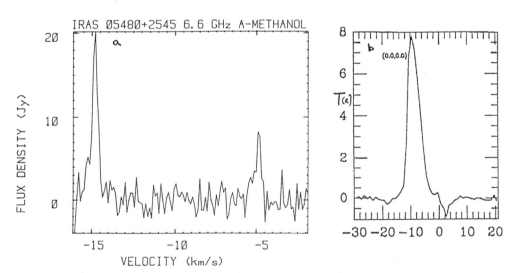

Figure 1. Spectra of the source IRAS 05480+2545 are shown. (a) The double-peaked 6.6-GHz A$^+$-methanol spectra. (b) The ^{12}CO (J=1→0) spectra.

CH IN TRANSLUCENT MOLECULAR CLOUDS

Loris Magnani
Department of Physics and Astronomy
University of Georgia
Athens, Georgia 30602

ABSTRACT

Under typical interstellar cloud conditions, the ground state hyperfine transitions of the methylidyne radical (CH) are anomalously excited. The resulting weak maser emission is particularly succesful in tracing low-density molecular gas. The CH ground state transitions at 3.3 GHz are routinely detected in diffuse clouds, translucent clouds, and dark molecular clouds. Recent observations of CH emission in translucent clouds are reviewed and a value for the $N(H_2)/W(CO)$ conversion factor appropriate for the translucent cloud MBM16 is derived on the basis of CO and CH data.

I. INTRODUCTION

The majority of the high-latitude molecular clouds identified in the Magnani, Blitz, and Mundy (1985) survey are translucent molecular clouds. Some of the properties of this class of interstellar cloud are described by van Dishoeck and Black (1988). Basically, a translucent cloud is a small molecular cloud with visual extinction in the 1-5 magnitude range in which both photoprocesses and gas phase reactions play an important role in the chemistry. Perhaps the most surprising property of the translucent high-latitude clouds is that they are not gravitationally bound and, in the absence of pressure confinement, are expanding on a time scale of 10^6 years. Clearly, this deduction depends critically on the mass of the clouds. A reliable method for determining the cloud mass is needed before the gravitational stability of the objects can be assessed.

The most common method for determining the mass of a given translucent cloud is to observe it in the CO(1-0) transition and then apply a conversion factor relating the integrated antenna temperature of CO [defined as $W(CO)$] to the column density of H_2. If the cloud distance is known, the column density of $N(H_2)$ leads immediately to a value for the cloud mass. Such $N(H_2)/W(CO)$ conversion factors are routinely used in determining the mass of giant molecular clouds and the derivation of the conversion factor is reviewed by Bloemen (1989) and van Dishoeck and Black (1987).

Magnani, Blitz, and Wouterloot (1988) used extinction data on high-latitude clouds and the Bohlin, Savage, and Drake (1978) color excess$-N(H_2)$ relation to calculate $N(H_2)/W(CO)$ conversion factors for several clouds. Their results ranged from 1 - 6 $\times$ 10^{20} cm^{-2} [K km s^{-1}]$^{-1}$ (hereafter, we will drop the units for brevity), consistent with the Bloemen (1989) value for the Galaxy of 2.3×10^{20}. Magnani, Blitz, and Wouterloot (1988) note that the range of results is probably real and indicates true variations in the conversion factor from cloud to cloud. An alternate derivation of the $N(H_2)/W(CO)$ ratio by de Vries, Heithausen, and Thaddeus (1987) relies on a detailed comparison between IRAS 100 μm cirrus data and CO and HI observations of a large cloud complex at high-latitudes. The values for the conversion factor obtained by the de Vries *et al.* group are an order of magnitude lower than the Magnani *et al.* values and thus lead to smaller estimates of the cloud masses. The smaller masses, in turn, exacerbate the disparity between the gravitational potential energy and the kinetic energy of the cloud. Both techniques contain some fairly severe shortcomings. The Magnani *et al.* work is forced to rely on star counts which contain notorious uncertainties in low-extinction regions. On the other hand, the de Vries *et al.* work calibrates the infrared emissivity of the molecular part of the clouds on the basis of the IR emissivity of the atomic portion. Boulanger (1989) questions the validity of the assumption that the atomic and molecular portions of an interstellar cloud have the same IR emissivity per hydrogen nucleon.

Another method for determining the $N(H_2)$ content of the clouds is required to resolve this controversy. In this paper we present preliminary results of a program to determine $N(H_2)$ from $N(CH)$ thereby obtaining independently of extinction or IR data the $N(H_2)/W(CO)$ ratio of a cloud.

II. CH AS A LINEAR TRACER OF MOLECULAR HYDROGEN

It is well-known that the efficient excitation properties of the CH ground state hyperfine transitions at 3.3 GHz and the subsequent weak maser behavior lead to detectable emission from interstellar clouds (*e.g.*, Rydbeck *et al.* 1976; Sandell and Johansson 1982; Federman and Willson 1982; Mattila 1986). The main line at 3335 MHz is readily detected in translucent clouds along virtually all lines of sight (Magnani *et al.* 1989) making this molecular transition an ideal tracer of the molecular gas in these objects. Moreover, the linear relationship between $N(CH)$ and $N(H_2)$ has been firmly established for visual extinctions less than 4 magnitudes both from theoretical considerations (Federman 1982) and various observational programs (for a review, see Mattila 1986). Observations of the CH main line thus lead immediately to values of $N(H_2)$ along the sampled line of sight which are reliable, in general, to a factor of two.

III. CH AND CO OBSERVATIONS OF MBM16

The CH observations which are used in this study are described in Magnani *et al.* (1989). The CH profiles from three positions (denoted as MBM16A, MBM16B, and MBM16C, respectively) at the edge of cloud 16 of the Magnani, Blitz, and Mundy (1985) compilation are compared to CO(1-0) data taken at the NRAO 12m telescope. The observing set up for the CO observations is described in Magnani and Onello (1992). The main difficulty in comparing the CH to the CO data is that the CH beam is 9' in size while the CO beam is only 1' in size. The CH beam was thus "synthesized" with 3 or 4 dozen CO observations per CH position.

IV. RESULTS AND DISCUSSION

The CH integrated antenna temperature is converted into a molecular hydrogen column density via equation 1 of Magnani *et al.* (1989) and the CH-$N(H_2)$ relation presented in Mattila (1986). The values for the $N(H_2)/W(CO)$ ratio which are obtained in this fashion range from 7 - 11 $\times 10^{20}$ with an average of 9×10^{20}. This latter value is consistent with the Magnani, Blitz, and Wouterloot (1988) determination for all high-latitude clouds (6.4×10^{20}) and it is an order of magnitude greater than the value quoted in de Vries, Heithausen, and Thaddeus (1987).

As pointed out by Magnani, Blitz, and Wouterloot, the conversion factor may not be constant from cloud to cloud. It is plausible that the value for the de Vries *et al.* molecular cloud complex is indeed smaller than for other high-latitude clouds. However, we note that all non-IR based conversion factors are in the several times 10^{20} range (see Blitz 1991) rather than in the 10^{19} range. Unless confirmation of the $N(H_2)/W(CO)$ ratio is available from independent observations, we caution against the use of IR-derived ratios in the 10^{19} range when determining translucent cloud masses. The method outlined in this paper using CH observations, when applied to a wide variety of translucent high-latitude clouds, may conclusively resolve this controversy.

REFERENCES

Blitz, L. 1991, in *Molecular Clouds*, ed. R.A. James and T.A. Miller (Dordrecht:Reidel), p. 49.

Bloemen, H. 1989, *Ann. Rev. Astr. Ap.*, **27**, 469.

Bohlin, R.C., Savage, B.D., and Drake, J.F. 1978, *Ap. J.*, **224**, 132.

Boulanger, F. 1989, in *The Physics and Chemistry of Interstellar Molecular Clouds*, ed. G. Winnewisser and J.T. Armstrong (New York:Springer), p. 127.

de Vries, H.W., Heithausen, A., and Thaddeus, P. 1987, *Ap. J.*, **319**, 723.

Federman, S.R. 1982, *Ap. J.*, **257**, 125.

Federman, S.R. and Willson, R.F. 1982, *Ap. J.*, **260**, 124.

Magnani, L., Blitz, L., and Mundy, L. 1985, *Ap. J.*, **295**, 402.

Magnani, L., Blitz, L., and Wouterloot 1988, *Ap. J.*, **326**, 409.

Magnani, L. *et al.*, 1989, *Ap. J.*, **339**, 224.

Magnani, L. and Onello, J.S. 1992, in preparation.

Mattila, K. 1986, *Astr. Ap.*, **160**, 157.

Rydbeck, O.E.H. *et al.* 1976, *Ap. J. Suppl.*, **31**, 333.

Sandell, G. and Johansson, L.E.B. 1982, in *Regions of Recent Star Formation*, ed. R.S. Roger and P.E. Dewdney (Dordrecht:Reidel), p. 429.

van Dishoeck, E.F. and Black, J.H. 1987, in *Physical Processes in Interstellar Clouds*, ed. G. Morfill and M. Scholer (Dordrecht:Reidel), p. 241.

van Dishoeck, E.F. and Black, J.H. 1988, *Ap. J.*, **334**, 771.

Ammonia Masers in the Interstellar Medium

T.L. Wilson P. Schilke

Max-Planck-Institut für Radioastronomie
Auf dem Hügel 69
W-5300 Bonn 1
Federal Republic of Germany

Summary

Since 1986, 10 ammonia masers have been found. Most of these are found in the sources W51D, W51e1/e2, NGC7538, DR21 and W33. Among them are a few masers or maser candidates, arising from metastable inversion lines. There are many more masers arising from non-metastable inversion lines. The most outstanding is intense maser emission from the (J,K)=(9,6) inversion line. Only in W51D are any masering non-metastable transitions of para-NH_3 to be found. Since the masering levels are more than 500 K above ground state, there are a large number of levels populated, and the excitation scheme must be complex. It is likely that there is no unique excitation scheme for all types of ammonia masers. Although there have been a few attempts to model the ammonia maser excitation, including excitation involving vibrationally excited levels, the quest for an all-encompassing ammonia maser excitation model is still going on.

Introduction

The ammonia molecule was first detected in the interstellar medium by means of inversion transitions, at a wavelength of 1.3 cm by Cheung et al.(1968). The first measurements were restricted to the metastable ($J = K$) lines, since the non-metastable ($J > K$) ammonia inversion lines are usually rather weak. The first detection of a non-metastable transition was made in 1972. With the use of large radio telescopes and sensitive receivers, the number of inversion-rotation transitions has risen steadily. The best source for detecting NH_3 inversion-rotation lines is Orion KL, for which the total number of NH_3 inversion-rotation lines found is 34 (Hermsen et al. 1988). The abundance of NH_3 in Orion KL is so large that 6 metastable transitions of $^{15}NH_3$ have been detected (Hermsen et al. 1985). All of the NH_3 inversion transitions found in Orion KL are consistent with Local Thermodynamic Equilibrium (LTE) conditions.

In Fig. 1 is shown an energy level diagram of NH_3 (taken from Wilson et al. 1990). The inversion levels marked with script "m" are those which seem to show maser emission in W51D. This region shows the most maser inversion lines. For most inversion lines, the evidence for masering rests on the spike-like line shapes, or time variability. The (11,9), (9,6) and (6,3) lines are markedly time variable; the brightness temperatures, T_{MB}, of the (9,6) and (9,8) lines are $> 10^{13}$ K (Pratab et al. 1991) and $> 10^8$ K (Wilson et al. 1991), respectively.

The details of the energy level structure of ammonia have been treated often (see Ho and Townes 1983). With one receiver one can measure inversion lines from 23 K (the (1,1) inversion line) to >1200 K (the (12,10) line) above the ground state (see, e.g. Mauersberger et al. 1986a, Wilson et al. 1990). Allowed rotational transitions in NH_3 follow the rule $\Delta K = 0$, $\Delta J = 0, \pm 1$, and parity change. The $\Delta J = 0$ transitions give rise to inversion lines. That is, a transition is allowed only within a K ladder. These transitions produce lines in the far IR. Only collisions can link different K ladders. The spin of the three H atoms separates energy levels into ortho or para-NH_3. The ortho- states are labelled in Fig. 1 in script. In the $K = 0$ ladder of ortho-NH_3, one of the inversion levels is missing because of the Pauli principle, which is not the case for any other K ladder. This asymmetry together with parity selection rules may be the crucial ingredient in masers found in the (3,3) inversion transition, since these levels are coupled by collisions. Transitions between ortho and para

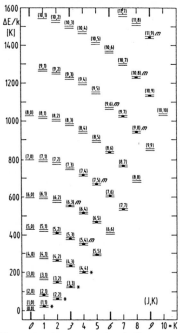

Figure 1: An energy level diagram of the vibrational ground state of NH_3. The vertical scale is energy above ground i K. The splitting of the (J, K) inversion levels is not to scale. The K quantum number in script is ortho-NH_3. The inversion levels with filled bars have been measured in W51d by Mauersberger et al. (1987). The doubletts with a star have been measured in W51d in $^{15}NH_3$. The levels marked with m show evidence for masering.

requires a spin flip in H. This transition is forbidden, so conversion can only occur via chemical reactions (e.g. by H_3^+) in the gas phase or on grains. For excitation considerations, ortho and para- NH_3 can be considered as separate species. There is hyperfine structure in NH_3 inversion lines. This is caused by the quadrupole moment of the ^{14}N nucleus; a finer splitting is caused by the magnetic interaction of the spins of the H nuclei and the rotation of the molecule. In dense molecular clouds, only the quadrupole hyperfine satellites can be separated from the main component.

In most sources, there is no evidence in metastable NH_3 inversion line spectra for deviations from Local Thermodynamic Equilibrium (LTE). This makes the measurement of metastable NH_3 lines very useful, since from the ratio of two metastable inversion lines, one can obtain an estimate of the kinetic temperature, T_{kin}, which does not depend on the beam filling factor (as with CO).

Metastable inversion transitions

Since one can measure many NH_3 lines, one simple but not entirely secure criterion for identifying masers is a line shape which is very different from other NH_3 lines in a source. (More secure but more involved are interferometric measurements of brightness temperatures which exceed the thermal equivalent of linewidth, and time variability.) We show an example, for NGC7538-IRS1, in Fig. 2.

This is a dense molecular cloud with NH_3 data showing a kinetic temperature, T_{kin}, of about 150 K to 200 K. This source shows the largest number of masers found so far. Here the (3,3) line of $^{15}NH_3$ has a very different shape from the (1,1) and (2,2) lines (Mauersberger et

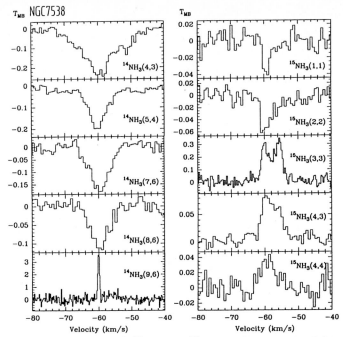

Figure 2: A set of selected ^{14}NH$_3$ and ^{15}NH$_3$-profiles toward NGC 7538-IRS1, by Schilke et al. (1991).

al. 1986b). Calculations by Walmsley and Ungerechts (1983) and Guilloteau et al. (1983) showed that maser emission of (3,3) inversion line, and only this line, could be produced. Thus it was natural to suspect that the (3,3) inversion line was caused by masering. Here the maser action was proved to be present by observations with the VLA (Johnston et al. 1989), but this is the only definitely known (3,3) inversion line maser to date. Gaume et al. (1991) have mapped the positions of the ^{15}NH$_3$ masers toward the continuum background. The radial velocities and positions indicate rotation-like motion, about a 30 to 60 solar mass object.

There are some difficulties with the theoretical inversion scheme for the (3,3) line. First, there is also emission in the metastable (4,4) and the non-metastable (4,3) inversion lines (Schilke et al. 1991), implying maser emission which is unaccounted for by the "standard" (3,3) pumping model. Furthermore, detailed calculations showed (Flower, Offer and Schilke 1991) that the parity selection rule which is responsible for the maser pumping and which is obeyed by He and para-H$_2$, appears *not* to be followed by collisions with ortho-H$_2$ Hence considerable amounts of ortho-H$_2$ quench the maser. Masering on the basis of the "standard" scheme would be expected only in special, non-LTE situations. Even in these favored situations, only low gain maser can be produced.

Other examples of line profiles which indicate maser action are the (5,5) inversion line in G9.62+0.19 (Cesaroni et al. 1992). As with the (4,4) inversion line in NGC7538, such a single maser line in para-NH$_3$ cannot be explained by any theory. Any explanation of masers from metastable lines of NH$_3$ must take into account the fact that many other metastable transitions would also be predicted to be masering. This is not found.

Non-Metastable Inversion Lines

Masers are much more commonly found in transitions between non- metastable levels. This is related to the fact that both IR fields and collisions influence the excitation of these levels, and also the level populations are small compared to metastable levels. Thus a small change in the level population can mean a large relative change (i.e. inversion) in the populations of the doublets. For example, sub-thermally excited metastable levels (J,K) can easily cause the populations of the non-metastable levels (J+1,K) to be inverted, because parity selection rules couple the lower of the (J,K) levels with the upper of the (J+1,K) levels. However, this mechanism can explain only some cases of inversion. Attempts to construct excitation models have been made by Schilke (1989) and Brown and Cragg (1991). The model of Brown and Cragg (1991) can explain some features of the (6,3) line emission, but no attempt was made to treat the more intense (9,6) inversion line masers. The models of Schilke (1989) used IR radiation from internal dust to populate the non-metastable levels and via excitation to the vibrational states. These models can invert populations of non-metastable levels, but these *cannot* selectively invert the (9,6) and (6,3) masers without also producing maser action in many other lines. Masers are found in the (9,6) line in a number of sources (Madden et al. 1986), but other NH_3 masers are not observed, except in W51D (Mauersberger et al. 1987). So, there must be some process favoring the (9,6) and (6,3) lines. In addition, the ortho and para-NH_3 masers in W51D have a very different time-variability behavior. For example, the (11,9), (9,6) and (6,3) lines vary rapidly, but the para masers hardly at all (Wilson and Henkel 1988). The (9,8) maser is found toward a compact continuum source (Wilson et al. (1991), and presumably amplifies the continuum radiation. The (9,6) maser has a brightness temperature of $> 10^{13}$ K, and is located near the ultra compact HII regions W51e1/e2 at present (Pratap et al. 1991). Before 1988, the strongest (9,6) emission was found toward W51D, but no precise position was determined then.

References

Brown R.D., Cragg D. 1991, ApJ, 378, 445
Cesaroni R., Walmsley C.M., Churchwell E.B. 1992, A&A, (submitted)
Cheung A.C., Rank D.M., Townes C.H., Thornton D.D., Welch W.J 1968, Phys. Rev. Lett., 21, 1701
Flower D.R., Offer A., Schilke P. 1990, MNRAS, 244, 4p
Gaume R., Johnston K.J., Nguyen H.A., Wilson T.L., Dickel H.R., Goss W.M, Wright M.C.H 1991, ApJ, 376, 608
Guilloteau S., Wilson T.L., Martin R.N., Batrla W., Pauls T.A. 1983, A&A, 124, 322
Hermsen W., Wilson T.L., Walmsley C.M, Batrla W. 1985, A&A, 146, 134
Hermsen W., Wilson T.L., Walmsley C.M., Henkel C. 1988, A&A, 201, 285
Ho P.T.T., Townes C.H. 1983, Ann. Rev. A&A, 21, 239
Johnston K.J., Stolovy S., Wilson T.L., Henkel C., 1989, ApJ, 343, L41
Madden S.C., Irvine W.M, Matthews H.E., Brown R.D., Godfrey P.D. 1986, ApJ, 300, L79
Mauersberger R., Henkel C., Wilson T.L., Walmsley C.M 1986a, A&A, 162, 199
Mauersberger R., Wilson T.L., Henkel C. 1986b A&A, 160, L13
Mauersberger R., Henkel C., Wilson T.L. 1987, A&A, 173, 352
Pratap P., Menten K.M., Reid M.J., Moran J.M., Walmsley C.M. 1991, ApJ, 273, L13
Schilke P. 1989, Diploma Thesis, Bonn Univ.
Schilke P., Walmsley C.M., Mauersberger R. 1991, A&A, 247, 516
Walmsley C.M., Ungerechts H. 1983, A&A, 122, 164
Wilson T.L., Henkel C. 1988, A&A 206, L26
Wilson T.L., Johnston K.J., Henkel C. 1990, A&A, 229, L1 Wilson T.L., Gaume R., Johnston K.J. 1991, A&A, 251, L7

6. OH MASERS IN STAR FORMING REGIONS

OH MASERS IN STAR-FORMING REGIONS :
INTERPRETATION OF OBSERVATIONS

M.D.Gray, K.N.Jones, R.C.Doel & D.Field
School of Chemistry
University of Bristol, Bristol BS8 1TS UK

R.N.F.Walker
Department of Physics
University of Bristol, Bristol BS8 1TL UK

ABSTRACT

OH masers are a potentially powerful probe of the physical conditions in massive star-forming regions, giving detailed information on the kinetic and dust temperatures, number densities of H_2 and OH, bulk flows and magnetic fields in the vicinity of massive young stellar objects. Using current theories and data, we describe how bright maser spots, as seen in VLBI at 18cm, 6 cm and 5 cm , may arise and we outline the physical conditions which may be associated with masers at different frequencies.

1. INTRODUCTION

We are concerned here specifically with the interpretation of bright OH maser spots , normally involving strong saturation of the OH inversions. The general question of OH inversion and anti-inversion, as a function of the ambient physical conditions, was dealt with in Gray, Doel & Field 1991, hereafter Paper I (see also Cesaroni & Walmsley 1991). The work reviewed here is described in considerably greater detail in Gray, Field & Doel 1992, hereafter Paper II. The important new element in the present work is the application of the semi-classical theory of radiation transport described in Field & Gray 1988 in conjunction with the model of Paper I. Our objective is to be able to interpret a spot on a VLBI map by stating that the presence of (say) 1665 MHz with this flux dictates that the physical conditions are so-and-so temperature, OH number density, bulk dynamics etc., or that 1612 MHz nearby, indicates that the medium is, for example, 15K colder than in the 1665 MHz zone. We are some way from achieving this objective but it is in sight. Moreover theories are now suggesting the need for specific new observations. In particular we are interested in regions in which the same column of gas may support more than one maser frequency at the same or at different velocities. We mention below a specific region in W3(OH) where 1665, 1720 and 4765 MHz (in J=1/2) are spatially superposed at a succession of velocity shifts.

2. THE MODEL

The initial requirement for any maser model is to calculate the unsaturated populations of the molecular energy levels. We use 48 hyperfine levels of OH corresponding to the first 12 rotational states up to an energy of 800 cm^{-1}. We employ a self-consistent but approximate set of rate coefficients for inelastic collisions between OH and H_2 for H_2(J=0) only. Kinetic and dust temperatures, T_K and T_d, are held equal and a dust emission law $I_{dust} = (\lambda/80)^{-2}$ x $B(\lambda, T_d)$ is used, where B() is the Planck function. To calculate the mean continuum and line intensities we use the LVG (or Sobolev) approximation. Line overlap, including the important phenomenon of multiple line overlap, is included (Doel, Gray & Field 1990). The maser radiation transport is semiclassical, treating the maser radiation field classically and the response of the molecules quantum-mechanically. The effects of saturation are coupled to the kinetic events of absorption, emission and collision in this model. A single maser ray, inducing emission at line centre, is considered. Problems of beaming, geometry and competition between rays are not considered. Velocity redistribution and polarisation are also both omitted (see Gray et al, this volume).

3. RESULTS

The range of physical conditions found is T_K = 30 to >150K, $[H_2]=10^6$ to 2 x 10^7 cm^{-3}, [OH]=10 to several hundred cm^{-3} and velocity shifts up to several kms^{-1}. Specific conditions apply for each maser line, see below. An interesting phenomenon to emerge is that of competitive gain (see Lewis, this volume and Field 1985) in which masers sharing common levels, or involving levels strongly coupled by radiation and collisions via other J states, compete for the same or strongly coupled populations as saturation becomes important. Competitive gain tends to lead to the emergence of one strong line, in the saturated regime.

We may summarise our major results as follows. Intense 1665 MHz spots are formed in gas supporting velocity shifts of > 1 kms^{-1}, T_K=50 to 75K, [OH] >20 cm^{-3}. An example is given in fig.1. The broad range of physical conditions coupled with short gain (amplification) lengths for 1665 MHz are not found for any other line and reveals why it is this frequency which most commonly forms bright maser spots (Gaume & Mutel 1987). The phenomenon of line overlap is a valuable probe of the bulk gas dynamics and indicates the presence of accelerating flows where 1665 MHz spots occur. 1612 MHz maser spots are also signposts of accelerating flows and of similar conditions to those for 1665 MHz except that T_K = 30-40K, [OH] can be as low as 10 cm^{-3}, and velocity shifts are >2 kms^{-1}. Intense1667 MHz maser spots may form under conditions similar to those of 1665 MHz with the important difference that no velocity shift or only a weak shift (e.g. <0.5 kms^{-1}) is present. Under a limited range of conditions 1667 MHz may however form in the presence of larger shifts. 1720 MHz masers form under similar conditions to 1665 MHz but now at velocity shifts between those of 1667 and 1665 MHz (e.g. around 1 kms^{-1}). The well-established observational correlation between 4765 and 1720 MHz is readily demonstrated. We can also locate the transition region in which the 1667/1720 MHz regime, [H$_2$]=<10^7 cm^{-3}, turns over into the 6035/6030 MHz dominated regime [H$_2$]>=2 x 10^7 cm^{-3}, in the absence of velocity shifts.

We have studied the region around the origin of the map of W3(OH) in Bloemhof, Reid & Moran, 1992, in which there is a strong clump of 1665 MHz masers at a Zeeman-corrected velocity of-46.3 kms^{-1}, a set of 1720 MHz masers peaking at -44.7 kms^{-1}(Fouquet & Reid) and a set of 4765 MHz masers peaking at -43.21 kms^{-1}(Baudry et al.1989). Observations strongly suggest that these are spatially superposed. It then follows from our calculations that this region is composed of an accelerating flow initiated at high pressure, T_K=125K, [H$_2$]=2 x 10^{7}cm^{-3}, and expanding adiabatically, into the intercloud medium, forming a lower pressure zone, T_K=63K, [H$_2$]=7.1 x 10^6 cm^{-3}. In the higher pressure zone, 4765 MHz rises and falls, in a manner that reproduces the observations. This is followed by the emergence at greater shifts of 1720 MHz. In the lower pressure zone, conditions yield strong 1665 MHz. The physical picture of an outflowing mass of material matches very well the findings of the proper motion studies for W3(OH) reported in Bloemhof et al.

4. CONCLUSIONS

The results reported here give the flavour of what could be achieved in the interpretation of OH maser data, if we can develop more reliable models. Such models require accurate rate coefficients for energy transfer, including collisions with H_2 $J>0$, replacement of the Sobolev approximation with Λ-iteration methods, inclusion of maser beaming, (complete) velocity redistribution and polarisation. Each of these developments constitutes a major project in its own right and reflects the need to place considerable emphasis on the analysis of the observational data. With regard to new data, we require more multi-wavelength observations to identify spatial superpositions. The VLBA would appear to be an excellent instrument for this purpose.

References

Baudry A., Diamond P., Booth R.S., Graham D. & Walmsley C.M., 1989, Astr. Astrophys., **201**, 105
Bloemhof E.E., Reid M.J. & Moran J.M., 1992, Astrophys. J., to appear
Cesaroni R. & Walmsley CM.M, 1991, Astr. Astrophys. **241**, 537
Doel R.C., Gray M.D. & Field D., 1990, Mon. Not. R.A.S. **244**, 504
Field D., 1985, Mont. Not. R.A.S. **217**,1
Field D. & Gray M.D., 1988, Mon. Not. R.A.S, **234**, 353
Fouquet J.E. 1 Reid M.J., 1982, Astron. J. Suppl.87, 691,
Gaume R.A. & Mutel R.L., 1987, Astrophys. J. Suppl. Ser. **65**,193
Gray M.D., Doel R.C. & Field D., 1991, Mon; Not. R.A.S. **252**, 30
Gray M.D., Field D. & Doel R.C., 1992, Astr. Astrophys., to appear

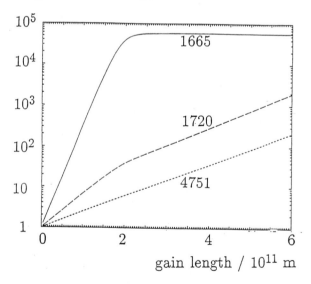

Fig.1 : Maser growth as a function of gain length for $T_K=50K$, [OH]=60 cm^{-3}, velocity shift =1.7 kms^{-1}

EXCITED OH 4.7 GHz MASERS ASSOCIATED WITH IRAS FAR-INFRARED SOURCES

M.R.W.Masheder(a), R.J.Cohen(b), J.L.Caswell(c), R.N.F.Walker(a) and M.Shepherd(b)
(a) Dept. of Physics, University of Bristol, Bristol, BS8 1TL, England
(b) N.R.A.L., Jodrell Bank, Macclesfield, Cheshire, SK11 9DL, England
(c) ATNF, CSIRO, PO Box 76, Epping 2121, NSW, Australia

Abstract

We describe the results of an all-sky search for maser emission from excited OH in the $^2\Pi_{1/2}$, J=1/2 state at 4765, 4750 and 4660 MHz, carried out at Jodrell Bank and at Parkes in 1989 and 1991. 129 sources were observed in all including all objects from the Cohen, Baart & Jonas (1988) (CBJ) sample of far infrared IRAS sources with 60 μm flux, F(60) > 4000 Jy for which OH 18-cm emission was already known. 18 objects were detected in all, including 7 new discoveries and a new maser region in W49. Three of these were also detected at 4750 MHz, including the first strong 4750 MHz maser (S252). Three objects were detected at 4660 MHz, including a new discovery seen in this line only. We found strong variations in seven sources.

Introduction

There are three hyperfine transitions in the Λ-doublet of the $^2\Pi_{1/2}$ J=1/2 excited state of OH : F=1-0 (4765 MHz), F=1-1 (4750 MHz) and F=0-1 (4660 MHz). These lines are often seen in broad-band 'quasi-thermal' emission from young compact HII regions and sometimes as maser emission, most often at 4765 MHz. (Gardner and Ribes 1971, Gardner and Martin-Pintado 1983). Evidently the pumping process is much less effective for masers in the excited states of OH than for those in the ground state. This provides an important key to the understanding of the pumping mechanisms at work by constraining the range of possible models. Field and Gray (1988) have carried out pumping and radiative transfer calculations for OH using a 48-level scheme in which they allow many masers to propagate simultaneously. (see also Gray et.al. 1991,1992). They find that there is considerable cross-coupling between the various masers, so the line strength ratios can be used to estimate the physical conditions in the active regions.

The present survey work was primarily motivated by the lack of statistically useful data on masers at 4.7 GHz. The selection of target objects was guided by surveys carried out by Cohen, Baart and Jonas (1988)(CBJ) at 18 cm. They searched objects which appeared in the IRAS Point Source Catalog with fluxes greater than 1000 Jy in both the 60 μm and 100 μm bands. There were 517 sources in this complete sample, of which 108 were known to be OH masers after their work was finished. Our goal is to search for 4.7 GHz maser emission in all of the powerful IRAS sources in their sample. So far we have made three surveys : at Jodrell Bank in 1989 (see Paper 1), at Parkes in 1991 and again at Jodrell Bank in 1991. The combined all-sky survey covered 129 objects with 18 detections and the report of that work is currently in preparation. The sources observed in the all-sky survey to date may be classified as follows:

a) All IRAS sources with F(60) > 4000 Jy for which 18cm OH maser emission was already known.

b) All IRAS sources with F(60) > 7500 Jy.

c) Some sources previously detected as 4.7 GHz masers. These were observed to check on variability.

Observations

Observations were made with the 76-m Lovell telescope at Jodrell Bank in 1989 and 1991 and with the 64-m telescope at Parkes NSW in 1991. The Jodrell Bank receiver was described in paper 1. The Parkes telescope had a beamwidth of 4.5 arcmin and was fitted with a dual polarisation receiver with $T_{SYS} = 65$ K on cold sky. Simultaneous 512 channel spectra for left and right hand circular polarisations were obtained using an autocorrelation spectrometer. The spectral resolution was 0.15 km/s. Spectra were taken in position referenced mode by taking reference spectra of regions known to be free of emission. Each object was observed for typically 15 minutes.

Spectra were processed offline on the CSIRO VAX system and the STARLINK VAX system at Jodrell Bank. The Spectral Line Analysis Package (SLAP) was used throughout (Staveley-Smith 1985). The data for the two polarisations were treated separately, although no polarisation was expected or observed. A first or second order polymomial baseline was first removed from each spectrum. The spectra for each position were then scaled and added according to the overall system gain and noise. The final spectrum for each source and each line was obtained as a weighted addition of the results for the left and right hand polarisations. This gave a final typical (3 σ) detection limit of 0.3 Jy.

Results

The table lists all the sources detected in the combined survey taken at Jodrell Bank in 1989, at Parkes in 1991 and at Jodrell Bank again in 1991, together with some results from Gardner and Martin-Pintado (1983), Rickard et.al. (1975) and Gardner and Ribes (1971). It is intended to be a complete list of detections so far, at least for maser emission. Column 1 gives the IRAS or other name; column 2 gives the flux given in the IPSC for the 60 μm band ; column 3 gives the velocity of a detection ; columns 4-6 give the peak flux density detected for each of the three lines or, for non-detections, the upper limit.

Seven of the sources listed are new discoveries: 06055+2039, 12073-6233, 16183-4958, 17175-3544, 17233-3606, 17271-3439 and 21413+5442. In addition, we discovered a new maser feature towards a region of the W49 complex previously unassociated with maser activity. We checked the positions of some other previously known 4.7 GHz masers and found that the emission ascribed to DR21(OH) actually comes from a position 1.5 arcmins North, DR21(OH)N, where Brebner (1988) found a weak secondary 18 cm maser source.

Two of our new discoveries are noteworthy inasmuch as they show 4.7 GHz masers but little or no emission from OH in the ground state. 06055+2039 (S252A) is unusual in showing its strongest maser at 4750 MHz as discovered in the paper 1 survey of 1989. It was not detectable with the Parkes telescope in 1991. The 0.4 Jy maser at 4765 MHz in 12073-6233 is a new discovery. This object shows absorption rather than emission in the ground state of OH.

OH 4.7 GHz Detections

IRAS Name	F60 (Jy)	Vel km/s	OH peak flux density (Jy) 4765	4750	4660	Object	Ref
02219+6152	12000	-37.3,-33.9	T=0.1,0.7			W3-IRS5	R,B
			2.2				GM
		-49.4	1.8				J2
02232+6138	9272	-44	3.9	<0.19	<0.32	W3(OH)	J1,J2,GM,R,GR
ORION-IR	-	8.4			T=0.25	Orion	R
		9.4,14.4			0.5,0.75		GM
06053-0622	13070	+10.8	0.53	<0.30	<0.26	Mon R2	P,J2,GM
06055+2039	1032	+9	0.25	1.0	<0.15	S252	J1
12073-6233	10670	+31.9	0.41	<0.28	<0.26		P
16183-4958	5150	-49.7	0.22				P
17175-3544	11370	-10.3	0.30	<0.18	<0.17		P
17233-3606	8788	2.0	<0.20	<0.21	1.06		P
17271-3439	7041	-20.7	1.70	<0.30	<0.32		P
17441-2822	12950	+60	1.4	Q 0.5	1.4	SgrB2	J1,GM,R,GR
18075-1956	9481	-1	0.4	<0.5	<0.8	W31	J1
			Q 0.15	Q 0.14			GM
18117-1753	2206	+35	Q 0.14	Q 0.13		W33	GM
18592+0108	10150	+42	Q 0.05			G35.2-1.8	GM
19078+0901	7566	+7	Q 0.4	Q 0.4	Q 0.7	W49	ALL
		+8.2	2.5			W49(SW)	P,J1,GM
		+2.2	1.5			W49	P,J1
W51	-	+57	0.42	<0.42	<0.33	W51	J1,GM
19598+3324	10590	-20	0.7	<0.20	<0.23	K3-50	J1,GM
20081+3122	1428	+24.1	2.2			ON1	GM
		+9.5	T=0.17				B
DR21	-	-5.7	Q 0.12	Q 0.06	Q 0.06	DR21	GM
DR21(OH)	-	0	3.8	<0.20	<0.20	DR21(OH)N	J1-2,GM
21413+5442	1143	-62	0.7	<0.17	<0.23		J1
23116+6111	7072	-59.2,-57.1	0.14,0.24			NGC7538	GM
		-58.8	0.2,0.9				GM,J2

Explanation: Q = Quasi-thermal, T = Antenna temperature (Rickard et.al. and Baudry)

References : J1 = Jodrell Bank survey 1989 (Paper 1) J2 = Jodrell Bank survey 1991
P = Parkes survey 1991   B = Baudry (1974)
GR = Gardner and Ribes (1971)   GM = Gardner and Martin-Pintado (1983)
R = Rickard et.al. (1975)

Discussion

The combined survey taken at Jodrell Bank in 1989, at Parkes in 1991 and again at Jodrell Bank in 1991 covers a complete sample of CBJ selected IRAS point sources with F(60) > 4000 Jy and 18 cm OH maser emission. It is also complete for all IRAS point sources with F(60) > 7500 Jy. It also includes a number of other known 4.7 GHz maser sources. Many of the objects in which maser emission was detected were observed a number of times, giving useful data of the variability of such sources. The table lists all the known detections to date found in our surveys and in previous work. Overall 18 maser sources have been detected at 4765 MHz, one at 4750 MHz and 4 at 4660 MHz. These numbers tend to confirm the theoretical findings of Field and Gray (1988), Gray et.al (1991 and 1992). They find that in a situation where masers are competing for gain, it is the masers at 4765 MHz that are expected to be seen. The unusual source in S252 discovered in the first part of the survey showed its strongest maser at 4750 MHz. This is the only maser ever seen on this line and is, interestingly, not accompanied by 18 cm emission. Gray et.al. in their 1991 paper find that this situation can be met by using models with a relatively large velocity gradient and scale length for the overlap between FIR lines coupling the rotational states. Sadly this source appears to be variable and was not visible at the time of our Parkes survey.

The aim of this work is to discover which types of objects are to be associated with maser activity of OH in this excited state. It is clear that the IPSC is a useful database on which to base a search programme. However it should be noted that two of our detections, DR21(OH) and W51 do not appear in the IPSC. Furthermore there do not seem to be any obvious characteristics of the FIR emission which distinguish between objects detected as 4.7 GHz masers and the other objects in the sample. There is a suggestion that maser emission is more readily detected in brighter IPSC objects, but the statistics are still too poor to make a firm statement. A simple statistical test based on the hypothesis of no dependence on $F(60)$ shows that a result at least as deviant as this would be expected with a probability of 18%. Two colour plots for 4.7 GHz masers and non-masers, using IRAS colours, do not show any significant separation of the two classes.

It is apparent that some, and possibly most, 4.7 GHz maser sources are variable over times of the order of a year. Of all the maser sources quoted in the table, there are six, W3(OH), DR21(OH), W49 (8.2 km/s), W51, K3-50 and SgrB2 which have changed little during the time that they have been known. Seven others, W3-IRS5, W31, W49 (2.2 km/s), ON1, S252, MonR2 and NGC7538, have been observed to vary by large factors within the times as short as six months. The remaining sources have not been sufficiently well observed to allow comment on variability at this stage. These include, of course, most of our new discoveries. It is not clear, as yet, what differences there are between the variable and non-variable sources. It is hoped that more light can be cast on this question by further observations of the known sources and by extending the survey of CBJ sources to lower FIR fluxes.

References

Baudry A. 1974, *Astr. Ap.*, **33**, 381
Brebner G.C. 1988 Ph.D Thesis, University of Manchester
Cohen R.J., Baart E.E. and Jonas J.L. 1988, *MNRAS* **231**, 205 ('CBJ')
Cohen R.J., Masheder M.R.W. and Walker R.N.F. 1991, *MNRAS* **250**, 611 ('Paper 1')
Field D. and Gray M.D. 1988, *MNRAS* **234**, 353
Gray M.D., Doel R.C. and Field D. 1991, *MNRAS* **252**, 30
Gray M.D., Field D. and Doel R.C. 1992, *Astr. Ap.* (submitted)
Gardner F.F. and Ribes J.C. 1971, *Astroph. Letters* **9**, 175
Gardner F.F. and Martin-Pintado J. 1983, *Astr. Ap.* **121**, 265
Rickard L.J., Zuckerman B. and Palmer P. 1975, *Ap.J.* **200**, 6
Staveley-Smith L.G. 1985, Ph.D. thesis, University of Manchester

THE MAGNETIC FIELD OF THE MILKY WAY:
OH MASER ZEEMAN RESULTS

Mark J. Reid and Karl M. Menten
Harvard-Smithsonian Center for Astrophysics
60 Garden Street, Cambridge, MA 02138

ABSTRACT

We report on observations with the VLA designed to map precisely the OH maser emission from many interstellar OH sources. The primary purpose of these observations is to locate oppositely polarized Zeeman-pairs and measure the strength and line-of-sight orientation of the magnetic field in the masing regions. The magnetic fields of interstellar OH masers appear to be of Galactic origin and can be used to map the global structure of the magnetic field of the Milky Way. Measurements of the line-of-sight direction of the magnetic field for 34 OH maser sources favor axi-symmetric (over bi-symmetric) models of the magnetic field of the Milky Way.

1. INTRODUCTION

OH maser emission occurs in the molecular material associated with ultra-compact HII regions. The Zeeman effect splits the OH transitions and gives rise to pairs of right (RCP) and left (LCP) circularly polarized lines. The magnetic (B) field strengths determined from Zeeman pairs for OH masers are typically about 5 milli-Gauss, enough so that the Zeeman splitting exceeds the maser line width. In this limit of large Zeeman splitting, the splitting is proportional to the *full magnitude* of the B-field, and not proportional to the line-of-sight component as in the case of small splitting (compared to the line width) usually encountered in 21-cm HI Zeeman measurements.

In addition to learning the full magnitude of the B-field, OH Zeeman measurements yield the *line-of-sight direction* of the field (i.e., whether it points toward or away from the observer). In fact simply noting whether the RCP line(s) are shifted to higher or lower frequency compared to the LCP line(s) indicates the line-of-sight direction of the B-field. Reid and Silverstein (1990) recently tested the suggestion of Davies (1974) that the magnetic field detected in OH masers follows galactic rotation. They searched the available literature of the past two decades looking for candidate Zeeman pairs in OH spectra and for interferometric results that would help identify them as originating from the same masing cloudlet and hence be true Zeeman pairs. They found 17 such sources. This sample more than doubled the number of OH masers with identified Zeeman pairs compared with those listed by Davies. Fourteen of the 17 sources had field directions that pointed in the direction of galactic rotation.

2. OBSERVATIONS

Recently we conducted a new survey of OH masers using the VLA in the A-configuration to search for Zeeman pairs in about 30 sources. The sources were observed in two OH lines and two polarizations with velocity coverage and resolution of 17 and 0.14 km s^{-1}, respectively. Most sources were observed at three sidereal times in order to obtain (u, v)-coverage better than for a simple snap-shot observation.

The interferometer data were first calibrated in a conventional manner. Since the standard phase calibration resulted in the incomplete removal of atmospheric phase fluctuations, at the level of nearly one radian, we "self-calibrated" the data from each transition in the following manner. We searched for a strong maser feature (in either polarization) for each transition by examining spectra formed by incoherent averaging ("amp-scalar") of all visibility data. Then we precisely located the position of this feature with a high resolution map, shifted the phase center of the interferometer to that position, determined the residual phases (primarily atmospheric in origin), and subtracted these residual phases from all visibility data for both polarizations of that transition. This procedure increased the dynamic range in the maps considerably.

Nearly 30,000 maps were generated by mapping all of the 128 (or 256) spectral channels from about 40 sites of maser activity in two transitions and two polarizations. The strongest maser spot in each map was located and its flux density and position were estimated by fitting an elliptical Gaussian brightness distribution. Candidate Zeeman pairs were then searched for by examining the fitted results for oppositely polarized maser components coincident to within 5×10^{15} cm on the sky (assuming a distance to the source). Our procedure required that emission in at least three spectral channels in both polarizations satisfied the spatial coincidence requirement. Finally, we plotted these data, generating spectra that remove all emission from outside the 5×10^{15} cm "beam", and arrived at Zeeman-pair identifications visually. An example of these spectra is shown in Figure 1.

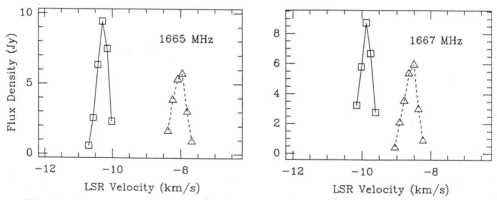

Fig. 1. Zeeman spectra from G351.8-0.5. Flux densities are from Gaussian fits in individual channel maps, and RCP and LCP are indicated with solid and dashed connecting lines, respectively. All emission shown is within a region of size 5×10^{15} cm. The center velocity and magnetic field for the two transitions are nearly identical.

3. RESULTS

We found 17 new maser sources for which one or more Zeeman pairs were identified. Following the grading criteria adopted by Reid and Silverstein (1990), the new sources were either of quality B or C. This doubles the number of maser sources with line-of-sight B-field directions measured from Zeeman shifts. These B-field "directions" are displayed on a schematic view of the Milky Way in Figure 2; also indicated in Figure 2 are the locations of spiral arms from the work of Georgelin and Georgelin (1976). It is important to note that even though the B-field direction arrows in the figure point towards (or away from) the Sun, all that is known is the that the B-field is within ±90° of that direction. Three of the sources measured in this study exhibited Zeeman pairs with opposite directions (signs) of the B-field across the masing region; these sources are plotted with lines perpendicular to the line-of-sight, an indication that the average B-field direction may be close to that direction. Sources with internal field reversals do not appear in the study of Reid and Silverstein, because their OH spectra were interpreted as yielding ambiguous B-field directions.

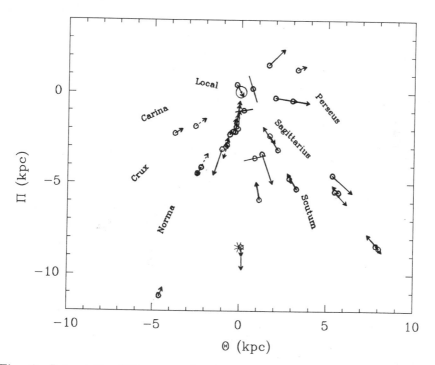

Fig. 2. Interstellar OH sources in the Milky Way with measured Zeeman-pairs. The Sun, indicated by a ⊙, is at the origin of the plot and 8.5 kpc from the Galactic Center, indicated by an *. The lengths and directions of the arrows indicate the magnitude and line-of-sight direction of the magnetic field. Dashed arrows indicate a very uncertain distance and perpendicular lines (to the Sun) indicate sources with internal field reversals. 23 line-of-sight field directions point clockwise (with galactic rotation) and 8 point counter clockwise.

In order to determine the locations of the maser sources in the Milky Way, we adopted a distance to each source, consistent with a kinematic model of the Milky Way assuming $R_0 = 8.5$ kpc and $\Theta_0 = 220$ km s^{-1}. For sources with a kinematic distance ambiguity, we attempted to choose the correct distance by consulting the literature for that star forming region. If insufficient information was available to resolve the distance ambiguity, we adopted the near distance and note this (with dashed B-field direction arrows) in the figure.

We find that 23 sources sources have field directions that point in the direction of galactic rotation, 8 sources point in the reverse direction, and 3 sources exhibit internal reversals. These Zeeman measurements seem to indicate that interstellar OH masers are sensitive to a magnetic field of **galactic** origin, and the galactic field has large scale features with the same line-of-sight direction. Further, the majority of B-field directions measured for the Perseus, Carina–Sagittarius, Crux–Scutum, and the Local Spiral Arm point in the direction of galactic rotation. Thus, our data favor axi-symmetric models for the galactic magnetic field rather than bi-symmetric field configurations (cf. Sofue, Fujimoto, and Wielebinski 1986). While uncertainties in the adopted source distances could place a small number of sources in the wrong spiral arm, and indeed the locations of the spiral are only approximately known, it seems unlikely that correcting these problems would change this general conclusion.

4. FUTURE WORK

While this paper has presented interferometric observations that double the number of OH maser sources with identified Zeeman pairs, it is only a progress report for a larger project that seeks also to provide interferometric maps for most strong OH maser sources. Using the VLA, and soon the VLBA, we are confident that we can locate Zeeman pairs in many more OH masers sources. Over the next few years we plan to at least double the number of measured sources. With such data, we hope to determine more clearly the magnetic field configuration in the spiral arms of the Milky Way.

REFERENCES

Davies, R. D. 1974, in IAU Symposium 60, Galactic Radio Astronomy, ed. F. J. Kerr and S. C. Simonson III (Dordrecht: Reidel), p.275)
Georgelin, Y. M. and Georgelin, Y. P., 1976, A&A, 49, 57
Reid, M. J. and Silverstein, E. 1990, ApJ, 361, 483
Sofue, Y., Fujimoto, M., and Wielebinski, R. 1986, Ann. Rev. Astr. Ap., 24,459

OH MASERS AND BIPOLAR MOLECULAR OUTFLOWS ASSOCIATED WITH MASSIVE STAR FORMATION

Douglas O. S. Wood

National Radio Astronomy Observatory

P.O. Box O, Socorro, NM 87801

ABSTRACT

I present observations of the OH maser emission from G5.89-0.39 (W28A$_2$), the most massive bipolar molecular outflow in the Galaxy. Together with recombination line and molecular line observations, these data suggest that a massive star is embedded in a dense disk of molecular gas. The star ionizes the inside of the disk and powers the bipolar outflow. OH masers form in the shocked regions just outside the ultra-compact H II region. G5.89-0.39 represents an important link between the phenomena of low and high-mass star formation as well as a chance to study the engine of a bipolar outflow in great detail.

I. Introduction

The phenomenon of bipolar molecular outflow is commonly associated with low-mass star formation. The most massive bipolar molecular outflow in the Galaxy, however, is associated with the high-mass star formation region G5.89-0.39 (W28A$_2$). Continuum observations by Wood and Churchwell (1988) and Zijlstra et al. (1989) have shown that G5.89-0.39 is a shell-shaped ultracompact H II region powered by an O5 star. Presumably the wind of the massive star compresses the H II region onto the surface of a sphere. Harvey and Forveille (1988) used ^{12}CO observations to estimate a mechanical luminosity of 1500 L$_\odot$ for the outflow. The CO line wings are an astounding 180 km s^{-1} broad. The outflow direction was very poorly determined in these observations but recently Cesaroni et al. (1992) have shown, using C^{34}S observations, that the outflow is oriented approximately north-south, with the red shifted molecular line emission in the north. Zijlstra et al. (1990) estimate, from H I absorption observations, that the distance to G5.89-0.39 is 4.0 kpc. These authors have also observed the OH maser emission of G5.89-0.39 but with lower resolution than the observations presented here.

I have been investigating the physical conditions of the ionized and molecular gas in G5.89-0.39 using the Very Large Array (VLA). Figure 1 shows a summary of the primary results. A complete presentation of this work will appear in a future publication. The 3.6 cm continuum emission shows a bright ring of emission and two, low surface-brightness extensions in the north and south through which ionized gas appears to be escaping. The H76α recombination line shows a remarkable velocity structure with red-shifted gas at *both* poles of the outflow. Using high resolution images of the NH$_3$ (3,3) and (2,2) transitions toward the source, I found very hot (~200K) gas surrounding the shell shaped H II region and a pronounced north-south outflow. The molecular gas is hottest and densest along the equator of the outflow. Along the line of sight to the continuum emission, NH$_3$ is in absorption and blue-shifted, indicating that the southern lobe of the outflow is pointed toward us while the northern lobe is pointed away.

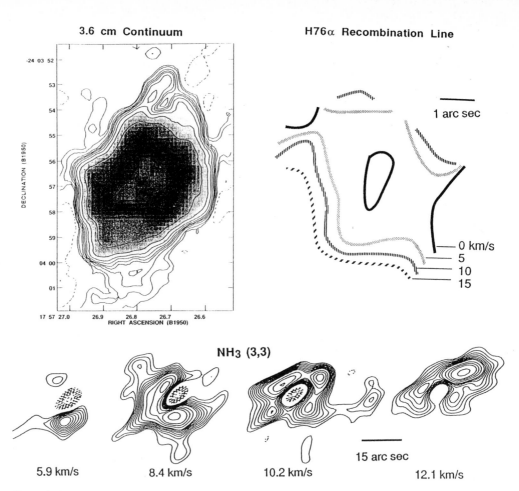

3.6 cm Continuum

H76α Recombination Line

NH₃ (3,3)

5.9 km/s 8.4 km/s 10.2 km/s 12.1 km/s

Figure 1 Continuum, recombination line and molecular line emission from G5.89-0.39.

II. Observations

I observed the 1665 and 1667 MHz transitions of OH in the A array of the VLA on 23 July 1991. At these frequencies the synthesized beam is $1\overset{\prime\prime}{.}4 \times 1\overset{\prime\prime}{.}0$, the highest possible with the VLA. The left and right circular polarizations were observed simultaneously in 127 spectral channels 0.137 km s^{-1} wide, centered at 10 km s^{-1}. Ten channels on each end of the band were averaged to produce a continuum image with a peak of 270 mJy beam^{-1}. After subtracting the continuum and cleaning, the line-only images had an r.m.s. of ~100 mJy beam^{-1}. I fitted two dimensional Gaussians to all emission above 3σ and examined the fit results from adjacent channels to determine where a particular spot reached a maximum. I considered spots in adjacent channels to be the same spot if they were separated by less than $0\overset{\prime\prime}{.}18$ (1/2 pixel). I found 67 spots at 1667 MHz and 54 at 1665 MHz. Figures 2 and 3 show the observed maser positions and intensities graphically. I have identified several possible Zeeman pairs and derived magnetic fields (assuming 1.96 kHz = 1 mG). I considered two spots to be a Zeeman

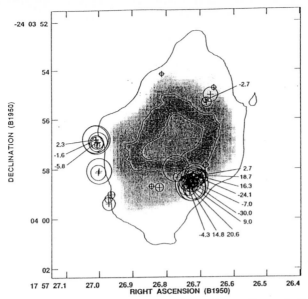

Figure 2 1665 Maser emission superimposed on a 3.6 continuum image. Crosses indicate the maser positions surrounded by a circle with diameter proportional to the log of the peak flux density. The brightest maser has a flux density of 10.1 Jy. The magnetic field strength (mG) from some possible Zeeman pairs is indicated.

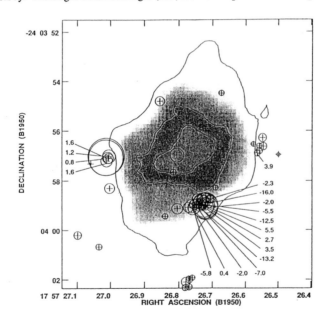

Figure 3 Same as Figure 2 except the 1667 Maser emission is shown.

pair if they were coincident to less than 0ʺ.2. Until higher resolution VLBI/VLBA observations are available, it is impossible to identify which of these spots are true Zeeman components.

III. Discussion

The main line maser emission from G5.89-0.39 is strongest at two positions that lie on either side of the southern opening in the H II region. The masers are clustered just outside the H II region, primarily along the equator of the outflow, avoiding the openings in the continuum emission. Although there is considerable scatter in the magnetic field measurements due to inadequate spatial resolution, the field tends to be negative (coming toward us) on the right and positive (going away from us) from us on the left. Although there are many field geometries that would reproduce this orientation of the field, any field that is symmetric about the axis of the out-flow cannot reproduce these observations.

These observations indicate that G5.89-0.39 is embedded in a thick disk of molecular material. The bipolar molecular outflow is perpendicular to this disk. This outflow is inclined by ~45° to the line of sight with the northern (red-shifted) lobe pointing away from us. At the center of the disk the wind of the massive star creates a cavity inside the ultracompact H II region. Molecular material from the disk is accreting onto the surface of the H II region, where it is ionized, producing the bright ring of continuum emission. The ionized gas flows from the equator to the poles along the surface of the sphere producing the remarkable velocity field see in the recombination line. Just outside the H II region, where the molecular gas is shocked and heated, OH masers and hot dense NH_3 emission are produced. The surrounding molecular gas participates in the outflow it and flows along with the ionized gas. Because the brightest maser emission is seen near the openings of the H II region, where the large bipolar molecular outflow originates, shocks must be present in the molecular gas to promote maser emission. Because the H II region is optically thick at 18 cm and the northern opening is obscured by the H II region, we do not see as many maser spots in the north.

It is clear that massive stars produce bipolar molecular outflows, an important link between low and high-mass star formation processes. Presumably high-mass stars produce more energetic outflows because they are more energetic than low-mass stars and because they are associated with more massive disks. One of the advantages of observing a high mass outflow such as G5.89-0.39 is that the energetic massive star provides several observational probes including a bright H II region, masers and strong molecular lines. G5.89-0.39 also presents an opportunity to study the physical engine that powers a bipolar outflow in detail.

References

Cesaroni, R., Walmsley, C. M., Kompe, C. and Churchwell, E., Astr. Ap. (Univ. Wisc. Preprint #401).
Harvey, P. M and Forveille, T., 1988, Astr. Ap., 197, L19.
Wood, D. O. S. and Churchwell, E., 1989, Ap. J. Suppl., 69, 831.
Zijlstra, A. A. et al., 1989, Astr. Ap., 217, 157.
Zijlstra, A. A., et al., M.N.R.A.S., 246, 217.

7. WATER MASERS IN STAR FORMING REGIONS

H₂O MASER SURVEY OF IRAS SOURCES
AT HIGH GALACTIC LATITUDE

P. Caselli[1,2], C. Codella[3], G.G.C. Palumbo[1,4], G. Pareschi[1,2]

(1) Dipartimento di Astronomia, Università di Bologna, Italy
(2) Istituto di Spettroscopia Molecolare, CNR., Bologna, Italy
(3) Istituto di Radioastronomia, C.N.R., Bologna, Italy
(4) I.Te.S.R.E., C.N.R., Bologna, Italy

ABSTRACT

A high galactic latitude ($|b| > 10°$, $\delta \geq -30°$) sample of 175 IRAS sources located within the core of molecular clouds has been searched for H_2O maser emission with the Medicina 32 m dish operated at 22 GHz . 17% of the sample previously searched by other authors contained only 4 detection. The search did not produce any new water maser emitter, while the previously known maser were seen again. The low value of 2% overall detection rate can be explained in terms of galactic distribution of massive cores.

1.THE SEARCH

As part of an extensive observational program aimed at detecting H_2O masers (6_{16}-5_{23}) in IRAS selected sources (1409 objects), H_2O emission has been searched for in two samples chosen using Emerson's criterium (Emerson 1987), that selects sources embedded in molecular cloud cores (Myers 1990). The first one is at $|b| > 10°$, $\delta \geq -30°$, and the other one at $|b| \leq 10°$, $\delta \geq -30°$. The radiotelescope used was the 32 m dish of Medicina (Bologna), operated at 22 GHz (1.3 cm) .

2.THE SAMPLES

Emerson's criterium for sources embedded in molecular cloud cores is:

[25-12] +0.4 to +1.0
[60-25] +0.4 to +1.3
[100-60] +0.1 to +0.7

where $[i–j]=\log(F_i/F_j)$, where F_i is the flux density in Jy, and i is the wavelenght in μm.

This selection produced 175 objects.

For comparison the same number of sources from the much larger sample of IRAS objects selected at $|b| \leq 10°$ was randomly extracted.

3.OBSERVATIONS

The $|b| > 10°$ sample was observed in May 1991. The $|b| \leq 10°$ comparison sample in December 1991 - January 1992.

Observations were made using the 1024 channels digital autocorrelator in the total power mode with an integration time of 5 min on the ON and OFF positions each. The bandwith was set to 25 MHz centered at zero velocity.

While data reduction is completed for the $|b| > 10°$ sources, for the others only preliminary results will be given.

1. The $|b| > 10°$ sample

30 of the 175 objects in the sample had already a measurement in the literature (Wouterloot and Walmsley 1986; Braz *et al.* 1989; Scappini, Caselli and Palumbo 1991). Four of these had reported water maser emission (Tab. 1). These are also CO outflow sources (Fukui 1989; Felli, Palagi and Tofani 1991).

Table 1. H_2O masers at $|b| > 10°$

IRAS	NAME	f_υ (Jy)	Δf_υ (Jy)
03259+3105	HH7-11 SSV13	31.9	1.6
05302-0537	Ori A-west	18.9	3.2
05445+0020	NGC2071	6521.0	33.9
06053-0622	Mon R2	557.8	1.8

2. The $|b| \leq 10°$ sample

These sources where chosen among those which had never been measured before. Our observations produced 8 new H_2O masers. The reduction of all spectra is in progress and will be published elsewhere. These detections will be used, in what follows, purely for statistical purposes. The detection rate in this sample is double than the one observed for the $|b| > 10°$ sample.

4.RESULTS

No new H_2O maser emission sources were discovered in the $|b| > 10°$ sample. The H_2O detection rate in high galactic latitude objects is therefore about 2%.

In the whole sample 15 CO outflow sources are present. We compared the IRAS sources FIR fluxes and the outflow mechanical luminosities between sources with and without maser emission taking the data from Felli, Palagi, and Tofani (1991). Let $<L_{m(CO)}>_{H2O}$, and $<L_{FIR}>_{H2O}$, be the averaged outflow mechanical luminosities and the FIR luminosities of the sources with H_2O masers respectively, and let $<L_{m(CO)}>$ and $<L_{FIR}>$ be the corresponding quantities for sources without H_2O masers (Table 4). We can see that:

$$<\mathbf{L}_{m(CO)}>_{H2O}/<\mathbf{L}_{m(CO)}> = \mathbf{3.1 \times 10^2}$$
$$<\mathbf{L}_{FIR}>_{H2O}/<\mathbf{L}_{FIR}> = \mathbf{7.6 \times 10^2}$$

Therefore H_2O detections are more frequent for sources with higher integrated FIR fluxes and higher mechanical luminosities of the outflows. This result is in agreement with what was found by Felli, Palagi and Tofani (1991) for a CO outflow sample.

5.DISCUSSION AND CONCLUSIONS

In order to understand the relevance of our selection criteria in the wider problem of H_2O emission from compact molecular cores we have plotted the two samples in Fig. 1 and 2. From the figures one notices that:

1) there are more sources in the compact HII (UC HII) regions area (Wood & Churchwell 1989) in the $|b| \leq 10°$ sample than in the $|b| > 10°$;

2) 75% of the detections in the $|b| \leq 10°$ sample lies in the HII regions area wherease only 2 out of 4 lie in the same area for the $|b| > 10°$ sample.

These results are summarized in Table 2.

Table 2. Statistical results

SAMPLE	UC HII (#)	UC HII (% of total)	detections in UC HII	detections in UC HII (%)	remaining sources (% of total)	detections (%)
$|b| > 10°$	58	33.1	2	3.4	66.9	1.7
$|b| < 10°$	67	38.3	6	9.0	61.7	1

From the above we conclude that: H_2O masers are preferably associated to massive cores which, in turn, tend to cluster in the galactic plane.

These cores are FIR bright and display high outflow mechanical luminosities.

REFERENCES

Braz, M.A., Scalise, E., Gregorio Hetem, J.C., Monteiro do Vale, J.L., Gailard, M.: 1989, *Astr. Ap.*, **77**, 465

Emerson, J.P.: 1987, in *IAU Symp. n.115, Star Forming Regions*, Peimbert-Jugaku eds., p.19.

Felli, M., Palagi, F., Tofani, G., 1992, *Astr. Astron.*, in press.

Fukui, Y.: 1989, proceedings of the *ESO Workshop on Low Mass Star Formation and Pre-Main-Sequence Objects*, Reipurth ed. (ESO, Garching), p.95.

Myers, P.C.: 1990, in *Molecular Astrophysics*, Hartquist ed. (Cambridge University Press), p.328.

Scappini, F.,Caselli, P.,Palumbo, G.G.C., 1991, *M. N. R. A. S.*, **249**, 763.

Wood, D.O.S.,Churchwell, E.: 1989, *Ap. J.*, **340**, 265.

Wouterloot, J.G.A.,Wasmeley, C.M., 1986, *Astr. Ap.*, **168**, 237.

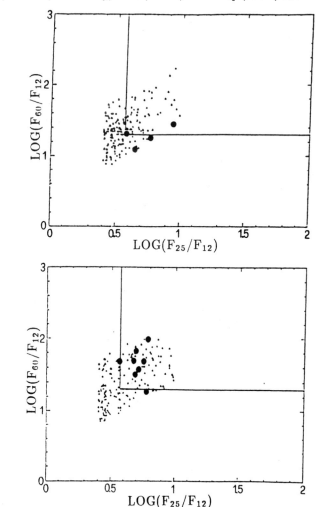

Fig.1

Fig.2

Fig.1. *Distribution of IRAS sources associated with* $|b| > 10°$ *cores. The box in the top right hand side bounds the region in which Wood and Churchwell (1986) colour criteria selects ultracompact HII regions associated with massive stars. The filled circles are detections.* **Fig.2.** *As in Fig.1 for a randomly selected sample of IRAS sources at* $|b| \leq 10°$ *satisfying cores colour criteria.*

WATER MASERS ASSOCIATED WITH COMPACT MOLECULAR CLOUDS AND ULTRACOMPACT HII REGIONS: THE EXTENDED SAMPLE

R. Cesaroni, G. Comoretto, M. Felli, F. Palla
Osservatorio Astrofisico di Arcetri
Largo E. Fermi 5, I-50125 Firenze, Italy

J. Brand, P. Caselli
Istituto di Radioastronomia
Via Irnerio 46, I-40100 Bologna, Italy

ABSTRACT

We present the results of a survey of water maser emission at 22.2 GHz towards a selected sample of IRAS-PSC sources which are believed to be associated with very young massive stars. The sample consists of 591 sources. The observations have been carried out using the Medicina 32-m radiotelescope, operated by the Istituto di Radioastronomia – C.N.R., Bologna.

Whereas previous searches for maser emission have been directed towards known HII regions, the aim of the present survey is to identify new sources in an even earlier evolutionary phase, corresponding to the development of an ultracompact HII region. It is shown that our sample contains a significant number of sources that could be considered good candidates of massive protostellar cores still in the main accretion phase.

1. IRAS SELECTION CRITERIA

The complete sample consists of 1259 sources with declination $\delta \geq -30°$, selected according to the color criteria suggested by Richards et al. (1987) to identify compact molecular clouds (CMC). These criteria are: (1) $0.61 \leq [60 - 25] \leq 1.74$; (2) $0.087 \leq [100 - 60] \leq 0.52$; (3) *no* upper limits at 25, 60, and 100 μm; (4) *no* coincidence with known HII regions; (5) $|b| \leq 10°$. In the following we define $[y - x] \equiv Log(F_y/F_x)$.

To further discriminate sources associated with ultracompact (UC) HII regions, we apply the color criteria of Wood and Churchwell (1989). Accordingly: (6) $[25 - 12] \geq 0.57$; (7) $[60 - 12] \geq 1.3$; (8) *no* coincidence with stars or galaxies.

The resulting number of candidate UC HII regions is **279** objects, hereafter referred to as "HIGH" in the discussion of the results. The remaining **980** sources, called "LOW", constitute an ensemble of sources whose classification remains to be clarified.

In order to better understand the nature of the LOW sources, we undertook a search for water maser emission towards the entire sample of HIGH sources and a similar number of LOW objects. The analysis of the detection properties of the two groups should give an indication of the fraction of HII regions among the LOW sources, and thus of their nature. A first account of the results of a subset of the whole sample has been given in Palla et al. (1991).

2. OBSERVATIONS

Using the Medicina 32-m antenna (see Comoretto et al. 1990 for a description of the instrument), we have observed 279 sources belonging to the HIGH sample, and 312 sources from the LOW sample, for a total of 591 objects. In Fig. 1 we show the [60–12] vs [25–12] color-color diagram of the entire sample. Also shown, are the limits set by the criteria of Richards et al. and Wood and Churchwell. The difference between the two samples is striking, with the HIGH one being characterized by a detection rate (18%) significantly higher than that of the LOW sources (5%). Furthermore, the spectra of the detected sources markedly differ in the two cases: complex for the former sample, weak and with a few lines in the latter.

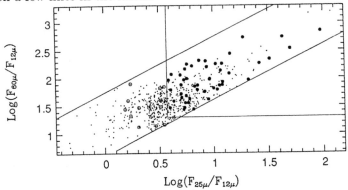

Fig. 1 Color-color plot of CMC sources non-detected (points) and detected (circles) in water. Filled and empty circles show detected sources of the HIGH and LOW samples respectively. The set of lines mark the boundaries of selected FIR colors

Next, we want to show that such a difference is real and see how it can enlighten us on the nature of the sources.

3. DISCUSSION: HIGH AND LOW SOURCES

A common claim is that UC HII regions must be strictly associated with H_2O maser emission. Indeed, Churchwell et al. (1990a) find a detection rate of 66% in a survey of known UC HII regions. This result is compatible with our lower rate (18%) for the HIGH sample when the different sensitivity of the Effelsberg and Medicina antennas is taken into account. In particular, 28 masers of the Churchwell et al. (1990a) survey have a flux greater than the Medicina 3σ detection limit, which leads to a "corrected" detection rate of 33%. This is still larger than our result, but the remaining discrepancy can be explained by the fact that we have taken out from our sample all sources with known associations with HII regions, unlike Churchwell et al. (1990a) that used the complementary criterium of selecting their targets from known UC HII regions. On this basis, we assume as a working hypothesis that HIGH sources are mostly associated with UC HII regions.

What about LOW sources? First of all, we want to show that an intrinsic difference between HIGH and LOW does indeed exist. For this purpose, we compare in Fig. 2 the distribution of the [25–12] color index of the whole sample with that of the detected sources only. The decrease of detections below the Wood and Churchwell threshold is quite sharp. A similar conclusion applies to the [60–12] color index distribution. In order to show that this trend is not a sensitivity effect, we also compare the detection rates of the LOW and HIGH samples as a function of the 60 μm flux. This is shown in Fig. 3. It must be stressed that the choice of $F_{60\mu m}$ is arbitrary, since the results would be qualitatively the same for any other of the IRAS fluxes. The conclusion is that at *any* flux HIGH sources show a detection rate much higher than LOW sources, with the exception of a couple of detections in the LOW sample below 10 Jy. However, one of this two sources lies very near to the Wood and Churchwell box of Fig. 1, while the other one satisfies the [60–12] criterium and has an upper limit on the flux at 12 μm that might shift it into the HIGH sample.

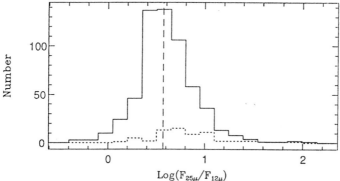

Fig. 2 Distribution of the color index [25–12] for the whole CMC sample (*continuous* histogram) and for detections only (*dashed* histogram). The vertical dashed line corresponds to the lower limit of the Wood and Churchwell criterium

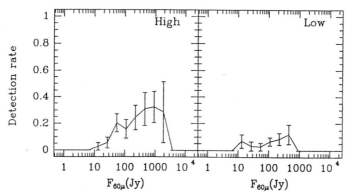

Fig. 3 Comparison between detection efficiencies of HIGH (*left*) and LOW (*right*) samples

We can state that the two samples are very different as far as water emission is concerned. However, the nature and/or evolutionary status of the LOW sources remains to be understood. Before trying any guess on their nature, let us show what they can*not* be. On the of the FIR colors, we can rule out a contamination from low and intermediate mass stars. In fact, FIR sources associated with T Tauri and Herbig Ae/Be stars have color indices that fall well outside the range of the LOW sources.

Moreover, it can be shown on statistical grounds that it is unlikely that most of the LOW sources are evolved HII regions. First of all, we have extracted from the Comoretto et al. (1990) survey of all known water masers associated with HII regions, those detected with the Medicina antenna *and* classifiable as either LOW or HIGH. The ratio LOW/HIGH is 53%, much higher than the relative detection rate of 33% of the present survey. This means that our LOW sources are not associated with the same type of objects that show H_2O maser emission, which are very likely to be related to HII regions. Secondly, we have considered all the IRAS sources satisfying Richards et al. criteria, but associated with known HII regions, and applied the distinction between LOW and HIGH. In this case the ratio LOW/HIGH is 29%, quite similar to 33% of our sample of candidate CMC detected in water.

We conclude that only LOW sources showing H_2O emission are those associated with HII regions. This is confirmed by the fact that the luminosity distribution of detected objects (computed from the IRAS fluxes and the kinematic distance) does not differ between the LOW and HIGH sample, both peaking around $10^4 \, L_\odot$.

4. SUMMARY AND CONCLUSIONS

In this work we have selected from the IRAS-PSC a sample of objects according to criteria that are presumed to identify high density molecular cores on the basis of their FIR color indices. We have then split this sample into two groups of sources using additional criteria to identify UC HII regions. We have found that H_2O masers are preferentially detected towards the sources of the HIGH sample and we interpret this as a proof that only a few sources of the LOW group are identifiable with evolved HII regions.

Our conclusions may thus be summarised as follows: (a) HIGH sources are UC HII regions, with low color temperature due to dust confined in a shell far from the central star, as suggested by the numerical models of Churchwell et al. (1990b). (b) LOW sources suffer only a little contamination of evolved HII regions, of the order of a few percent, and are dominated by very young massive stars, possibly still in the accreting phase; the high color temperature that characterizes these sources may be due to dust distributed in a protostellar envelope near to the central hot core (e.g. Scoville and Kwan 1976).

We therefore conclude that the LOW sample contains good targets for searching high mass protostellar cores still in the accreting phase. We plan to observe other molecular lines (NH_3, CS) to determine the main physical parameters of these regions in order to test our hypothesis.

REFERENCES

Comoretto, G., et al. 1990, A&AS, 84, 179
Churchwell, E., Walmsley, C.M., Cesaroni, R. 1990a, A&AS, 83, 119
Churchwell, E., Wolfire, M.G., Wood, D.O.S. 1990b, ApJ, 354, 247
Palla, F., Brand, J., Cesaroni, R., Comoretto, G., Felli, M. 1991, A&A, 246, 249
Richards, P.J., Little, L.T., Toriseva, M., Heaton, B.D. 1987, MNRAS, 228, 43
Scoville, N.Z., Kwan, J. 1976, ApJ, 206, 718
Wood, D.O.S., Churchwell, E. 1989, ApJ, 340, 265

MOLECULAR OUTFLOWS AND H2O MASERS: WHAT TYPE OF CONNECTION?

M. Felli[1], F. Palagi[2], G. Tofani[1]

[1] Osservatorio Astrofisico di Arcetri

[2] C.N.R., Gruppo Nazionale di Astronomia

Largo E. Fermi 5, I-50125 Firenze, Italy

ABSTRACT

From the search of water maser emission in a sample of 142 molecular outflow sources, we find a relationship between the integrated water maser luminosity and the mechanical luminosity of the outflow (Felli et al. 1992). This relationship is also expected by the model of masers excited by shocks produced by stellar winds.

The association between masers and molecular outflows is characterized by a detection rate which increases: i) with the mechanical luminosity of the outflow, ii) with the total CO line width and iii) with the FIR luminosity of the central object.

The possibility that the stellar wind from luminous Young Stellar Objects (YSO) is the common source of energy for the maser and the outflow, is reinforced by the analysis of the radio continuum emission found in a subsample of sources.

1. INTRODUCTION

We have investigated in a quantitative way the observational evidence that molecular outflows and water masers are often found associated. The results of the analysis have been related to the EHM model (Elitzur et al. 1989), which proposes shock excitation of the maser emission in star forming regions.

Water masers were observed with the Medicina radiotelescope at 22.2 GHz. Emission (above a sensitivity limit of 3 Jy) has been found in 56 molecular outflows out of a sample of 142 known sources north of $\delta \geq -30°$ (Fukui, 1989). Most of the outflows in the sample have an associated IRAS Point Source.

2. THE MASER MODEL AND LUMINOSITY CORRELATIONS

The isotropic luminosity of a water maser generated in a layer of thickness d and aspect ratio a in the EHM model is given by:

(1)

$$L_{H_2O} / L_\odot = 4.2 \ 10^{-6} \ a_1^3 \ d_{13}^2 \ \Delta v_5 \ N_c,$$

where $a_1 = a/10$, $d_{13} = d/10^{13}$ cm, Δv_5 is the half power width of the line in $10^5 \ cms^{-1}$ and N_c is the number of masers beamed in the direction of the observer.

The EHM model gives also an expression for the isotropic mechanical luminosity of the shocks exciting the masers:

(2)
$$L_{mH_2O}/L_\odot = 14.3 \; n_7 \; v_{s7}^3 \; a_1^4 \; d_{13}^2 \; N_c.$$

From the two relations we obtain a link between L_{H_2O} and L_{mH_2O} :

(3)
$$Log(L_{H_2O}/L_\odot) = -6.5 + \; Log(L_{mH_2O}/L_\odot) + Log(\Delta v_5/n_7 \; v_{s7}^3 \; a_1).$$

In eq. (3) we may substitute the unknown shock luminosity L_{mH_2O} with the mechanical luminosity of the CO molecular outflow. In fact, for momentum driven outflows the molecular outflow luminosity is a fraction of the wind luminosity ($L_{mCO} = L_{wind}$ v_{max}/v_w). We also assume that the shock luminosity is a fraction ϵ of L_{wind}, ($\epsilon \sim 0.1$). Then:

(4)
$$Log(L_{H_2O}/L_\odot) = -6.5 + \; Log(L_{mCO}/L_\odot) + Log(\Delta v_5/n_7 \; v_{s7}^3 \; a_1).$$

Within these approximations, the EHM model predicts that L_{H_2O} should correlate with L_{mCO} and should be at least 6.5 order of magnitude smaller than L_{mCO} .
For sources where both masers and outflows are detected, we derive the following correlations between the computed luminosities:

(5)
$$L_{mCO} = 1.1 \; 10^{-3} \; (L_{FIR})^{1.1},$$

(6)
$$L_{H_2O} = 1.12 \; 10^{-9} \; (L_{FIR})^{1.02},$$

(7)
$$L_{H_2O} = 1.8 \; 10^{-6} \; (L_{mCO})^{0.54}.$$

The agreement between the observations and the model suggests that stellar winds may represent the common source of energy for masers and outflows.

3. DETECTION RATES

The distributions of sources with and without H_2O as a function of L_{mCO} are shown in fig. 1. While the non-detections are found at lower L_{mCO} , detections are associated with larger luminosities.

Masers are found only above a threshold mechanical luminosity of $\sim 10^{-2} L_\odot$. It is also interesting to note that the peak of the distribution for the detections is at $\mathrm{Log}(L_{mCO}) \sim 0.0$, close to the nominal value of L_{mCO} for one maser component ($N_c = 1$) beamed in the observer direction.

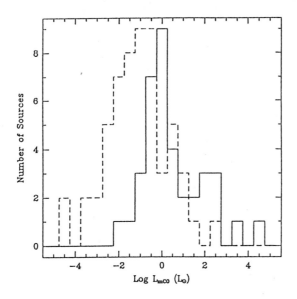

Fig. 1 The distribution of 94 outflows as a function of the mechanical luminosity L_{mCO}. Full line are sources with H_2O maser, dashed without.

The distribution of L_{mCO} with H_2O detections decreases rather sharply towards lower luminosities. This effect may depend on the geometrical configuration of the maser emission. For L_{mCO} greater than the peak value one would expect many maser components beamed in our direction and a proportionality between L_{mCO} and L_{H_2O}. For L_{mCO} less than the nominal value, there may be too few masers to beam in all directions, and the detection rate goes down. The peak of L_{mCO} should indicate that there is at least one maser in any direction. Sources with lower L_{mCO} may have few saturated masers, beamed into $\Omega \le 4\pi$, and the detection rate goes down.

Since the ratio between L_{mCO} and L_{FIR} is $\sim 10^{-3}$, the peak of the L_{mCO} distribution implies that only YSOs with relatively high luminosity ($\mathrm{Log}(L_{FIR}/ L_\odot) \ge 2.5$) will be able to produce H_2O masers.

4. STELLAR WINDS

If the YSO radiation is converted in to a ionized stellar wind, we should observe a radio continuum emission with spectral index $\alpha \sim 0.6$ from a pointlike source (≤ 0.2 arcsec).

In the literature we find that a small (19) number of the outflow sample have been detected with high resolution continuum observations.

The quantity $S_\nu d^2$ gives an estimate of the mass loss rate of an optically thick stellar wind. This parameter is shown in fig. 2 as a function of the bolometric luminosity of the YSO.

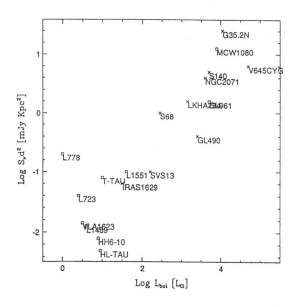

Fig. 2 The radio continuum normalized flux density $S_\nu d^2$ as a function of the bolometric luminosity. * are sources with H_2O masers, squares without.

It is difficult to establish if all these sources are real stellar winds, for the lack of spectral and spatial informations. The normalized radio flux increases almost linearly with the bolometric luminosity as expected by the fact the powerfull YSO will deliver more energy to the surrounding ambient in the form of a stellar wind or of an ionized gas envelope. In fig. 2 there is also a marked separation between outflow sources with and without H_2O masers, the first ones being preferentially associated to the more luminous objects.

REFERENCES

Elitzur, M., Hollenbach, D.J., McKee, C.F. 1989, ApJ, 346, 983
Felli, M., Palagi, F., Tofani, G. 1992, A&A, 255, 293
Fukui, J. 1989, ESO workshop on "Low mass star formation and pre-main sequence objects", Ed. B. Reipurth, 95

A SHOCK ORIGIN FOR INTERSTELLAR H_2O MASERS

David Hollenbach
NASA Ames Research Center
Moffett Field, CA 94035

Moshe Elitzur
University of Kentucky
Lexington, KY 40506

Christopher F. McKee
University of California
Berkeley, CA 94720

ABSTRACT

We present a comprehensive model for the powerful H_2O masers observed in star-forming regions. In this model the masers occur behind dissociative shocks propagating in dense regions (preshock density $n_o \sim 10^6 - 10^8$ cm^{-3}). This paper focuses on high-velocity ($v_s \gtrsim 30$ km/s) dissociative shocks in which the heat of H_2 reformation on dust grains maintains a large column of $\sim 300 - 400$ K gas, where the chemistry drives a considerable fraction of the oxygen not in CO to form H_2O. The H_2O column densities, the hydrogen densities, and the warm temperatures produced by these shocks are sufficiently high to enable powerful maser action, where the maser is excited by thermal collisions with H atoms and H_2 molecules. A critical ingredient in determining the shock structure is the magnetic pressure, and the fields required by our models are in agreement with recent observations. The observed brightness temperatures (generally $\sim 10^{11} - 10^{14}$ K) are the result of coherent velocity regions which have dimensions in the shock plane that are 5 to 50 times the postshock thickness.

1. INTRODUCTION

Interstellar H_2O masers often appear to be individual clumps, streaming away from some center of activity at velocities up to 200 km/s. Individual features have apparent sizes of 10^{13} cm and brightness temperatures usually in the range $T_b \sim 10^{11} - 10^{14}$ K (Genzel 1986). The isotropic luminosity of individual maser spots range from $\lesssim 10^{-6}$ to 0.08 $L_\odot$ in the Galaxy (Walker, Matsakis & Garcia-Barreto 1982). Pumping by an external source of radiation is ruled out by observations (e.g., Genzel 1986) and an internal source of pump energy, such as produced in a shock, seems required. The development of powerful shocks in maser regions is inevitable in light of the high velocities observed in the sources, and the H_2O maser luminosity correlates with the mechanical luminosity in the observed outflows (Felli, Palagi & Tofani 1992), as would be expected in a shock model.

We (Hollenbach, McKee & Chernoff 1987, Elitzur, Hollenbach & McKee 1989, hereafter EHM) propose that fast ($v_s \gtrsim 30 - 50$ km/s) dissociative J shocks incident on preshock hydrogen nuclei densities of $n_o \simeq 10^6 - 10^8$ cm^{-3} provide a natural site for H_2O maser action. In such shocks the molecules are first completely dissociated by the extremely high postshock temperatures ($\sim 10^5$ K) immediately behind the shock front. Further downstream, where the material cools down, H_2 molecules reform on the dust grains and are ejected to the gas phase with sizable internal energies which provide a source of heating for the gas. This heating maintains a nearly isothermal column of $T_p \sim 300 - 400$ K gas–warm enough to drive all oxygen not locked in CO to form H_2O and to collisionally populate the maser levels which lie ~ 600 K above ground. The hydrogen column density N_p of the heated region can be as large as $\sim 10^{22-23}$ cm^{-2}, and the H_2O column as high as 3×10^{19} cm^{-2}. This "H_2 formation plateau", or temperature plateau, is an ideal site for H_2O masers.

2. SHOCK STRUCTURE

We have used the J shock code described by Hollenbach & McKee (1989, hereafter HM) to probe the (n_o, v_s) parameter space likely to produce intense H_2O masers. The code is a 1D, steady-state code which follows the chemistry, heating and cooling, and radiative transfer in the shocked gas. The numeric results are well approximated by analytic formulae derived in HM and EHM, which we summarize here.

The preshock magnetic field B_o is parameterized by the Alfven speed v_A in the ambient gas, which is observed to be ~ 2 km/s in molecular clouds (Heiles et al 1992): $B_o = 0.54 v_{A5} n_o^{0.5}$ μGauss, where $v_{A5} = v_A/1$ km s^{-1} and n_o is in cm^{-3}. The importance of the preshock field is that it limits the compression of the warm postshock gas. The hydrogen gas density in the masing plateau is given

$$n_{maser} = 1.4 \times 10^9 n_{o7} v_{s7} v_{A5}^{-1} \text{ cm}^{-3}, \tag{1}$$

where $n_{o7} = n_o/10^7$ cm^{-3} and $v_{s7} = v_s/100$ km s^{-1}. The magnetic field in the masing region is independent of the preshock field; since magnetic pressures dominate in the masing region, the magnetic pressure is set by the ram pressure of the shock, or

$$B_{maser} \simeq 200 n_{o7}^{0.5} v_{s7} \text{ mGauss}, \tag{2}$$

No simple analytic formula can be derived for T_p, but the numerical results show that 300 K$\lesssim T_p \lesssim 400$ K for 10^5 cm$^{-3} \lesssim n_o \lesssim 10^9$ cm^{-3} and $v_s \gtrsim 30$ km/s. For $n_o \lesssim 10^5$ cm^{-3}, there is no "H_2 formation plateau" because the energy of H_2 formation is radiated away by the newly formed molecules before it can be collisionally transferred to heat. T_p is moderately sensitive to the rate coefficient γ for H_2 formation on warm ($T_{gr} \sim 30 - 100$ K) dust grains. This coefficient is not well determined, but variation of γ from $10^{-18} - 3 \times 10^{-17}$ cm^3 s^{-1} produces a range of only 200 K$\lesssim T_p \lesssim 600$ K.

The hydrogen column in the plateau is given by

$$N_p \simeq 7 \times 10^{21} v_{A5} \gamma_{-17}^{-1} \ \text{cm}^{-2}, \tag{3}$$

where $\gamma_{-17} = \gamma/10^{-17} \ \text{cm}^3 \ \text{s}^{-1}$. The thickness of the plateau (which we identify with the maser spot size) is given

$$d = 5 \times 10^{12} v_{A5}^2 \gamma_{-17}^{-1} n_{o7}^{-1} v_{s7}^{-1} \ \text{cm}. \tag{4}$$

The numerical results, which use the Hollenbach and McKee (1979) formulation for γ as a function of T_{gr}, can be fitted by the following formula for $v_{A5} = 1$:

$$d \sim 1 \times 10^{13} n_{o7}^{-0.5} \ \text{cm}, \tag{5}$$

accurate to a factor of 2 for $10^6 \lesssim n_o \lesssim 10^8 \ \text{cm}^{-3}$ and $30 \lesssim v_s \lesssim 160 \ \text{km/s}$.

The H_2O abundance in the plateau is $\gtrsim 10^{-4}$ in this same parameter space, thereby producing an H_2O column of order $\gtrsim 10^{18} \ \text{cm}^{-2}$ at a temperature of 300-400 K. The column is velocity coherent, in the sense that the thermal line width far exceeds any velocity gradient in the flow. In an ideal 1D shock, the velocity coherence is also maintained to an infinite distance in the shock plane (i.e., perpendicular to the column or the flow). However, the scale length l for coherence in the shock plane is probably set by the size scale for density homogeneity (clumps) in the preshock gas. We estimate this to be of order $10^{14-15} \ \text{cm}$, which then produces velocity coherent disks with aspect ratios $a = l/d \sim 10 - 100$.

3. H_2O MASER SLAB MODELS APPLIED TO SHOCK RESULTS

We have separately modeled the H_2O maser emission from an isothermal, isodensity, velocity-coherent disk of thickness d and diameter l (EHM, Elitzur, Hollenbach & McKee 1992). We have calculated the H_2O level populations for the lowest 40 levels of ortho-H_2O and the resultant brightness temperatures in the 22 GHz maser transition for the range of parameters which occur behind interstellar J shocks. In the range 10^6 $\text{cm}^{-3} \lesssim n_o \lesssim 10^8 \ \text{cm}^{-3}$ and $30 \ \text{km/s} \lesssim v_s \lesssim 160 \ \text{km/s}$, we find that the disks consist of an unsaturated core of radius $\sim d/2$ with a saturated region extending radially from $d/2$ to $l/2$. Such planar masers beam the maser radiation radially in the plane of the disk. The observed spot size is $\sim d$ (the observer only sees rays which pass through the unsaturated core, which exponentially amplifies the intensities along the rays). The observed brightness temperature is given by

$$T_b = T_o \left(\frac{l}{d}\right)^3 \equiv T_o a^3. \tag{6}$$

In the same parameter range, the numerical results for $v_{A5} = 1$ can be fitted to give

$$T_o \simeq 10^9 \frac{n_{o7}^{1.5} v_{s7}}{1 + v_{s7}^2 n_{o7}^2} \ \text{K}, \tag{7}$$

which is accurate to a factor of 2. In the shock parameter range specified, the observed T_b of H_2O masers in the ISM can be matched with $a \sim 5 - 50$. Outside this range, T_b rapidly falls. For $n_o \gtrsim 10^8$ cm^{-3}, the maser is quenched; for $n_o \lesssim 10^6$ cm^{-3}, the entire disk is unsaturated and T_b becomes exponentially smaller.

4. SUMMARY AND DISCUSSION

Figure 1 summarizes the J shock parameter space which produces 22 GHz H_2O masers. For $v_s \lesssim 30-40$ km/s, shocks may not be J type but instead C type (Hollenbach, Chernoff & McKee 1989); C shocks have yet to be studied for possible H_2O maser production. To the left and below the solid line, there is no H_2 formation plateau. Above the dash-dot line, the maser is quenched in J shocks. Below the dashed line, the maser is unsaturated and weak. J shocks in the range $v_s \gtrsim 30$ km/s and 10^6 cm$^{-3} \lesssim n_o \lesssim 10^8$ cm^{-3} produce strong, saturated, beamed 22 GHz H_2O masers.

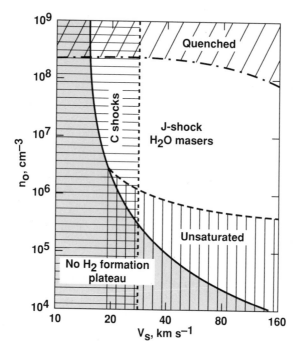

Figure 1. The range of preshock density n_o and shock velocity v_s which produces strong J shock H_2O masers, assuming $v_A \sim 1$ km/s.

Genzel (1986) reviews the observations of H_2O masers and we summarize here the shock model explanation of the data. Their space velocities are often $\gtrsim 30$ km/s; this is the velocity required to produce the H_2 formation plateau. Their densities are inferred to be $\gtrsim 10^9$ cm^{-3}; the shock compresses gas of density $\sim 10^7$ cm^{-3} to this density. Their spot sizes are about 10^{13} cm; the shock thickness and unsaturated core size are of this order for $v_{A5} \sim 1$ and the (n_o, v_s) range in Figure 1. Their observed isotropic luminosities range from $10^{-7} - 10^{-1}$ $L_\odot$ and their brightness temperatures range from $10^{11} - 10^{14}$ K; the shock models in the parameter range shown in Figure 1 produce these numbers with $a \sim 5 - 50$ or, equivalently, coherence lengths of 10^{14-15} cm. Lower radial velocity masers tend to have higher T_b; in shocks, lower radial velocities mean that the line of sight is close to the shock plane, resulting in higher a and T_b. H_2O masers are associated with star formation and outflows, the H_2O luminosity correlating with the mechanical luminosity seen in the CO outflow; the mass loss produces the shocks which, in turn, produce the masers. Millimeter observations, the newly detected 321 GHz H_2O maser (Neufeld & Melnick 1990), and the observation that there are not enough external photons to pump the maser all point to warm 300-400 K gas; the shock produces a large column of H_2O at 300-400 K, where collisions can pump the maser. Finally, Fiebig & Güsten (1989) have observed the Zeeman splitting of the 22 GHz H_2O maser in W49 and estimated the component of the B field along the line of sight to be about 100 mG; the shock model predicts exactly this sort of field in the masing region (Eq. 2).

5. REFERENCES

Elitzur, M., Hollenbach, D.J., & McKee, C.F. 1989, *Ap. J.*, **346**, 983(EHM).

Elitzur, M., Hollenbach, D.J., & McKee, C.F. 1992, *Ap.J.*, in press.

Felli, M., Palagi, F., & Tofani, G. 1992, *Astr. Astrophys.*, in press.

Fiebig, D. & Güsten, R. 1989, *Astr. Ap. Letters*, **214**, 333.

Genzel, R. 1986, in *Masers, Molecules and Mass Outflows in Star Forming Regions*, ed. A.D. Haschick (Westford, MA:Haystack Observatory), p233.

Heiles, C., Goodman, A.A., McKee, C.F., & Zweibel, E.G. 1992, in *Protostars and Planets III*, ed. E. Levy, J. Lunine & M. Matthews (Tucson: Univ. of Arizona Press), in press.

Hollenbach, D.J., Chernoff, D., & McKee, C.F. 1989, in *Infrared Spectroscopy in Astronomy*, ed. B. Kaldeich, ESA SP-290, p245.

Hollenbach, D.J. & McKee, C.F. 1979, *Ap. J. Suppl.*, **41**, 555.

Hollenbach, D.J. & McKee, C.F. 1989, *Ap. J.*, **342**, 306(HM).

Hollenbach, D.J., McKee, C.F., & Chernoff, D. 1987, in *Star Forming Regions*, ed. M. Peimbert & J. Jugaku (Dordrecht: Reidel), p334.

Neufeld, D. & Melnick, G. 1990, *Ap. J. Letters*, **352**, L9.

Walker, R.C., Matsakis, D.N., & Garcia-Barreto, J.A. 1982, *Ap. J.*, **255**, 128.

WATER MASER EMISSION TOWARD ULTRACOMPACT HII REGIONS

Peter Hofner and Ed Churchwell

Washburn Observatory, University of Wisconsin-Madison

475 North Charter Street, Madison, Wisconsin 53706

ABSTRACT

We present preliminary results of a recent VLA observation of the 6_{16}-5_{23} masing transition of interstellar H_2O toward the ultracompact HII regions G29.96 − 0.02 and G5.89 − 0.39. The positions of the maser spots relative to the ionized gas are given and their radial velocities are compared with velocities obtained from molecular line and hydrogen recombination line observations.

I. INTRODUCTION

Ultracompact (UC) HII regions are manifestations of newly formed massive stars which are still embedded in their natal molecular cloud. The 1.3 cm water maser occurs with a high frequency in the neighborhood of these objects as was clearly demonstrated in a survey of the 1.3 cm H_2O maser transition toward UC HII regions with the MPIfR 100 m telescope by Churchwell et al., 1990, who detected water maser emission in 67 % of the sources.

The main goal of this project is to determine absolute positions and velocities of the H_2O maser sources in order to investigate the kinematics of the molecular gas in the neighborhood of newly formed, massive stars and to study the role of water masers in the massive star formation process. It is of particular interest to investigate whether the H_2O maser sources participate in the bipolar, molecular outflows which have been detected toward some ultracompact HII regions.

We have recently obtained VLA observations of 21 UC HII regions taken from the sample of Churchwell et al., 1990. Here we present preliminary results for two sources. The observations are described in section II and the results are given in section III. Section IV contains a summary.

II. OBSERVATIONS

Observations of the 6_{16}-5_{23} transition of interstellar H_2O were made on December 14, 1991 with the VLA in its B-configuration. The synthesized beam FWHM of our observations was about 0.4″. We used a total bandwith of 3.125 MHz with 128 channels which resulted in a velocity coverage of 42 km/sec and a channel separation of 0.329 km/sec. The typical integration time was 8.5 minutes.

III. RESULTS

a) G29.96 − 0.02

G29.96 − 0.02 is a UC HII region with a cometary morphology at a distance of 7.4 kpc (Churchwell et al., 1990). Single dish observations of several molecular lines ($C^{34}S$,

^{13}CO, CH$_3$CN, NH$_3$) have shown, that the molecular gas in which G29.96 − 0.02 is embedded has an LSR velocity between 97 and 99 km/sec. The dynamics of the ionized gas has been investigated by Wood and Churchwell, 1991 who found a steep, systematic velocity gradient across the source with velocities increasing from about 85 km/sec along the leading edge to about 98 km/sec along the ridge of maximum brightness, reaching a maximum value of about 105 km/sec 2″ behind the arc.

We detect water maser emission from two distinct locations. Figure 1 shows the positions and spectra of the masers overlayed on a 2 cm continuum map taken from Wood and Churchwell, 1989. Numerical values are given in Table 1.

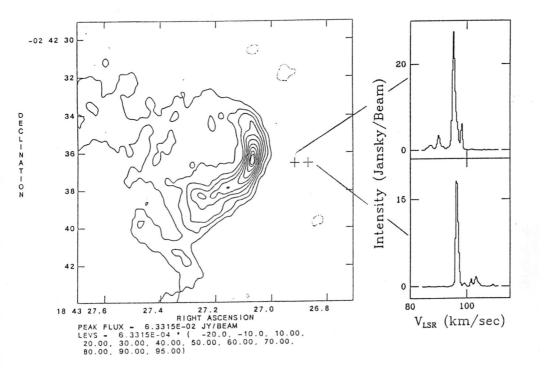

Figure 1. 2 cm VLA continuum map of G29.96 − 0.02. The crosses indicate the position of the H$_2$O maser features. The right panel shows the spectra obtained at the maser positions.

The maser emission occurs close to the ionized gas, in front of the cometary arc. Comparing the spectra at the two locations we note an interesting symmetry: At both maser spots the strongest emission occurs at velocities close to the velocity of the molecular gas. However, the maser spot which appears closest to the cometary arc shows additional features which are mostly blue shifted whereas only additional red shifted features are found at the other maser spot. Thus it appears that the maser spectra reflect a well ordered motion of the molecular gas around G29.96 − 0.02.

TABLE 1

OBSERVED WATER MASER FEATURES

Source	α (1950) (h min sec)	δ (1950) (° ′ ″)	V_{LSR} (km s^{-1})	S_ν (Jy)
G5.89−0.38	17 57 26.815	−24 03 56.11	20.2	36.1
	26.814	56.07	18.2	8.8
	26.814	56.16	15.9	1.1
	17 57 26.698	−24 04 02.59	−3.2	26.5
	26.698	02.59	−4.2	7.6
	26.694	02.28	8.4	1.6
	26.698	02.59	11.0	1.9
	17 57 26.741	−24 04 00.01	10.3	2.3
	17 57 26.671	−24 03 59.86	8.7	10.9
	17 57 26.472	−24 04 00.70	8.7	9.3
	17 57 26.703	−24 03 48.27	78.4	47.2
G29.96−0.02	18 43 26.917	−02 42 36.64	95.3	27.4
	26.917	36.63	98.3	6.1
	26.916	36.63	90.0	3.4
	26.916	36.64	87.4	1.1
	18 43 26.871	−02 42 36.70	96.6	20.0
	26.866	36.39	103.2	3.5
	26.880	36.72	101.6	1.9
	26.870	36.69	109.1	0.3

b) G5.89 − 0.39

G5.89 − 0.39, also known as W28 A2 is a strong radio continuum source with a shell morphology at a distance of about 4 kpc (Zijlstra et al., 1990). A massive bipolar outflow has been detected toward this source by Harvey and Forveille, 1988. Both OH and H_2O masers as well as NIR emission are associated with this source.

In Figure 2 we show the locations and spectra of the water masers on a 1.3 cm map from the same observation. The high velocity feature at 78 km/sec which has previously been detected in single dish spectra is seen about 10″ to the north of the radio continuum. We detect one maser feature which is projected onto the maximum of the radio continuum at the NE part of the shell. The position of the NIR source IRS1 (Moorwood and Salinari, 1981) is coincident with this maser spot within about 1″. A velocity gradient along the North–South axis is apparent in the spectra. The sense and orientation of this velocity gradient is consistent with the bipolar molecular outflow as mapped in $C^{34}S$ by Cesaroni et al., 1991.

IV SUMMARY

Although it is too early to draw any conclusion, a preliminary look at our data suggests that water maser emission in the vicinity of UC HII regions often occurs at spatially distinct positions with individual features spanning a wide range of velocities. It will be

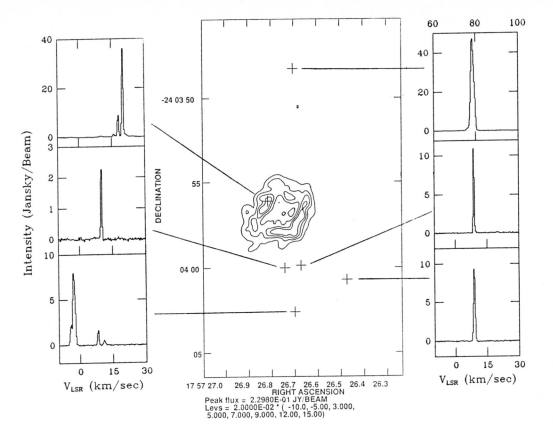

Figure 2. 1.3 cm VLA continuum map of G5.89 − 0.39. The crosses indicate the position of the H_2O maser features. The spectra obtained at the maser positions are displayed on both sides of the continuum map. Note the different velocity scale for the spectrum in the uppermost panel on the right hand side.

interesting to see whether the observed positions and velocities can constrain existing models of the dynamics of the molecular gas around newly formed massive stars.

REFERENCES

Cesaroni, R. et al. (1991). *Astr. Ap.*, **252**, 278.
Churchwell, E. et al. (1990). *Astr. Ap. Suppl.*, **83**, 119.
Harvey, P. M. & Forveille, T. (1988). *Astr. Ap.*, **197**, L19.
Moorwood, A. F. M., & Salinari, P. (1981). *Astr. Ap.*, **94**, 299.
Wood, D. O. S. & Churchwell, E. (1989). *Ap. J. Suppl.*, **69**, 831.
Wood, D. O. S. & Churchwell, E. (1991). *Ap. J.*, **372**, 199.
Zijlstra, A. A. et al. (1990). *M.N.R.A.S.*, **246**, 217.

INTERACTING H_2O MASERS IN STAR-FORMING REGIONS

Nikolaos D. Kylafis & Konstantinos G. Pavlakis

University of Crete, Physics Department
714 09 Heraklion, Crete, Greece
and
Foundation for Research and Technology-Hellas
P.O. Box 1527, 711 10 Heraklion, Crete, Greece

ABSTRACT

We have studied the interaction of H_2O masers in star forming regions as a physical mechanism for the explanation of the very strong H_2O maser sourses. We have carried out detailed numerical calculations for both saturated and unsaturated masers and have derived approximate analytic expressions for the expected brightness temperature from interacting masers. We have found that the interaction of two low or medium power masers can in principle lead to the appearance of a very strong one.

1. INTRODUCTION

An interesting idea regarding the powerful Galactic H_2O masers was proposed by Deguchi & Watson (1989). They demonstrated that two medium power masers, separated by distances characteristic of the size of star-forming regions and aligned to within the angular size of the beam of one maser, can result in reduced beam size and therefore enhanced brightness temperature. Elitzur, McKee, & Hollenbach (1991) extended the work of Deguchi & Watson and proposed that the two giant bursts of H_2O maser emission in W49 and Orion were the result of interacting masers.

In this paper we present a qualitative discussion of interacting (both saturated and unsaturated) masers. The results of our numerical calculations appear in a lengthier publication (Kylafis & Pavlakis 1992).

2. QUALITATIVE DISCUSSION

Consider a maser region in the form of a cylinder of diameter d and length l (see Figure 1). Its maser radiation is emitted mainly through the bases of the cylinder and into solid angles Ω along the axis of the cylinder. Let the brightness temperature of the maser along the axis of the cylinder be T_{b0}. Now consider in addition a background source (in Figure 1 it is shown as a similar maser) at a distance $D \gg l$ from the maser and an observer on the other side of it. When the radiation of the background source enters the maser, and if its intensity is high, it causes the maser radiation to be emitted in the solid angle $\Omega_s \approx (d/D)^2 \ll \Omega$ (Deguchi & Watson 1989; for the case of dissimilar sources see Elitzur, McKee, & Hollenbach 1991). The observer on the right, who detects this radiation and does not know its origin, calculates an *isotropic* luminosity which is orders of magnitude larger than what would be inferred if the maser were alone. In what follows we explain qualitatively and quantitatively the effects of the background source on the observed intensity.

Let ϕ_i be the net rate (i.e., stimulated emission minus absorption) of maser photon production per unit volume due to the internal radiation of the maser and ϕ_s the corresponding rate due to the radiation of the background source. The photons produced with rate ϕ_i are emitted on either side of the maser and into solid angle 2Ω, while those produced with rate ϕ_s are emitted toward the observer and into solid angle $\Omega_s \approx (d/D)^2 \ll \Omega$.

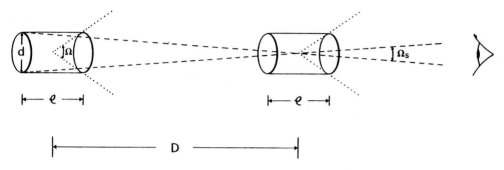

Fig. 1. Schematic representation of two similar masers interacting with each other.

The brightness temperature or the intensity seen by the observer can be written as

$$T_{obs} \approx (|T_x| + T_s) \exp(|\tau|) \qquad \text{or} \qquad I_{obs} \approx \frac{\phi_i}{2\Omega} + \frac{\phi_s}{\Omega_s} , \qquad (1)$$

where T_x is the excitation temerature of the maser line, T_s is the brightness temperature of the background source and τ is the optical depth of the maser line.

For the sake of this qualitative discussion let us treat the intensity of the background source as a parameter. When the mean intensity of the photons that are produced with rate ϕ_s is much smaller than the mean intensity of those produced with rate ϕ_i, then clearly the effect of the source on the maser is negligible. An equivalent condition is $T_s \ll |T_x|$. For collisionally pumped H_2O masers the excitation temperature is typically in the range $10^2 \lesssim |T_x| \lesssim 10^3$ K and $|\tau| \lesssim 25$ (Kylafis and Norman 1991).

As T_s increases, a point is reached where $T_s \approx |T_x|$ or $\phi_s/\Omega_s \approx \phi_i/2\Omega$. At this point, the effect of the background source on the observed intensity begins to become important. Note however that the equation $T_s \approx |T_x|$ implies $\phi_s \ll \phi_i$. That is, the radiation from the background source has a significant contribution to the observed intensity *despite the fact that it causes the production of a negligible number of photons.*

As T_s increases further, the observed intensity grows essentially linearly with T_s (see eq. [1]). This is because with $\phi_s \ll \phi_i$, the background radiation has a negligible effect on the populations N_1 and N_2 and therefore on τ and T_x.

The linear increase of T_b will continue until $\phi_s \approx \phi_i$. This occurs when $T_s \approx (2\Omega/\Omega_s)|T_x|$. At this point, equal number of photons are emitted into solid angles 2Ω and Ω_s. A further increase of T_s can have different effects on the observed intensity depending on whether the maser is saturated or unsaturated.

2.1. Saturated Masers

If the maser is saturated, then the photon production rate per unit volume ϕ_i has essentially its maximum possible value. That is, for every pumping event a photon is emitted. As T_s increases, ϕ_s increases, but only at the expense of ϕ_i. Thus, more and more photons that would be emitted into solid angle 2Ω are emitted into Ω_s and this last increase of T_s can at most increase the observed brightness temperature by a factor of two.

In summary, to within a factor of two, the brightness temperature T_b^f of a saturated maser *in the direction of the observer* as a function of the brightness temperature T_s of the background source is given by

$$T_b^f \approx \begin{cases} T_{b0}, & \text{if } T_s \lesssim |T_x|; \\ (T_{b0}/|T_x|)T_s, & \text{if } |T_x| \lesssim T_s \lesssim |T_x|(2\Omega/\Omega_s); \\ T_{b0}(2\Omega/\Omega_s), & \text{if } |T_x|(2\Omega/\Omega_s) \lesssim T_s . \end{cases} \tag{2}$$

The background source affects not only the forward brightness temperature T_b^f, but also the brightness temperature T_b^b of the maser *in the direction of the source*. It is straightforward to show (Kylafis & Pavlakis 1992) that the background brightness temperature T_b^b of the maser is given by

$$T_b^b \approx \begin{cases} T_{b0}, & \text{if } T_s \lesssim |T_x|(2\Omega/\Omega_s); \\ T_{b0}(|T_x|/T_s)(2\Omega/\Omega_s), & \text{otherwise} . \end{cases} \tag{3}$$

Similarly the behavior of the optical depth $|\tau|$ as a function of T_s can be written as

$$|\tau| \approx \begin{cases} |\tau_0|, & \text{if } T_s \lesssim |T_x|(2\Omega/\Omega_s); \\ |\tau_0| + 2.3 \log\left[\frac{|T_x|(2\Omega/\Omega_s)}{T_s}\right], & \text{otherwise} , \end{cases} \tag{4}$$

where τ_0 is the optical depth of the maser when it is unaffected by the background source.

If the background source is a maser (call it 2) identical to the one under study (call it 1), then T_s cannot be arbitrary. The radiation of maser 2 affects maser 1, but also the radiation of maser 1 affects maser 2. Self-consistency is obtained when the output of maser 2 is the input to maser 1 and visa versa (Deguchi & Watson 1989). This means $T_s = T_b^b$. Equating T_s with expression (3) we find that self-consistency requires

$$T_s = T_b^b \approx \left(T_{b0}|T_x|\frac{2\Omega}{\Omega_s}\right)^{1/2} . \tag{5}$$

2.2. Unsaturated Masers

If the maser is unsaturated, then the photon production rate per unit volume ϕ_i is not the maximum possible. Therefore ϕ_s and consequently T_b^f can continue increasing linearly with T_s, while ϕ_i remains constant. This linear increase will stop when the maser saturates. For a collisionally pumped maser, saturation occurs roughly when the radiative rate becomes comparable to the typical collision rate $C_{ji} \approx 10^{-2}(n_{H_2}/10^9 \text{cm}^{-3})$ s^{-1} (Palma *et al.* 1988). From then on ϕ_s can increase only at the expense of ϕ_i and only a small increase in T_b^f can occur with increasing T_s.

According to the above discussion, one can write approximate analytic expressions similar to those for saturated masers. Thus, for the forward brightness temperature we have

$$T_b^f \approx \begin{cases} T_{b0}, & \text{if } T_s \lesssim |T_x|; \\ (T_{b0}/|T_x|)T_s, & \text{if } |T_x| \lesssim T_s \lesssim T_s^*; \\ (T_{b0}/|T_x|)T_s^*, & \text{if } T_s^* \lesssim T_s, \end{cases} \qquad (6)$$

where T_s^* is a brightness temperature of the background source at which the net radiative rate becomes comparable to the typical collision rate. For the backward brightness temperature we have

$$T_b^b \approx \begin{cases} T_{b0}, & \text{if } T_s \lesssim T_s^*; \\ T_{b0}(T_s^*/T_s), & \text{otherwise}, \end{cases} \qquad (7)$$

and for the gain

$$|\tau| \approx \begin{cases} |\tau_0|, & \text{if } T_s \lesssim T_s^*; \\ |\tau_0| + 2.3\log(T_s^*/T_s), & \text{otherwise}. \end{cases} \qquad (8)$$

3. SUMMARY AND CONCLUSIONS

In agreement with previous work on interacting H_2O masers, we have found that the interaction of two low or medium power masers can in principle lead to the appearance of a very strong one.

The interaction of a very weak and a medium power maser can also lead to a very strong one.

We find it tempting to suggest that probably all H_2O masers in star-forming regions with observed brightness temperature $T_b \gtrsim 10^{13}$ K are the result of interacting masers. The reason for this is the following: To obtain brightness temperatures $T_b > 10^{13}$ K one needs aspect ratios (i.e., ratio of length to width) $a > 10$ if the maser region is static (Elitzur, Hollenbach, & McKee 1989) and $a \gtrsim 100$ if the maser region has velocity gradients (Kylafis & Norman 1991). Since velocity gradients are probably present in star-forming regions, it makes more sense to think of the masers with $T_b \gtrsim 10^{13}$ K as interacting masers rather than as very long cylinders.

We thank W. Watson and J. Caswell for useful discussions. We are indebted to N. Anderson for giving us the energy levels and dipole moment matrix elements for water. We also thank S. Green for sending us the collision rate coefficients of Palma *et al.* (1988) by electronic mail.

REFERENCES

Deguchi, S., & Watson W. D. 1989, ApJ, 340, L17

Elitzur, M., Hollenbach, D. J., & McKee, C. F. 1989, ApJ, 346, 983

Elitzur, M., McKee, C. F., & Hollenbach, D. J. 1991, ApJ, 367, 333

Kylafis, N. D., & Norman, C. 1991, ApJ, 373, 525

Kylafis, N. D., & Pavlakis, K. G. 1992, ApJ, submitted

Palma, A., Green, S., DeFrees, D. J., & McLean, A. D. 1988, ApJS, 68, 287

ORIGIN OF WATER MASERS IN W49N

Mordecai-Mark Mac Low

Space Science Division, Mail Stop 245-3, NASA Ames Research Center

Moffett Field, CA 94035 USA

and

Moshe Elitzur

Department of Physics and Astronomy, University of Kentucky

Lexington, KY 40506 USA

VLBI observations by Gwinn, Moran, & Reid (1992, also the current proceedings) of the proper motions of H_2O masers in W49N show that they have an elongated distribution expanding from a common center. Features with high space velocity only occur far from the center, while low velocity features occur at most distances, as shown in Figure 1. If the maser distribution is interpreted as a biconical expansion of individual blobs of material from a single common center, it requires a rather sudden acceleration to velocities of $\sim$ 200 km s^{-1} at the outer edges of the maser cluster, at $\sim 10^{17}$ cm.

We propose instead that H_2O masers in star-forming regions occur early in the expansion of thin shells swept up by high-velocity winds from young, massive stars, as outlined in Figure 2 (Mac Low & Elitzur 1992). Stellar winds of the strength that we require have been inferred in young stellar objects from the mapping of molecular outflows much larger and older than the region that we are examining (Scoville et al. 1986). The fast stellar wind ends in a strong shock, producing a region of high temperature ($T > 10^6$ K) gas that sweeps up the shell and drives a shock into the ambient gas (Weaver et al. 1977).

In W49N, confinement of the shell by a density distribution with an axial cavity can explain both the velocity field and the shape of the maser distribution. We use a modified version of the thin-shell code described by Mac Low & McCray (1988), including radiative cooling, to dynamically model the expanding shell. We describe the cavity with a density distribution of the form

$$n(\rho, z) = \min \left[n_0 (1 + \exp[\frac{\rho - \rho_0}{H}]), \; n_m \right], \tag{1}$$

where ρ is the cylindrical radius, ρ_0 and H are free parameters that determine, respectively,

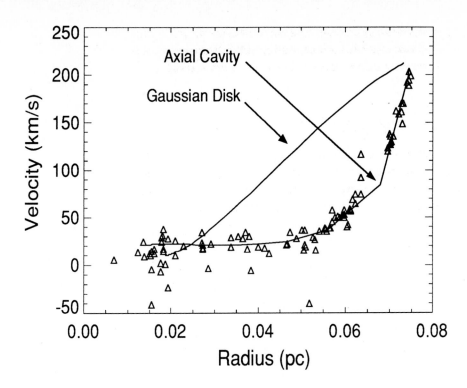

Figure 1: Space velocities and distances from the center of expansion of H_2O masers in W49N. Crosses are data points from Figure 5 of GMR. Solid lines show the results of full dynamical calculations of wind-driven expanding shells. The Axial Cavity line corresponds to the model described by equation (1) and subsequent discussion. The Gaussian Disk model has $n(z) = n_0 \exp[-(z/H)^2]$ with $n_0 = 10^9$ cm^{-3} and $H = 2.8 \times 10^{16}$ cm, and a wind characterized by $\dot{M} = 3.0 \times 10^{-3}$ $M_\odot$ yr^{-1}, and $v_w = 1.6 \times 10^8$ cm s^{-1}. The plotted curve corresponds to 800 years after wind onset.

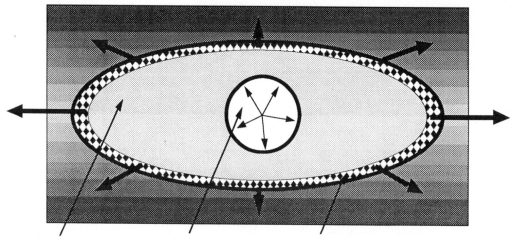

hot shocked wind   strong stellar wind   swept-up shell with masers

Figure 2: Structure of our proposed model for H_2O masers. A strong stellar wind gets stopped and heated at a termination shock. The resulting hot gas sweeps up a shell from the ambient gas. The wind termination shock and the outer shock, in which the masers form, are shown as thicker lines. If the density is higher to the sides than towards the poles, the poles will decelerate more slowly, producing higher velocities at larger radii, as observed in W49N.

the radius of the cavity and the steepness of its walls, and n_0 and n_m are, respectively, the density along the cavity axis and far away from it (at $\rho - \rho_0 \gg H$). In Figure 1 we show a model with stellar mass loss rate of 3×10^{-3} $M_\odot$ yr^{-1}, and wind velocity 1.4×10^8 cm s^{-1}. The cavity is characterized by $H = 10^{15}$ cm and $\rho_0 = 1.5 \times 10^{16}$ cm, and we selected $n_0 = 2 \times 10^6$ cm^{-3} and $n_m = 2 \times 10^8$ cm^{-3} to fit the low and high velocities observed. The temperature of the ambient medium is 30 K. The model is shown at $t = 250$ years after the wind onset. A simple plane-stratified Gaussian disk can explain the low and high velocities observed, but badly fails for intermediate velocities.

Our model fits the observations well and suggests that this system is of order 300 years old. Our proposal explains the cause and location of the shocks needed to produce water masers in the model of Elitzur, Hollenbach, & McKee (1989). The very short lifetime suggested by our model, about 10^{-4} of the lifetime of an O star, is consistent with W49 being a unique object in our Galaxy. These observations may show the very first stages of the collimation of a jet from a young stellar object. Icke et al. (1992) propose a similar toroidal density distribution denser at the equator to confine jets.

References

Elitzur, M., Hollenbach, D. J. & McKee, C. F. 1989, ApJ, 346, 983.

Gwinn, C., Moran, J., & Reid, M. 1992, ApJ, submitted.

Icke, V. Mellema, G., Ballick, B., Euldrink, F. & Frank, A. 1992, Nature 355, 524.

Mac Low, M.-M. & McCray, R. 1988, ApJ, 324, 776.

Mac Low, M.-M. & Elitzur, M. 1992, ApJ (Letters), submitted.

Scoville et al. 1986, ApJ 303, 416.

Weaver, R., McCray, R., Castor, J., Shapiro, P. and Moore 1977, ApJ 218, 377.

VLBI OBSERVATIONS OF THE H_2O SUPERMASER OUTBURST IN

THE ORION NEBULA

L. I. Matveyenko
Institute for Space Research, Moscow, Russia

P. J. Diamond
NRAO, Socorro, NM 87801, USA

D. A. Graham
MPIfR, Auf Dem Hügel, Bonn, Germany

W. Junor
NRAO, Socorro, NM 87801, USA

ABSTRACT

We have conducted a campaign of annual VLBI observations of the H_2O supermaser region in the Orion-KL. Most of our observations have been conducted with a small number (~ 3) of antennas. Here we report on a 7 station observation taken in 1985. We have produced an image of the supermaser region. Our data indicate that the masers lie along an approximately E-W line, and that there is a significant velocity gradient across the maser region. We suggest that the masers originate in several thin rings in a rotating, expanding disc of gas and dust.

1. INTRODUCTION

From 1979 to 1986 the H_2O maser in Orion-KL has been observed to have the highest average flux density of any known H_2O maser. Within this period there have also been three bursts of even stronger emission; in 1980.3, 1983.7 and 1984.9 the maser reached flux density levels of between $2.5 \times 10^6 \rightarrow 7 \times 10^6$ Jy. The profile of this supermaser is simple and smooth with low or high velocity tails, depending on the epoch. The velocity and overall shape of the profile are also observed to change slowly with time and this has been attributed to the blending of different components (Garay et al., 1989). The average velocity of the supermaser was 7.9 km s^{-1} and the average linewidth was $0.3 \rightarrow 0.7$ km s^{-1}. The emission was also strongly linearly polarized (Abraham et al., 1986; Matveyenko et al., 1988; Garay et al., 1989)

2. OBSERVATIONS and RESULTS

We have conducted a VLBI campaign to monitor the H_2O supermaser emission region (Matveyenko et al., 1988). We report here on observations performed in 1985.8.

These were the most extensive of our series; we used antennas at Simeiz (22m), Onsala (20m), Effelsberg (100m), Haystack (37m), Green Bank (43m), VLA (25m) and Owens Valley (40m). The angular resolutions we obtained varied from 0.3 → 0.6mas (0.15 → 0.3AU at the distance of Orion), and enabled us to image the complex region of the H_2O supermaser. The data were correlated on the MkII correlator of the NRAO at Socorro. The calibration and imaging were performed within the AIPS package.

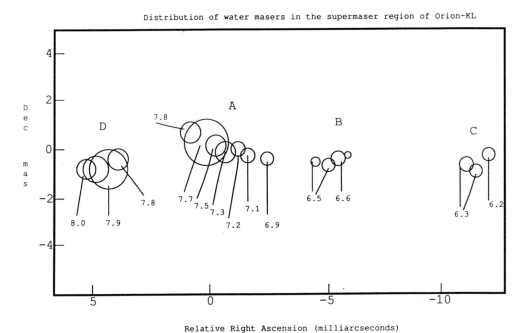

Fig. 1. The total intensity VLBI map of the H_2O supermaser region. The synthesised beam is 1.0×0.3mas. The circles represent the logarithm of the flux density of a component. The total flux density observed from the supermaser region is 2×10^6 Jy. Each component is labelled with its velocity with respect to the LSR. The position of the brightest maser component is RA $05^h32^m46\overset{s}{.}64$, Dec $-05°24'29\overset{''}{.}8$ (B1950.0).

The structure of the supermaser region is shown in Fig.1. It consists of four or five groups of compact components lying along an approximate EW line 20mas in length. The total flux density of this region is ∼ 2×10^6 Jy. The main group (group A in Fig.1) consists of 7 compact components distributed along an arc ($\chi \sim 60°$). The component sizes are ∼ $0.2 \rightarrow 0.3$mas ($0.1 \rightarrow 0.15$AU). The distances between the individual components vary between 0.4 and 0.8 mas ($0.2 \rightarrow 0.4$AU). The total size of group A is approximately 4 mas (2 AU). Other smaller groups of components (B, C and D) lie between 5 and 10 mas from Group A.

There is an evident velocity gradient running across the supermaser region. In Fig.2 we show the velocity as a function of position. The overall gradient is 0.27 km s^{-1} mas^{-1} (0.54 km s^{-1} AU^{-1}).

The linewidths of the individual compact components are very narrow and lie between 0.1 → 0.13 km s^{-1} depending on the component.

The Simeiz antenna has been used to study the linear polarization of the supermaser with a frequency resolution of 15kHz ($\Delta V \sim 0.2$ km s^{-1}). These data (Matveyenko et al., 1988) show that the degree of polarization and its position angle vary across the supermaser profile. It can be seen from these data that each spectral component has a specific position angle; the low-velocity components ($V \sim 7$ km s^{-1}) have a position angle $\chi \sim -40°$, while the high velocity features ($V \sim 8$ km s^{-1}) have $\chi \sim -15°$. This corresponds to a gradient of polarization position angle $\Delta\chi/\Delta L \sim 13°$ AU^{-1}).

The brightness temperatures of the compact components are $T_B \sim 10^{16-17} K$.

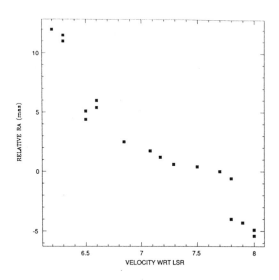

Fig. 2. A plot of the velocity of the maser components as a function of RA offset (mas) from the reference feature. Note the distinctive velocity gradient of 0.27 km s^{-1} mas^{-1}.

3. DISCUSSION

From our data, the most plausible model to explain the supermaser characteristics for groups A, B and C is:

a) The compact maser components lie in a thin, edge-on, rotating and expanding disc. The disc has a radius of ~ 5AU, a rotational velocity of ~ 5 km s^{-1}, an expansion velocity of ~ 3.8 km s^{-1} and is oriented at a position angle of 60°. The strongest maser components lie along the edge of the disc where the longest pathlengths are available.

b) The disc is composed of a series of thin rings. Each ring generates one maser component of narrow linewidth e.g. the 7 components visible in group A (Fig.1).

c) A magnetic field of strength $\sim$ 30mG (Garay et al., 1989) lies parallel to the disc.

This model would then explain naturally the velocity gradient that is observed. The thin ring hypothesis would explain also the multiple components observed in group A. Furthermore, this model explains the change in the polarization position angle, since the velocity vector of each compact component when combined with the magnetic field vector (which is constant over the disc) would cause a change in position angle across the disc. This model would also explain naturally the fact that the compact maser components vary independently (Matveyenko et al., 1988), since the variations would be caused by the changes in the optical depths in each ring.

Garay et al. (1989) suggest that the supermaser is saturated because the observed T_B are greater than the canonical saturation temperature of H_2O masers (Reid and Moran, 1981). However we suggest that the H_2O masers are highly beamed ($\Omega < 10^{-4}$) and that they are only partially saturated. The short time scale and high flux density variations can then be explained by a relatively small ($\frac{\Delta \tau}{\tau} \sim 3\%$) change in optical depth caused by a non-uniform distribution of H_2O in the rings.

We also suggest that the strong ($P > 80\%$) linear polarization observed is due to the pumping of the masers by an anisotropic mechanism (Varsalovich, 1971; Western & Watson, 1983, 1989).

REFERENCES

Abraham, Z., Vilas Boas, J.W.S. and del Campo, L.F., 1986, *Astr. Ap.*, **167**, 311.

Garay, G., Moran, J. M. and Haschick, A. D., 1989, *Ap. J.*, **338**, 244.

Matveyenko, L. I., Graham, D. A. and Diamond, P. J., 1988, *Sov. Astron. Letters*, **14**, 468.

Reid, M. J. and Moran, J. M. , 1981, *Ann. Rev. Astr. Ap.*, **19**, 231.

Varshalovich, A., 1971, Uspechi Fiz. Nauk, **13**, 429.

Western, R. and Watson, W. D. , 1983, *Ap. J.*, **275**, 195.

Western, R. and Watson, W. D. , 1989, *Ap. J.*, **285**, 158.

THE ZEEMAN EFFECT OR LINEAR BIREFRINGENCE?

— VLA Polarimetric Spectral Line Observations of H_2O Masers

Jun-Hui Zhao, W. M. Goss, P. Diamond

National Radio Astronomy Observatory
P.O. Box O, Socorro, New Mexico 87801

ABSTRACT

We present line profiles of the four Stokes parameters of H_2O masers at 22 GHz observed with the VLA in full polarimetric spectral line mode. With careful calibration, the instrumental effects such as linear leakage and the difference of antenna gain between RCP and LCP, can be minimized. Our measurements show a few percent linear polarization. Weak circular polarization was detected at a level of 0.1 percent of the peak intensity. A large uncertainty in the measurements of weak circular polarization is caused by telescope pointing errors. The observed polarization of H_2O masers can be interpreted as either the Zeeman effect or linear birefringence.

1. INTRODUCTION

H_2O masers are thought to be associated with an early stage of star formation (Moran, 1990). The density of maser regions requires 10^{9-11} cm^{-3} to give rise to H_2O maser amplification. Strong magnetic fields up to 100 mG might then be expected in the dense circumstellar gas clumps. The only existing method for measuring magnetic field strengths in molecular clouds is the Zeeman effect in radio frequency spectral lines. Circularly polarized radiation from astrophysical masers has been observed by Fiebig and Güsten (1989, hereafter FG), using the Effelsberg 100 m antenna. The observed antisymmetric variation of the difference about the center of the spectral line can be explained by Zeeman splitting. The typical measured circular polarization, $V/I \sim 0.001$, is extremely small and magnetic field strengths up to 50 mG were inferred for the component parallel to the line of sight in a number of H_2O maser regions.

If linear polarization is present at a level of a few percent, the circular polarization could be caused by linear birefringence (Heiles *et al.*, 1991). This phenomenon has been discussed by a number of authors (Kylafis and Shapiro, 1983; Deguchi and Watson, 1985). Calculations indicate that the circular polarization generated by linear birefringence is antisymmetric about the line center. This antisymmetric profile mimics, and can easily be mistaken for, the normal Zeeman effect. However, Nedoluha and Watson (1992) show that the 22 GHz H_2O maser feature is a blend of the three strongest hyperfine components and the profile for the net circular polarization is quite asymmetric. Observations of all four Stokes components are necessary to clarify which effect dominates the observed circular polarization in H_2O masers.

In this paper we discuss preliminary results of VLA observations of H_2O masers obtained by simultaneously measuring the four correlations RR, LL, RL and LR. The observations were made on 2 April 1989 (C array) and 29 March 1991 (D array), and 17 January 1992 (B-array) with the VLA at a rest frequency of 22.235 GHz for the H_2O maser transition. There are some fundamental limitations in circular polarization measurements. The details of the calibrations and corrections for the instrumental effects, such as beam squint and pointing errors, are discussed elsewhere (Zhao *et al.*, 1992).

2. RESULTS

The observations presented here are an attempt to use the VLA to confirm the results of FG. From their sample, we selected four sources (S140, W3(2), Orion-KL and W49N) with an isolated single velocity profile. However, our observations show that many of the components are blended with nearby velocity features (*e.g.*, W3(2) at -40.5 km s^{-1}), and some have faded away (*e.g.*, Orion-KL at -49.7 km s^{-1}). W49N shows a very complex spatial and spectral structure at the velocities 11.4 and -52.7 km s^{-1}. In the following, we will concentrate on the main component A of S140 at RA(1950) $= 22^h\ 17^m\ 41.10^s$ and DEC(1950) $= 63°\ 03'\ 41.0''$.

a) Linear Polarization

Linear polarization spectra of S140 are shown in Figure 1a. The instrumental effects caused by the polarization leakage (the D terms) have been corrected. Thus, the typical uncertainty in the degree of linear polarization is about 0.1 percent. The degree of linear polarization and position angle of the E-vector of S140-A are shown in Figure 1b. The degree of linear polarization of S140-A is about 1.5 percent. The position angle appears to change by 6° across the profile (from -15.8 to -14.8 km s^{-1}).

b) Circular Polarization

In addition to the corrections for linear leakage, further correction for the instrumental circular polarization caused by the gain differences between RCP and LCP (the α term) must be applied to the V spectra. Reconstruction of the actual V spectrum requires an accurate determination of the instrumental gain factor, α. Figure 2 shows the V profile which is obtained by applying the instrumental gain corrections with different α determined by two independent methods. First, we assume that the actual V profile is antisymmetric; the ratio of the integrated instrumental V profile to the integrated I profile indicates $\alpha = -0.0075 \pm 0.0005$. The quoted error is due to the r.m.s. noise of the spectra.

Secondly, α can be independently determined from the observations of 3C 84. Adopting the average value of $< \alpha > = -0.009$, the residual V spectral profile is no longer antisymmetric (see Figure 2, dashed line in the middle panel). The uncertainty of this technique is mainly caused by the instability of the instrumental gains as a function of time. The origin of this instability is thought to be telescope pointing errors (see Zhao *et al.*, 1992). To minimize the uncertainty due to the time variation of the instrumental gain factor requires frequent observations of a strong calibrator near the target source during an observing run.

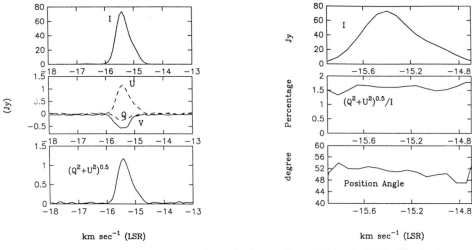

Figure 1: a. Full polarimetric spectra of S140-A observed on 29 March 1991. The instrumental effects due to linear leakage (the D terms) have been corrected in Stokes Q, U, and V, but the factor (α) of gain difference between RCP and LCP which dominates the Stokes V spectra has not been applied. b. Degree of linear polarization and position angle of the E-vector.

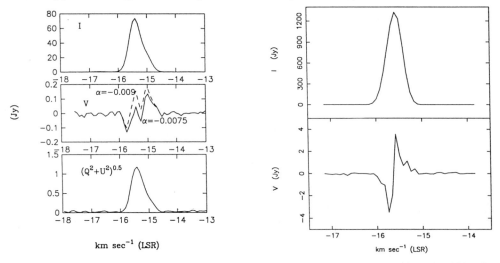

Figure 2 (left): The true V profile (middle panel) of S140-A, observed on 29 March 1991, after correcting the instrumental effect by applying $\alpha = -0.0075$ (solid line) and $\alpha = -0.009$ (dashed line). The gain factor α was determined by two independent methods: (1) The assumption is that the actual V profile is antisymmetric; the ratio of the integrated instrumental V profile to the integrated I profile indicates $\alpha = -0.0075$. (2) The factor $\alpha = -0.009$ was independently determined from the observations of 3C 84. The Stokes I (top panel) and linear polarization $\sqrt{Q^2 + U^2}$ (bottom panel) are presented.

Figure 3 (right): The V profile (bottom panel) of S140-A, observed on 2 April 1989, after correcting the instrumental effect by applying $\alpha = 0.01$ based on the assumption of an anti-symmetric V profile. The Stokes I (top panel) is also shown. Stokes Q and U were not observed on this date.

Table 1: Measurements of Polarization for S140-A

Obs. date	$\Delta V^\dagger$ (Jy)	I_{max} (Jy)	$\Delta V/2I_{max}$	$\sqrt{Q^2 + U^2}/I$
29Mar91	0.28 ± 0.05	73 ± 0.1	0.0019 ± 0.0004	0.016 ± 0.002
02Apr89	$7.0 \pm 1.3^\ddagger$	1325 ± 0.5	0.0026 ± 0.0005	$-$

$^\dagger \Delta V = V_{max} - V_{min}$, where V_{max} and V_{min} are the maximum and minimum values in the Stokes V spectra, respectively. ‡The error is mainly due to the uncertainty of the leakage of linear polarization into the Stokes V.

Nevertheless, both independent methods provide good agreement in the determination of α. The instrumental circular polarization gain is $\sim 1\%$. After correcting for this instrumental effect, the difference $\Delta V = V_{max} - V_{min}$ remains at a level of 0.3 Jy and is insensitive to the uncertainties in the determination of the α term.

In addition, a strong, circularly polarized signal of $\Delta V = 7$ Jy from S140-A was detected in our earlier VLA observations (on 2 April 1989) in which only the correlations RR and LL were observed (see Figure 3). The instrumental gain factor $\alpha = 0.01$ was determined by assuming that the actual V profile is antisymmetric. The correction for instrumental linear leakage term in this observation was not possible.

The determinations of the polarization properties for S140-A obtained from the two VLA observations are summarized in Table 1. The ratios ($\Delta V/2I_{max}$) of 0.0026 and 0.0019 were observed on 2 April 1989 and 29 March 1991, respectively. The degree of circular polarization between the two dates has not changed significantly while the peak flux density of the maser has decreased by a factor of 17. This ratio $\Delta V/2I_{max}$ observed by FG was about a factor of four less at an earlier period (during December 1987 to February 1988) when the flux density was at a level of a several hundred Jansky (Jy).

3. DISCUSSIONS AND IMPLICATIONS

Using the VLA, a few percent linear polarization and a few tenths percent circular polarization from H_2O masers have been detected. Origins of the polarization from astrophysical maser sources have been discussed by many authors (Goldreich, Keeley, and Kwan 1973; Deguchi and Watson 1986; Nedoluha and Watson 1990a). Maser polarization characteristics can be related to the interaction of the molecules with a magnetic field (Goldreich, Keeley, and Kwan 1973). For nonparamagnetic molecules such as H_2O, the Zeeman splitting is much smaller than the typical linewidths of 50 kHz but can be larger than the stimulated rates (~ 1 s^{-1}). Under these circumstances, linear polarization can be produced (Deguchi and Watson, 1986). Calculations indicate that the Zeeman splitting can produce a weak circular polarization (FG, Nedoluha and Watson 1990b). Detailed calculations of polarization radiation of 22 GHz H_2O masers have been carried out by Nedoluha and Watson (1992). These authors find that the 22 GHz maser feature is a result of blending of the three strongest hyperfine components and the profile for the net circular polarization (V profile) is asymmetric. For most observed line widths, the degree of linear polarization from the 22 GHz maser radiation tends to be smaller than ten percent .

Circular polarization from molecules can result from propagation effects such as birefringence (Kylafis and Shapiro 1983; Deguchi and Watson 1985). These authors have shown that if the particles are partially aligned, magneto-rotation converts linear polarized radiation into circularly polarized radiation. The peak circular polarization tends to be only about one-tenth of the linear polarization in the emitted radiation. The line profile of Stokes V caused by linear birefringence is antisymmetric about the line center.

Thus, the only characteristic that distinguishes between the Zeeman effect and linear birefringence appears to be the shape of the line profile. Any significant deviation from antisymmetry in observed profiles for Stokes V would indicate that the circular polarization is dominated by the Zeeman effect. Then, the strength of the magnetic field along the line-of-sight can be inferred by measuring the degree of circular polarization and the linewidth of the spectral line (FG; Nedoluha and Watson 1992). Based on the current observations, it is not possible to distinguish between the two phenomena.

ACKNOWLEDGEMENTS

We are grateful to Jim Moran, Dick Crutcher, Gerald Nedoluha, and Carl Bignell for helpful discussions. The National Radio Astronomy Observatory is operated by Associated Universities, Inc., under a cooperative agreement with the National Science Foundation.

REFERENCES

Deguchi, S. and Watson, W. D. 1986, *Ap. J.*, 302, 750.

Deguchi, S. and Watson, W. D. 1985, *Ap. J.*, 289, 621.

Fiebig, D. and Güsten, R. 1989, *Astron. Astrophy.*, 214, 333.

Goldreich, P., Keeley, D. A., and Kwan, J. Y., 1973, *Ap. J.*, 179, 111.

Heiles, C., Goodman, A. A., McKee, C. F., and Zweibel, E. G., in *Fragmentation of Molecular Clouds and Star Formation*, Eds. E. Falgarone *et al.*, (IAU Symp. 147, Kluwer Academic, Dordrecht, 1991), p43.

Kylafis, N. D. and Shapiro, P. R. 1983, *Ap. J.*, 272, L35.

Moran, J. M., in *Galactic and Extragalactic Magnetic Fields*, eds. R. Beck *et al.*(IAU Symp. 140, Kluwer Academic, Dordrecht, 1990), p301.

Nedoluha, G. E. and Watson, W. D. 1992, *Ap. J.*, 384, 185.

Nedoluha, G. E. and Watson, W. D. 1990a, *Ap. J.*, 354, 660.

Nedoluha, G. E. and Watson, W. D. 1990b, *Ap. J.*, 361, L53.

Zhao, J.-H., Goss, W. M. and Diamond, P. 1992, in preparation.

8. METHANOL MASERS IN STAR FORMING REGIONS

THE ARCETRI METHANOL VLBI RECEIVER

M. Catarzi
Osservatorio Astrofisico di Arcetri
largo E.Fermi 5, I-50125 Florence, Italy

L.Moscadelli
Dipartimento di Astronomia
largo E.Fermi 5, I-50125 Florence, Italy

1. INTRODUCTION

From 1970 so far various methanol lines have been observed towards star-forming regions both in emission and in absorption. At the present time at least 14 transitions are known be masing. The marked separation between the detection probability of different transitions in different objects has led to the definition of two classes of methanol masers. The prototype of class I masers is Orion-KL, while W3(OH) is that of class II masers. In no case a class I methanol maser emits in a class II maser line and viceversa.

Class II methanol masers show several velocity components spread over a few km/s, while class I masers have few components confined to a velocity interval of less than 1 km/s. Sometimes class I spectra show narrow components of 0.3-0.7 km/s superimposed to a broad (quasi-) thermal emission. Transitions of the class II masers show comparable velocity ranges, but the resemblance of spectra from different transitions is not as close as that between class I maser spectra.

While class II methanol masers are observed near compact HII regions, IR sources and H_2O and OH masers, class I masers are found at distances of about 1 pc from these objects. Several observational data (Plambeck, 1990) suggest that class I masers originate in shock fronts associated to powerful molecular outflows: class I masers can therefore be used to investigate the low mass star formation, where molecular outflows are supposed to play a dominant role. On the contrary class II masers, being probes of the physical and kinematic conditions of the hot and dense molecular cores around compact HII regions, are very useful to study the formation and the first evolution of massive stars.

Before 1991 the strongest ever observed methanol maser was that at 12.2 GHz (Batrla, 1987): in this year Menten (1991) has observed maser emission still strongest at 6.6 GHz. Both masers are of class II.

In order to add the Medicina and Noto radiotelescopes to the VLBI network and to provide frequent monitoring of the maser spectra, a 12.2 GHz uncooled low noise receiver has been realized. The system is equipped with commercial frontend for satellite broadcasting composed by a low noise HEMT amplifier followed by a conventional hetherodyne system.

The resulting noise temperature is included in the range 140-160 K with a bandwidth of 400 MHz.

The system has been tested on the Medicina radiotelescope by observing maser sources quoted in the literature.

2. THE DESIGN AND LABORATORY TESTS OF THE RECEIVER

According to the present intensities of the 12.2 GHz line, a sensitivity of 1 Jy seemed to be a reasonable compromise for the detection of galactic methanol masers.

Single dish observations are possible with the Medicina 32 m radiotelescope, which has an antenna efficiency of 0.12 K/Jy. With a spectral resolution of 24 KHz/channel and a typical integration time of 5 min., the required T_{sys} is of the order of 180-200 K.

Furthermore for polarization analysis, the receiver is equipped with two channels of opposite circular polarizations.

The required system temperature and bandwidth can be easily satisfied using commercial low noise satellite receiver with some minor modifications.

Since the receiver must be used also for VLBI observations, the signals of the local oscillator are obtained by a direct multiplication of the tone generated by a Rodhe & Schwarz synthetizer locked to a hydrogen maser standard.

We have measured a 3 dB bandwidth of about 250 MHz with a ripple lower than 2 dB and a total gain of 73 dB.

The noise temperature of the receiver at the feed aperture is 154 ± 14 K measured by the Y method.

Using the nitrogen load, the calibration noise signal generator of the receiver has been tested and measured.

During our observations, the system noise temperature was between 174 and 191 K. The zenith radiation temperature at this frequency is about 10 K and the extimated antenna temperature is about 20 K; these values, added to the receiver noise temperature, give a system temperature in a good agreement with field results.

3. OBSERVATIONS AND FIRST RESULTS

The receiver has been installed at the Medicina radiotelescope in the period April 1990- November 1991.

The antenna is a 32 m dish in Cassegrain configuration with an angular resolution of 2.5" at 12.2 GHz. The IF signal from the receiver is filtered and amplified by the VLBI MARKIII terminal; the output frequency is then converted via a SSB mixer to a 0-50 MHz bandwidth suitable for the 1024 channels digital autocorrelator. The velocity resolution can be selected from 0.018 km/s to 1.23 km/s in binary steps.

Our observations have been performed in total power mode with a right ascension offset of 60" and a typical integration time of 5 min.

In each on and off spectra the system temperature is measured using a standard noise generator; the antenna efficiency is derived from DR 21 and 3C286 measurements.

Spectra are analyzed by the Toolbox software package. The reduction steps are: 1) removal of the baseline 2) gaussian fitting of the profile 3) computation of the integrated flux

In order to test the performance of the receiver, we observed all sources already presented in literature. The comparison with the literature shows a good agreement in terms of spectral profile and flux density differences less than 25 % .

The selfconsistency of our measurements has been verified by observing the same sources in different periods and comparing the results, given the stability suggested in the literature. The sample is composed of 14 sources with flux densities between 10 Jy and 1000 Jy.

In three observing runs we obtained about 150 spectra. The main results of our observations are:

(1) For each source the spectra profile is stable (e.g. for W3(OH) the ratio between the main and the minor components changes within 3%).

(2) The flux density variation is less than our measurement error of 20% .

For 8 sources a comparison between the mean peak flux densities into two different runs are reported in Fig.1. Error bars express our 20% error estimation.

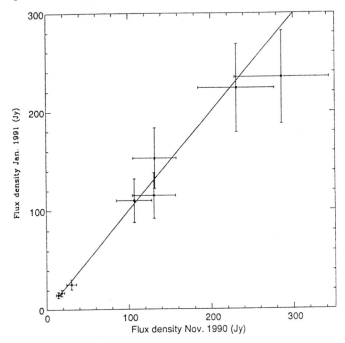

4. CONCLUSIONS

We have designed and built a 12.2 GHz receiver to observe the $2_0 - 3_{-1}$ methanol transition. Laboratory tests and first spectral observations show that its stability and sensibility are satisfactory for our purposes. With the highest possible velocity resolution of our spectrometer (0.018 km/s, about 10 times less than the width of the narrowest spectral feature), the sensitivity is 4-6 Jy.

Our future plans concern both single dish and VLBI observations.

Contrary to H_2O masers no accurate study of the variability of CH_3OH has been performed till now. Single dish observations at time intervals of days and months will permit to estabilish if 12.2 GHz methanol masers are really stable as suggested from yearly observations, now available in the literature.

By the study of proper motion of 12.2 GHz maser spots, we can investigate the kinematics of the dense and hot molecular cores around compact HII regions. With respect to ground state OH masers, 12.2 GHz methanol masers have the advantage of:

(1) Highest frequency (about 7 times higher resolution).

(2) No confusion due to Zeeman components.

(3) A greater time stability.

REFERENCES

Batrla W., 1987, Nat 326, 49

Menten K.M., 1991, Proceedings of the Third Haystack Observatory Meeting, ed. A.D. Haschick & P.T.P. Ho, p.119, San Francisco: Astronomical Society of the Pacific

Plambeck R.L., Menten K.M., 1990, ApJ 364, 555

A STUDY OF METHANOL MASERS AT 36 AND 44 GHz
AND 48 GHz THERMAL EMISSION AROUND THEM.

S. V. Kalenskii, I. I. Berulis, I. E. Val'tts, V. I. Slysh
Astro Space Center
Profsoyuznaya 84/32, 117810, Moscow

R. Bachiller, J. Gomez-Gonzalez, J. Martin-Pintado, A. Rodriguez-Franco
Centro Astronomico de Yebes, O. A. N.
Apartado 148, E-19080, Guadalajara

ABSTRACT

An extensive search for new methanol sources were made. Three new masers in the 7_0-6_1A$^+$ transition and twenty-four new thermal sources in the 1_0-0_0A$^+$ transition were detected. Our results support the idea that high- velocity flows may increase the abundance of methanol and that methanol masers arise in methanol-rich sources.

1. INTRODUCTION

A search for 7_0-6_1A$^+$ methanol maser emission and 1_0-0_0A$^+$ thermal emission at 44 and 48 GHz was made with 14-meter radio telescope of Centro Astronomico de Yebes near Guadalajara in Spain. The masers were searched toward cold IRAS sources - young, highly embedded in dust envelopes stellar objects. To look for a relation between maser and thermal methanol emission, we observed the majority of these and some additional sources in the 1_0-0_0A$^+$ transition at 48 GHz.

We detected three new masers toward cold luminous IRAS sources. These masers were also observed in the 4_{-1}-3_0E (36 GHz) transition with the 22-meter radio telescope of Radio Astronomical Station of Astro Space Center in Pushino near Moscow. Toward two of them 36 GHz masers were found. We detected also 24 new thermal sources and determined, where possible, methanol abundance.

2. NOTES ON INDIVIDUAL SOURCES

GGD 27. A strong narrow maser line was observed in the $7_0-6_1A^+$ tran sition toward IRAS 18162-2048 (GGD 27 IRS2). The five-point mappin showed that the source is unresolved with 2' beam and approximate 20" offset to the north-east from IRS2. At 36 GHz we also detected narrow line, approximately at the same radial velocity as 44 GHz lin 48 GHz emission was not found.

L 379. A strong asymmetrical line was detected in the $7_0-6_1A^+$ tran sition toward cold source IRAS 18265-1517 (L 379 IRS3). The lack symmetry and the different shape of the lines toward different dire tions suggest a presence of several components. The observed profil can be explained by assuming that two components are present: "narro and "broad" components with the parameters given in the Table 1.

In the $4_{-1}-3_0E$ transition toward IRAS 18265-1517 a double - peak line with a narrow ($\Delta V = 1$ km s^{-1}), probably, maser feature at 20.4 s^{-1} and a broader component ($\Delta V = 1.9$ km s^{-1}) was observed. In t $1_0-0_0A^+$ transition thermal emission toward IRAS 18265-1517 was dete ted.

IC 1396N. A narrow ($\Delta V = 0.6$ km s^{-1}) maser line was detected in t $7_0-6_1A^+$ transition at -0.5 km s^{-1} toward cold source IRAS 21391+58 in the bright-rimmed globule IC 1396N. 36 and 48 GHz emission was n found.

Table 1. Parameters of Newly Detected Masers.

Source Name	R.A. 1950 Dec. 1950	Transition	T_A K	V_{LSR} km s^{-1}	$\Delta V(FWHM)$ km s^{-1}	S_ν Jy
GGD 27	$18^h16^m13^s8$	$7_0-6_1A^+$	1.50	13.5	0.8	135
	$-20°48'31"$	$4_{-1}-3_0E$	0.43	14.0	<1	17
L 379	18 26 32.9	$7_0-6_1 A^+$	0.70	18.0	1.1	63
	-15 17 51		0.50	19.0	3.3	45
		$4_{-1}-3_0E$	0.88	18.0	1.9	36
			0.96	20.4	1.2	38
IC 1396N	21 39 10.3	$7_0-6_1 A^+$	0.14	-0.5	0.6	15
	58 02 29	$4_{-1}-3_0E$	<0.15			<6

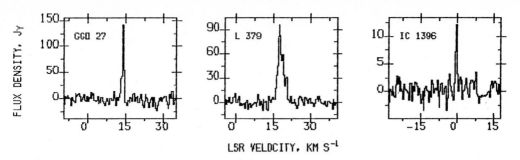

Fig. 1. 44 GHz spectra of newly detected masers.

3. DISCUSSION

Three new class A methanol masers were detected toward cold, deeply embedded in dust IRAS sources. All these IRAS sources are centers of high- velocity outflows and may be surrounded by dense discs (Fukui et al., 1989 and references therein). This result supports the idea of Plambeck and Menten (1990), who suggested a relationship between methanol masers and high - velocity outflows. They suggested that the interaction of high-velocity flows with ambient gas leads to enhanced abundance of methanol and that methanol masers may arise in these methanol-rich regions.

Comparison of known 44 GHz masers (Haschick et al, 1990; Bachiller et al., 1990) with the catalog of bipolar flows (Fukui et al., 1989) revealed 14 high - velocity flows, connected with methanol masers. The driving sources are cold, deeply embedded in dust objects; the majority of them are strong at 60 and 100 μm (usually $S_{100} \geq 1000$ Jy). This fact suggests a possible correlation between far-infrared and maser luminosities and explains low detection rate of new masers (the majority of the IRAS sources from our list are not luminous and has $S_{100} < 1000$ Jy).

We use our 48 GHz observations to check (i) if the methanol abundance is enhanced in the environments of high-velocity flows; and (ii) if the methanol maser emission arises in the regions with high methanol abundance. Histogram (a) on fig. 2 shows the methanol abundance distribution. The sources were divided into four groups: Group I contains sources with the abundances lower than 2×10^{-9}; group II is for the abundances between 2×10^{-9} and 10^{-8}; Group III is for the abundances between 10^{-8} and 10^{-7}; and Group IV is for the abundances above 10^{-7}. The histogram contains all sources with known methanol abundances, or with upper limits of 2×10^{-9}; to our data Ori-KL (Menten et al.,1988)

193

and Sgr B2 (Morimoto et al.,1985) were added. The dashed region corresponds to the sources with the high-velocity outflows (velocity extend greater than 20 km s^{-1}). The histogram shows that the highest abundance occurs mostly in sources with the high-velocity outflows. But the statistics is poor and further observations of regions with high - velocity outflows are very desirable.

The histogram (b) is similar to histogram (a). The dashed portion of the histogram corresponds to the sources with methanol masers. The histogram shows that indeed the masers are observed in regions with high methanol abundance. Unfortunately, the statistics is poor again and further observations are necessary to check this conclusion.

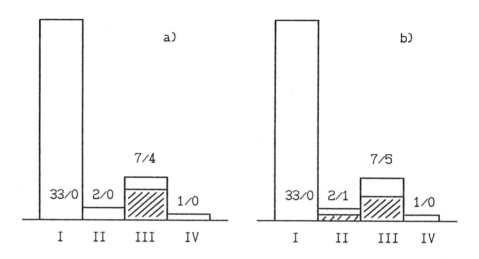

Fig. 2. The methanol abundance distribution. See the text for explanations. The fractions on the histograms means total number of sources in the group / number of sources with outfows (a) or masers (b).

REFERENCES

Bachiller R., Menten K.M., Gomez-Gonzalez J., Barcia A. (1990), Astron. & Astrophys., 240, 116.
Fukui Y. in: ESO Workshop on Low Mass Star Formation and Pre-Main Sequence Objects, (1989), 11-13 Jul., ed. R.Reipurth, 95-117.
Haschick A.D., Menten K.M., Baan W.A. Astrophys. J., (1990), 354, 556.
Menten K.M, Walmsley C.M., Henkel C., Wilson T.L. Astron. & Astrophys., (1988), 198, 253.
Morimoto M., Ohishi M., Kanzava T. Astrophys. J., (1985), 288, L11.
Plambeck R.L., Menten K.M. Astrophys. J., (1990), 364, 555.

Methanol masers in the southern Galaxy

G. C. MacLeod and M. J. Gaylard
Hartebeesthoek Radio Astronomy Observatory, South Africa

Abstract

We have detected 6.6-GHz A^+-methanol masers towards all nineteen 12.2-GHz E-methanol masers at declinations below $-42°$. We also searched all known OH masers in the longitude range $260°$ to $345°$ and detected 52 of 88. The detection rate of 6.6-GHz methanol masers is four times higher in the Norma area of the Galaxy than in the Carina arm. Although this may be partly due to sensitivity limitations, it may also be a product of the radial metallicity gradient in the Galaxy.

Fourteen 6.6-GHz masers with $S_{6.6} \geq 100$ Jy were searched for associated 12.2-GHz E-methanol masers and four new masers were detected. The statistics of the non-detections suggest that there may not be a one-to-one correspondence of 6.6- to 12.2-GHz methanol masers.

We have reobserved five 12.2-GHz E-methanol masers, originally observed in 1987. We found that $\sim80\%$ of the maser components have varied by less that 10% over the four year period, implying that these masers are generally saturated. In several sources maser components have shifted in velocity.

Introduction

A strong masing transition, 6.6-GHz $5_1 \rightarrow 6_0$ A^+-methanol, was recently found by Menten (1991b) to occur widely in Galactic star-forming regions. The southern limit of his search was $-42°$. We have searched for 6.6-GHz masers towards all 1.6-GHz OH masers and 12.2-GHz E-methanol masers south of his limit.

In this paper we examine the detection statistics of our search. We compare the fluxes of the 6.6-GHz masers with those of the 12.2-GHz methanol, 1.6-GHz OH and the 22-GHz H_2O masers, as well as the IRAS $60\mu m$ fluxes of the star-forming regions. Finally we discuss the results of our search for new 12.2-GHz masers towards 6.6-GHz masers and the variability of known 12.2-GHz masers.

Results and Discussion

All observations were made with the 26-m Hartebeesthoek telescope, which is equipped with 6.6- and 12.2-GHz receivers. We detected 6.6-GHz masers towards all 12.2-GHz methanol masers south of −42° declination (MacLeod, Gaylard & Nicolson 1992), corroborating the results of Menten (1991b). All known OH masers not observed by Menten and which were visible from HartRAO (dec. ≤ +45°) were also searched (MacLeod & Gaylard 1992). In total, we found 60 new 6.6-GHz masers. In the longitude range 260° to 345°, where we searched the complete sample of 88 known OH masers, the detection rate was 59%. The 5 Jy detection limit of our observations was ten times higher than that of Menten (1991b). If we limit the detections in Menten (1991b) to those with $S_{6.6} \geq 5$ Jy, thereby excluding eleven masers, then his detection rate would have been 63%, which is similar to ours.

In total, 143 6.6-GHz methanol masers are now known, of which 60 are found in the first and 72 in the fourth quadrant. Even with the limited sensitivity of our observations in the fourth quadrant, the distribution of the 6.6-GHz masers matches the north/south asymmetry reported by Robinson et al. (1988) for the integrated CO intensity in the two quadrants.

In general the detection rate of 6.6-GHz to OH masers is similar throughout the Galaxy, with the exceptions of the Norma and Carina regions. In the Norma region (335°–340°) the rate of detection is 80%, while in the Carina arm (280°–305°) it is only 20%. This may simply be a sensitivity problem; some of the sources in the Carina arm have velocities which imply that they are at large kinematic distances. Several of the Carina OH sources have relatively low infrared fluxes ($S_{60\mu m} \leq 1000$ Jy) or no identifiable IRAS counterpart. However, for ∼25% of the 60 6.6-GHz masers which we detected, the corresponding $60\mu m$ flux is less than 1000 Jy. Another effect to be considered is the relative deficiency of metal atoms in the outer Galaxy reported by Shaver et al. (1983). This would reduce the formation of methanol molecules in the molecular clouds of the Carina arm. Methanol molecules contain two metal atoms while hydroxyl and water each contain only one. This implies that the effect of the radial metallicity gradient should be more readily seen in the galactic distribution of methanol masers than hydroxyl or water masers.

For the complete set of 6.6-GHz masers we have compiled the data on the fluxes of the associated 12.2-GHz methanol, 1.6-GHz OH, and 22-GHz H_2O masers. On average the 6.6-GHz masers are ∼6 and ∼10 times brighter than the 12.2-GHz methanol and hydroxyl masers respectively, and almost as bright as the water masers.

There is no correlation of detection rate of 6.6-GHz maser emission with IRAS colour, with the exception of sources with a very large $S_{25\mu m}$ to $S_{12\mu m}$ ratio, in which the rate is higher than average. For those star-forming regions with usable 12- and 25-μm IRAS fluxes, the rate of detection for the 110 sources with $\log(S_{25}/S_{12}) \leq 1.3$ is 67%, while for the sources with $\log(S_{25}/S_{12}) \geq 1.3$ the detection rate is ten out of twelve, or 83%. However, given the small numbers in the second category, the difference may not be significant.

It has been shown by comparison of the fluxes of the OH and 12.2-GHz methanol masers in star-forming regions with the $60\mu m$ IRAS fluxes that it is feasible that these masers could be radiatively pumped by mid-infrared photons (Cohen, Baart & Jonas 1988; Kemball, Gaylard & Nicolson 1988). However, several 6.6-GHz methanol and H_2O masers have fluxes greater than that of the $60\mu m$ flux, implying that radiative pumping by mid-infrared photons alone cannot be responsible for their excitation. Collisional pumping is currently favoured as the excitation mechanism for H_2O masers. The strength of the 6.6-GHz masers suggests that it is the most likely mechanism for the class II methanol masers, although it is also thought to be responsible for the class I masers (Menten 1991a).

We have searched 14 of the strongest 6.6-GHz masers ($S_{6.6} \geq 100$ Jy) for associated 12.2-GHz maser emission. We detected new 12.2-GHz masers towards G023.01-0.41, G035.19-0.74, V645 Cyg, G322.16+0.63 and possibly Mon R2.

The distribution of the logarithmic ratios of the fluxes of the 6.6- and 12.2-GHz masers in the 49 star-forming regions in which both are detected is well fitted by a Gaussian profile. We used this to estimate the confidence level that 12.2-GHz emission above our detection level would be observed towards a 6.6-GHz maser with a given peak flux. The fitted Gaussian has a logarithmic mean of 0.75 and a standard deviation of 0.55. The brightest 6.6-GHz source which we observed at 12.2-GHz was G305.20+0.21, where $S_{6.6} = 400$ Jy, and we achieved an upper limit of $S_{12.2} \leq 3$ Jy. Hence $\log(S_{6.6}/S_{12.2}) \leq 2.1$, which is almost three standard deviations from the mean. This source runs counter to the general one-to-one occurence of 6.6-GHz and 12.2-GHz methanol masers.

McCutcheon et al. (1988) reported little or no variation over an eight month period in the 12.2-GHz E-methanol masers which they observed. In December 1991 we reobserved five sources that were originally observed at Hartebeesthoek in 1987. Over the four year period the average internal variation in the spectra relative to the brightest peak was not more that 10%. In one case where the rms noise was small we measured a variation of 5±2%. This behaviour implies that in general these 12.2-GHz masers appear to be saturated, in contrast to hydroxyl and water masers in star-forming regions. However we did find two components that varied significantly in this period. In NGC 6334F one of the components increased by ~100%, while one component in G345.01+1.79 vanished.

The very bright, complex 12.2-GHz E-methanol maser profile in G323.77-

0.21 has narrowed over the last four years (see Figure 1). The overall velocity range has decreased by 0.2 km s^{-1}, the velocity resolution of these observations being 0.06 km s^{-1}. It is possible that this has been caused by retardation of the outward expansion of the masering region or by infalling masering material encountering a denser medium.

References

Cohen, R.J., Baart, E.E. & Jonas, J.L., 1988. *Mon. Not. R. astr. Soc.*, **231**, 205.

Kemball, A.J., Gaylard, M.J. & Nicolson, G.D., 1988. *Astrophys. J.*, **331**, L37.

MacLeod, G.C., Gaylard, M.J. & Nicolson, G.D., 1992. *Mon. Not. R. astr. Soc.*, **254**, 1P.

MacLeod, G.C. & Gaylard, M.J., 1992. *Mon. Not. R. astr. Soc.,* in press.

McCutcheon, W.H., Wellington, K.J., Norris, R.P., Caswell. J.L., Kesteven, M.J., Reynolds, J.E. & Peng, R.-S., 1988. *Astrophys. J.*, **333**, L79.

Menten, K.M., 1991a. *Skylines,* Proc. Third Haystack Observatory Meeting, p. 119, Haschick, A. D. & Ho, P. T. P., Astronomical Society of the Pacific, San Francisco.

Menten, K.M., 1991b. *Astrophys. J. (Lett.)*, **380**, L75.

Robinson, B.J., Manchester, R.N., Whiteoak, J.B., Otrupcek, R.E. & McCutcheon, W.H., 1988. *Astr. Astrophys.*, **193**, 60.

Shaver, P.A., McGee, R.X., Newton, L.M., Danks, A.C. & Pottasch, S.R., 1983. *Mon. Not. R. astr. Soc.*, **204**, 53.

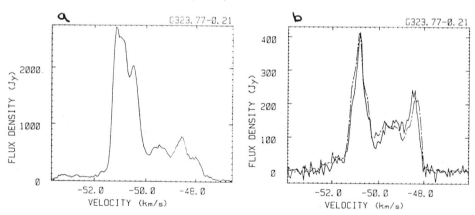

Figure 1. (a) The 6.6-GHz spectra of G323.77+0.21. (b) The 1987 (dashed line) 12.2-GHz spectra of G323.77-0.21 superimposed on its 1991 (solid line) 12.2-GHz spectra.

INTERSTELLAR METHANOL MASERS

Karl M. Menten
Harvard-Smithsonian Center for Astrophysics
60 Garden Street, Cambridge MA 02138, USA

ABSTRACT

The observational status of methanol maser research is summarized with an emphasis on recent developments. The main properties of Class I and II methanol masers are discussed. In particular, a collisional pumping mechanism that is capable of inverting all known Class I methanol masers is explained. Finally, a short summary of the observational properties of the newly detected very strong and widespread 6.6 GHz Class II maser transition is given.

1. INTRODUCTION – CLASS I AND II METHANOL MASERS

To date about two dozen transitions of interstellar methanol (CH_3OH) have been found to show strong maser emission toward numerous star-forming regions. All methanol maser sources can be divided in two classes (Batrla *et al.* 1987, Menten 1991a): *Class I* methanol masers have Orion-KL as their prototype and show maser action in the 25 GHz $J_2 \rightarrow J_1 E$ ($J = 2, 3, 4, ...$) lines and a number of millimeter-wave transitions. They are frequently found offset, by up to 1 pc, from ultracompact HII regions and OH or H_2O maser centers. *Class II* methanol maser sources show maser action in the 12 GHz $2_0 \rightarrow 3_{-1} E$ transition, and in the extremely strong and very widespread 6.6 GHz $5_1 \rightarrow 6_0 A^+$ line. Toward a few sources maser emission in a number of other, higher frequency, lines has been detected, which in all cases is much weaker than the 6.6 and 12 GHz emission. Interferometric observations have shown that Class II masers are located in the dense envelopes of ultracompact HII regions, the same environment that gives rise to interstellar OH masers. None of the transitions that is masing toward Class I sources is also masing toward Class II sources and vice versa. In particular, the most prominent Class II maser transitions, the 6.6 and 12 GHz lines show enhanced absorption toward many Class I regions.

A review of the properties of interstellar methanol masers has recently been presented by Menten (1991a). In the following we shall concentrate on recent developments, which include the discovery of the 6.6 GHz $5_1 \rightarrow 6_0 A^+$ line.

2. THE EXCITATION OF CLASS I METHANOL MASERS

One remarkable property of Class I methanol masers is the close resemblance between spectra of different maser transitions taken toward the same source (see

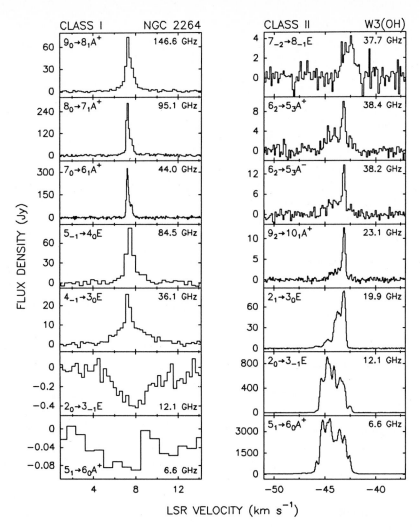

Figure 1. Spectra of various methanol maser transitions observed toward the Class I source NGC 2264 (*left hand panels*) and the prototypical Class II source W3(OH) (*right hand panels*). In the left part of the figure, the top five panels show spectra of various Class I CH_3OH maser transitions observed toward NGC 2264. Note that in the 36.1 GHz line the maser spike is superposed on broad (quasi-)thermal emission that covers a velocity range comparable to that of the absorption observed in the 6.6 and 12.1 GHz lines (bottom two panels). Toward Class II sources like W3(OH), the latter two transitions always show the strongest maser emission. Maser action 19.9, 23.1, 38.2, and 38.4 GHz lines has only been detected toward W3(OH) and NGC 6334-F.

Fig. 1). This leads one to suspect that all the maser transitions are inverted by the same mechanism. We also note that for all known Class I maser lines from the E-symmetry species, except for the $J_{k=2} \to J_{k=1}E$ transitions, the upper energy levels have k quantum numbers of -1, while all Class I masers from A-type methanol have upper levels with $K = 0$. Lees (1973) points out that one might expect inversion in these lines as a natural consequence of basic properties of the methanol molecule. The observed masers might occur because molecules that have been collisionally excited into high J_k states decay radiatively into the lowest energy or "backbone" k ladder, which is the $k = -1$ ladder for the E-species and the $K = 0$ for A-type methanol. Once there, they funnel down this ladder and overpopulate the lower levels relative to levels in neighboring ladders. This is because $\Delta k = 0$ collisions are strongly preferred over $|\Delta k| = 1$ and radiative transitions out of the "backbone" ladder are much slower than transitions down this ladder. This produces, in the case of E-type methanol, the $4_{-1} \to 3_0 E$ and $5_{-1} \to 4_0 E$ masers at 36 and 84 GHz, respectively, and, for the A-species, the $7_0 \to 6_1 A^+$, $8_0 \to 7_1 A^+$, and $9_0 \to 8_1 A^+$ masers at 44, 95, and 146 GHz, respectively. This mechanism *also* produces enhanced absorption ("overcooling") in the 12 GHz $2_0 \to 3_{-1} E$ line and, as predicted by Menten (1991a), in the 6.6 GHz $5_1 \to 6_0 A^+$ transition (see Fig. 1). It does not straightforwardly explain the 25 GHz $J_2 \to J_1 E$ masers. However, the latter transitions can be inverted, if, following Johnston *et al.* (1992), one makes the *ad hoc* assumption that $|\Delta k| = 3$ collisions occur at one-tenth the rate of $|\Delta k| = 0$ collisions. This is because the "weakly" allowed $|\Delta k| = 3$ collisions transfer population from levels in the $k = -1$ backbone ladder into the $k = 2$ ladder, causing an overpopulation of the J_2 relative to their neighboring J_1 levels.

The collisional pumping mechanism described above is capable of inverting the observed Class I masers without the presence of energy sources close to the maser region. Indeed, as mentioned, Class I methanol masers frequently are found far away from known infrared sources. Plambeck & Menten (1990), motivated by their high resolution observations of the 95 GHz $8_0 \to 7_1 A^+$ line toward the DR21 molecular outflow source, propose that Class I methanol masers arise from interface regions between high velocity outflows and dense clumps in the ambient molecular medium. The elevated temperatures created by the passing shock could enhance the methanol abundance in the maser region by evaporating solid methanol ice from dust grain mantles, thus creating methanol column densities sufficient for high gain maser amplification. Recent VLA observations of the 25 GHz $J_2 \to J_1 E$ masers in the Orion-KL region suggest that these are also located at the interface of a high-velocity outflow and surrounding dense gas (Johnston *et al.* 1992). High resolution observations of both Class I masers and tracers of the outflowing gas, such as shock-excited H_2 and high-velocity CO emission are highly desirable (see also Pratap & Menten, these proceedings). Furthermore, more extensive and more sensitive searches will certainly lead to the detection of more Class I sources (S. V. Kalenskii *et al.*, these proceedings).

3. THE 6.6 GHz CLASS II MASER LINE

The strength and ubiquity of the maser emission observed in the $5_1 \to 6_0 A^+$ line at a frequency of 6.668518 GHz make this line a particularly useful probe of the molecular envelopes of ultracompact HII regions (Menten 1991b). To date, maser emission in the $5_1 \rightarrow 6_0 A^+$ line has been reported toward $\approx$ 150 regions, all of which have associated hydroxyl masers (Menten 1991b; MacLeod, Gaylard & Nicolson 1992; MacLeod & Gaylard 1992 and these proceedings). In fact, 6.6 GHz CH_3OH masers have been detected toward about 70% off all interstellar OH masers searched. Since typical detection limits of the existing 6.6 GHz surveys are 5–10 times higher than those of the OH surveys that produced the target lists, more sensitive methanol surveys would be expected to increase the number of 6.6 GHz maser detections.

The 6.6 GHz line has been detected toward all sources with known masers in the 12 GHz $2_0 \rightarrow 3_{-1} E$ transition. In all cases the 6.6 GHz photon luminosity, $L_{6.6}$, is greater, and in many cases much greater, than the 12 GHz photon luminosity, L_{12}, assuming that both lines are equally beamed. In the extreme case of W51, $L_{6.6}/L_{12} \approx 200$. In general, 6.6 and 12 GHz spectra cover similar velocity ranges, which are also comparable to the velocity ranges covered by OH maser emission. In some cases, such as W3(OH), individual velocity components in the 12 GHz spectra seem to have counterparts at identical velocities in the 6.6 GHz spectra (Fig. 1). Moreover, recent interferometer observations of the 6.6 GHz line (R. P. Norris, these proceedings) indicate that some 6.6 GHz maser spots coincide spatially with 12 GHz maser spots emitting at identical velocities. Therefore, maps of the same source in both methanol lines should be able to provide powerful constraints on the excitation mechanism of Class II methanol masers and, eventually, on the physical conditions of the masing region.

Future interferometric studies of the 6.6 GHz transition will yield important information on the dynamics of the gas surrounding young massive stars. Moreover, measurements of *absolute* proper motions of CH_3OH maser sources relative to background quasars, using VLBI phase-referencing techniques, seem feasible. Such measurements could potentially lead to an accurate determination of the galactic rotation parameters.

REFERENCES

Batrla, W., Matthews, H. E., Menten, K. M., Walmsley, C. M. 1987, Nature, 326, 49
Johnston, K. J., Gaume, R., Stolovy, S., Wilson, T. L., Walmsley, C. M., Menten, K. M. 1991, ApJ, 385, 232
Lees, R. M. 1973, ApJ, 184, 763
MacLeod, G. C., Gaylard, M. J. 1992, submitted to MNRAS
MacLeod, G. C., Gaylard, M. J., Nicolson, G. D. 1992, MNRAS, 254, 1p
Menten, K. M. 1991a, in *"Skylines"*, *Proceedings of the Third Haystack Observatory Meeting*, ed. A. D. Haschick & P. T. P. Ho, p. 119, San Francisco: Astronomical Society of the Pacific
Menten, K. M. 1991b, ApJ, 380, L75
Plambeck, R. L., and Menten, K. M. 1990, ApJ, 364, 555

SYNTHESIS IMAGES OF 6.7-GHZ METHANOL MASERS

R.P.Norris, J.B.Whiteoak, J.L.Caswell, and M. Wieringa

Australia Telescope National Facility, CSIRO,

PO Box 76, Epping, NSW 2121, Australia

ABSTRACT

We have produced images of a number of galactic methanol sources in the 6.7 GHz (5_1 - 6_0 A^+) transition, and compared them with earlier maps of the 12.2-GHz (2_0 - 3_{-1} E) methanol masers. We find that, in several cases, the 6.7-GHz and 12.2-GHz maser positions are coincident to within 20 milliarcsec, placing a tight constraint on pumping mechanisms. We also confirm the 12.2-GHz result that the methanol masers tend to be located along lines, perhaps indicating jets, shock fronts, or edge-on protoplanetary discs.

1. INTRODUCTION

Interstellar masers are potentially extremely powerful tools for studying the kinematics of star formation, although only modest advances have resulted from the three decades of intensive study of interstellar OH and H_2O masers. The recently discovered strong methanol maser transitions at 6.7 (Menten 1991) and 12.2 GHz (Batrla et al. 1987) promise to breathe new life into this area of research. However, to date there have been a number of surveys but very little mapping. This is largely because both methanol maser transitions have non-standard radio-astronomy frequencies which are not generally available on synthesis arrays.

Over the last few years, some of the Australian VLBI antennas have been equipped with 12.2-GHz receivers to study these masers. This has resulted in a number of VLBI experiments to study the 12.2-GHz masers, from which the first images are now available (Norris et al. 1992a, in preparation).

In addition, the 6-km Australia Telescope Compact Array is already equipped with receivers which operate in the 6-GHz region, although their nominal specification does not extend to 6.7 GHz. However, because of the enormous maser intensities, even the relatively poor performance of the receivers at this frequency is adequate for studying the masers. Consequently, several sources have been mapped, and relative positions of maser spots measured to an accuracy of ~0.02 arcsec. Their absolute positions have also been measured to an accuracy of about 0.5 arcsec. Here we present a few of the first results from this program, and compare the distribution of maser spots at 6.7 GHz to that at 12.2 GHz. A full account of this work, and details of the observations and analysis, will be published elsewhere (Norris et al. 1992b, in preparation).

2. RESULTS

G351.42+0.64 (NGC6334F)

The 6.7-GHz and 12.2-GHz maps and spectra of G351.42+0.64 are shown in Fig. 1. This source demonstrates an important result which is also common to other sources. The spectra at the two transitions are similar, and several maser spots at the same velocity in the two transitions are coincident within tens of milliarcsec.

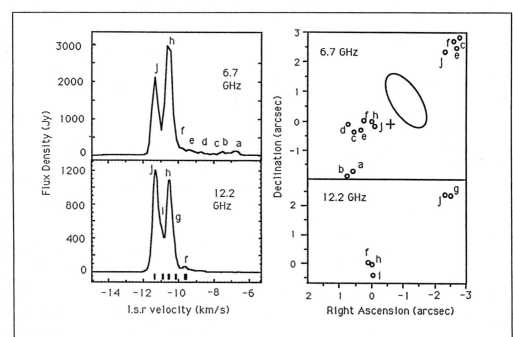

Fig. 1 *Map and spectra of the 6.7- and 12.2-GHz maser emission from G351.42+0.64 (NGC6334F). Letters on the maps correspond to the lettered features on the spectra, and the same letters are used in both transitions for corresponding maser features. Relative positional uncertainty is estimated at 0.02 arcsec. The 12.2-GHz data are adapted from Norris et al. (1988). The B1950 origin of both maps is at RA 17:17:32.35, Decl. -35:44:04.2, with an estimated uncertainty of 0.5 arcsec. The cross represents the position of the continuum peak, and the ellipse the approximate position of the OH masers, as measured by Gaume and Mutel (1987).*

We see this effect in several sources, the most dramatic being in G345.01+1.79, where the 6.7-GHz spectrum is nearly identical to that at 12.2 GHz except for the addition of one group of features. The resulting maps are also nearly identical for the two transitions, so that we can compare the relative positions of the masers in the two transitions. We find unambiguous coincidence of several of the components to within 20 milliarcsec.

Despite this important coincidence of some features in G351.42+0.64, two other features (A & B) appear in the 6.7 GHz transition but are absent from the 12.2 GHz transition, and no 12.2 GHz masers appear from this location.

This is the only source of our sample for which comparable continuum and OH data exist. Comparison of the absolute position with those of OH masers and continuum emission (Gaume & Mutel, 1987) shows the methanol masers to be close to, but not coincident with, the OH masers. The methanol masers also appear to occupy a broader range of radii from the centre of the HII region than that occupied by the OH masers.

G339.88-1.26

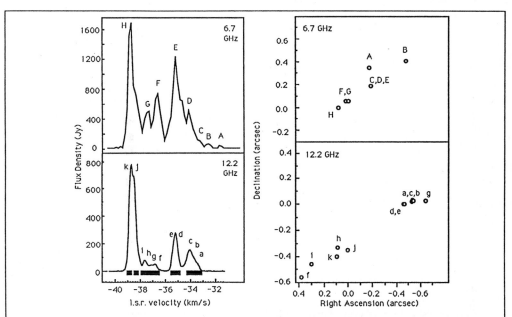

Fig. 2 *Map and spectra of the 6.7- and 12.2-GHz maser emission from G339.88-1.26. Details are as in Fig. 1, except that the labelling at 6.7 GHz does not correspond to that at 12.2 GHz, as in this case detailed correspondence between the two transitions is limited. The B1950 origin of the 6.7-GHz map is at RA 16:48:24.76, Decl. -46:03:33.9.*

The 6.7-GHz and 12.2-GHz map and spectra of G339.88-1.26 are shown in Fig. 2. This source demonstrates a further result which is also common to other sources. The maser positions, in both transitions, appear to follow a line. We see this effect in several (but by no means all) of the other sources, sometimes with a clear velocity gradient along the line. In the case of G339.88-1.26, the velocity gradient at 6.7 GHz is monotonic along the line with the exception of one feature (labelled A); at 12.2 GHz the gradient is less clear.

The 6.7 GHz and 12.2 GHz transitions show similar but not identical spectra, and the measured positions of common features in the two transitions are similar. However, in this case only a coarse similarity is found, with maser components lying in clusters at similar positions.

3. CONCLUSION

Comparison of the 6.7-GHz maps shown here with the earlier maps produced at 12.2 GHz produces four main results

i) In several cases, the 6.7-GHz and 12.2-GHz masers are coincident to within 20 milliarcsec. This implies that one maser spot is masing in both transitions, which is surprising given that the two transitions are from completely different excitation species. We believe this is the first time this phenomenon has been observed in interstellar masers, and should place a tight constraint on pumping mechanisms.

ii) Both the 6.7-GHz and the 12.2-GHz masers tend to be located along lines. This suggests a geometry involving jets, shock fronts, or edge-on protoplanetary discs.

iii) The methanol masers are located close to, but are not coincident with, the OH maser clusters, implying that slightly different conditions are required for the two masers.

iv) There is a tentative indication that the 6.7-GHz masers sample a wider range of radii than either the 12.2-GHz methanol masers or OH masers, and therefore should prove particularly useful for kinematical studies.

REFERENCES

Batrla, W., Matthews, H. E., Menten, K. M., Walmsley, C. M., 1987, *Nature*, **326**, 49.

Gaume, R. A., & Mutel, R. L., 1987, *Ap. J. Suppl.*, **65**, 193.

Menten, K.M., 1991, *Ap. J. (Letters)*, **380**, L75.

Norris, R. P., McCutcheon, W. H., Caswell, J. L., Wellington, K. J., Reynolds, J. E., Peng, R.-S., & Kesteven, M. J., 1988, *Nature*, **335**, 149.

Production of 12.2 GHz CH$_3$OH Masers
— Preliminary Results of an Excitation Study

R. Peng[1], J. B. Whiteoak[2]

[1] Department of Astronomy, University of Illinois at Urbana-Champaign
   1002 W Green St., Urbana, IL61801, U.S.A.
[2] CSIRO Australia Telescope National Facility
   P.O.Box 76, Epping, NSW 2121, Australia

Abstract

Excitation of the $2_0 \rightarrow 3_{-1}$E transition of CH$_3$OH at 12.2 GHz has been examined in a statistical equlibrium study. The results show that maser emission can be produced in a molecular cloud with a bright HII region as immediate background. For a cloud of kinetic temperature of 30 K, velocity gradient of 5 km s^{-1} pc^{-1}, H$_2$ density of $\sim 10^5$ cm^{-3} and methanol abundance of 10^{-6}, and a continuum background of 50 K, the maser emission can reach a brightness temperature of $\sim$ 1300 K. This value would be much higher with smaller velocity gradient, where bright maser emission is usually expected.

Introduction

12.2 GHz maser emission in the $2_0 \rightarrow 3_{-1}$E transition of methanol (CH$_3$OH) has been widely detected in star forming regions throughout the Galaxy (Batrla et al. 1987, Norris et al. 1987, Kemball et al. 1988). Interferometry observations (Norris et al. 1988, Menten et al. 1988) have revealed that the maser component has a brightness temperature of over 10^{10} K. The transition has also been observed in absorption towards many HII regions in the Galaxy (Peng & Whiteoak, 1991), and even against the 2.7 K microwave background (Walmsley et al. 1988, Whiteoak & Peng 1989). Peng and Whiteoak (1992) studied the peculiar excitation conditions of the transition in dark clouds. As an extension of this study, we have explored the excitation pattern of the transition at various continuum background temperatures.

The model and results

Statistical equilibrium calculations were carried out involving the lowest 69 rotational levels of E-species CH$_3$OH (J up to 11). The escape probability technique, which makes use of the large scale velocity gradient in molecular clouds, was employed to decouple the line radiation transfer equations and statistical equilibrium equations. Molecular excitation procedures include collisional excitation with H$_2$ as prime colliding partner, radiative excitation by the 2.7 K microwave background and extended continuum background, and photon trapping in the cloud.

Radiation transitions of CH$_3$OH are mainly dipole type, bound by the selection rules $|\Delta J| \leq 1$, $|\Delta K| \leq 1$. Collisional transitions, on the other hand, are much less restricted. Based on the experimental studies by Lees (1973) and Lees and Haque (1974), a set of empirical formulae were derived to obtain collissional transition rates between various energy levels. Photon trapping in the cloud is incorporated by the use of the escape probaility technique. Refer to Peng and Whiteoak (1992) for further details.

Calculations were carried out over an H_2 density range of 10^2–10^7 cm^{-3} for a cloud with a kinetic temperature T_K of 30 K and a velocity gradient of 5 km s^{-1} pc^{-1}. The methanol abundance of the cloud relative to H_2 was varied between 10^{-6} and 10^{-8}. The brightness temperature of the extended continuum background (excluding the 2.7 K microwave background) T_c was varied over the range 0-50 K.

Fig.1 shows the variation of excitation temperature T_{ex} and line temperature T_L of the $2_0 \rightarrow 3_{-1} E$ transition with the density of the cloud for various background temperatures.

Discussion

When $T_c < T_K$ (fig. 1a, b and c), the excitation temperature is near the background temperature at low H_2 densities ($\leq 10^2$ cm^{-3}), and maximizes at the kinetic temperature at high H_2 densities. In these cases, one would only see absorption, sub-thermal or thermal emission from clouds with different H_2 densities. Note that the absorption in fig. 1a occurs against the 2.7 K microwave background.

When $T_c = T_K$ (fig. 1d), there is little change in excitation temperature, which remains close to the cloud's kinetic temperature. The transition occurs generally in absorption, which maxiumizes at 2.7 K at high H_2 densities ($> 10^6$ cm^{-3}).

When $T_c > T_K$ (fig. 1e, f), the variation of excitation temperature shows large discountinuities, indicating sharp transformation of excitation states from superthermal to masing, or from masing to superthermal. With increasing H_2 density the transition undergoes a progression of superthermal, masing, superthermal and thermal excitation. In the masing state, the excitation temperature remains essentially constant. The highest maser brightness temperature occurs towards the high density end of this state. The density range of the masing state becomes wider in clouds of lower methanol abundances. The maximum maser intensity decreases drastically with methanol abundance. With the increasing effects of molecular collisions, the transition undergoes rapidly the narrow phase of transformation from masing to thermal excitation, and the cloud appears in absorption.

Note that the H_2 density range covered in the transformation from maser emission to absorption is rather small (≤ 0.6 magnitude for clouds with a methanol abundance of 10^{-6}). Such a small density range can often be accommodated in molecular clouds, and explains why maser emission and absorption occur in same molecular clouds. Since absorption occurs at higher densities than those of maser emission, in a cloud with an inward density gradient, the absorption would come from the dense core, and maser emission from outer part of the cloud.

Compared with the observed maser brightness temperature of $\geq 10^{10}$ K (Norris et al. 1988), the maximum temperature of 1300 K obtained in fig. 1f is rather low. The situation would improve remarkably in clouds of smaller velocity gradients. Under similar conditions ($T_K=30$ K, $T_c=50$ K and methanol abundance=10^{-6}), a test calculation obtains a brightness temperature of 1.2×10^7 and 2.7×10^9 K at an H_2 density of 5×10^4 cm^{-3} with a velocity gradient of 1 and 0.5 km s^{-1} pc^{-1} respectively. Note that the parameters used do not necessarily represent the most favorable conditions for producing strong maser emission.

In summary, the maser production scheme described here requires that the extended continuum background temperature be greater than the kinetic temperature of the cloud. This is in general compatible with observations: 12.2 GHz CH_3OH masers are mostly detected towards HII regions and star forming regions, yet not all the clouds with significant background continuum emission exhibit maser emission. The observed maser brightness temperature may be obtained in clouds of small velocity gradients and high methanol abuncances.

Conclusions

The study of molecular excitation of the $2_0 \rightarrow 3_{-1}E$ transition of CH_3OH in molecular clouds associated with extended background continuum emission has shown that
- In clouds where the kinetic temperature is greater than the background continuum temperature, only absorption, subthermal and thermal emission processes are presented.
- Maser emission can be produced in clouds where the kinetic temperature is lower than the continuum background temperature. Clouds of this configuration can produce maser emission, maser or superthermal emission plus absorption, and thermal absorption.
- A cloud with an H_2 density of 10^4–10^5 cm^{-3}, a velocity gradient of ≤ 1 km s^{-1} pc^{-1} and a methanol abundance of 10^{-6} is capable of producing CH_3OH maser emission with a brightness temperature comparable with measured values. On the other hand, a methanol abundance of 10^{-8} or less is thought to be too small to produce any strong maser emission.

Further study of molecular excitation in clouds with small velocity gradient should provide a clearer picture of the maser process.

Reference

Batrla, W., Matthews, H. E., Menten, K. M. and Walmsley, C. M., 1987, *Nature*, **326**, 49

Kemball, A. J., Gaylord, M. J. and Nicholson, G. D., 1988, *Ap. J.* (Letters) **331**, L37

Koo, B-C., Williams, D. R. W., Heiles, C. and Backer, D. C., 1988, *Ap. J.* **326**, 931

Lees, R. M., 1973, *Ap. J.* **184**, 763

Lees, R. M. and Haque, S. S., 1974, *Can. J. Phys.* **52**, 2250

Menten, K. M., Reid, M. J., Moran, J. M., Wilson, T. L., Johnston, K. J. and Batrla, W., 1988, *Ap. J.* (Letters) **333**, L83

Norris, R. P., Caswell, J. L., Gardner, F. F. and Wellington, K. J., 1987, *Ap. J.* (Letters) **321**, L159

Norris, R. P., McCutcheon, W. H., Caswell, J. L., Wellington, K. J., Reynolds, J. E., Peng, R. S. and Kesteven, M. J., 1988, *Nature*, **335**, 149

Peng, R. S. and Whiteoak, J. B., 1991, *Mon. Not. R. Astr. Soc.* in press

Peng, R. S. and Whiteoak, J. B., 1992, *Mon. Not. R. Astr. Soc.* submitted

Walmsley, C. M., Batrla, W., Matthews, H. E. and Menten, K. M., 1988, *A. Ap.* **197**, 217

Whiteoak, J. B. and Peng, R. S., 1989, *Mon. Not. R. Astr. Soc.* **239**, 677

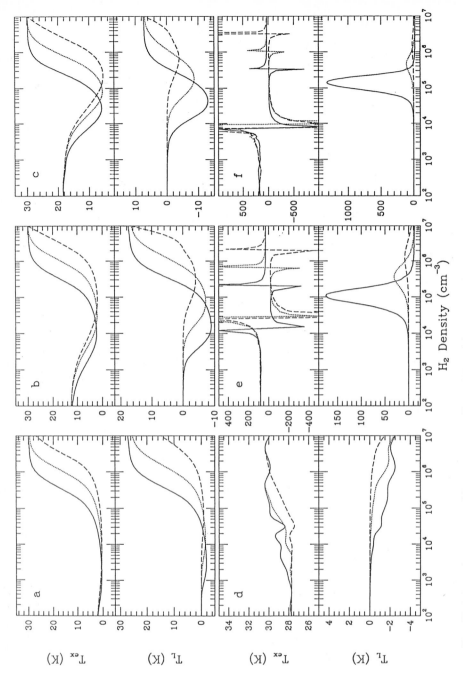

Fig. 1 Variations of excitation and line temperature, T_{ex} and T_L, of the $2_0 \rightarrow 3_{-1}E$ transition of CH_3OH with H_2 density in a cloud associated with the 2.7 K microwave background and extended continuum background of temperature T_c. The cloud has a kinetic temperature T_K of 30 K, a velocity gradient of 5 km s^{-1} pc^{-1}. Solid, dotted and dashed lines represent a methanol abundance of 10^{-6}, 10^{-7} and 10^{-8}, respectively. a. $T_c = 0$ K; b. $T_c = 10$ K; c. $T_c = 20$ K; d. $T_c = 30$ K; e. $T_c = 40$ K; f. $T_c = 50$ K.

210

INTERFEROMETRIC OBSERVATIONS OF 95 GHz METHANOL MASERS

Preethi Pratap and Karl Menten
Harvard-Smithsonian Center for Astrophysics
60 Garden Street
Cambridge, MA 02138

ABSTRACT

High resolution observations of the 95 GHz Class I methanol maser transition and the CS $J = 2 \rightarrow 1$ line toward the W33 and W51 star-forming regions were made with the BIMA array. In both cases, the methanol maser positions are distinctly offset from continuum sources and water maser centers but agree with positions of known masers in other methanol lines. In W33 the 95 GHz masers arise from several spots that are close to a high density clump marked by CS emission. Strong thermal emission from methanol is seen toward the W51 e1/e2 ultracompact HII regions.

1. INTRODUCTION

Methanol (CH_3OH) masers are generally found toward molecular cloud cores containing sites of active high-mass star formation. Multi-transition studies indicate that methanol masers can be divided into two distinct classes (Batrla et al. 1987, Menten 1991): Class I masers which are found to be offset from HII regions, infrared sources, and centers of OH and H_2O maser activity, and Class II masers which are seen toward compact HII regions and are coincident with OH masers. Class I masers arise from the $4_{-1} \rightarrow 3_0 E$, $7_0 \rightarrow 6_1 A^+$, $5_{-1} \rightarrow 4_0 E$, and $8_0 \rightarrow 7_1 A^+$ lines frequencies near 36, 44, 84 and 95 GHz, respectively, as well as the series of $J_{k=2} \rightarrow J_{k=1} E$ ($J = 2, 3, 4, ...$) transitions near 25 GHz. The strongest Class II masers arise from the $5_1 \rightarrow 6_0 A^+$ and the $2_0 \rightarrow 3_{-1} E$ lines at 6.6 and 12 GHz.

Qualitatively, the excitation of the Class I masers is easy to understand. In the case of the E-type symmetry species, the lower levels of the $k = -1$ energy ladder can be overpopulated relative to its neighboring ladders over a wide range of physical conditions. This leads to maser emission in the 36 and 84 GHz lines. Similarly, the $K = 0$ ladder of A-type methanol is overpopulated relative to $K = 1$, leading to the 44 and 95 GHz masers. To further understand the physical mechanisms that cause this population inversion, it is necessary to have accurate positional information to see if masers from different transitions do indeed arise from the identical positions. It is also useful to have an idea of the relationship between the masers, the dense molecular gas and the HII regions. Finally, accurate positions can be useful in making sensitive searches for deeply embedded infrared sources or other phenomena that could possibly be coincident with the masers.

In the past, high resolution observations of the 95 GHz lines have been made toward three sources. Toward Orion-KL, Plambeck & Wright (1988) find a narrow 95 GHz maser feature to be coincident with one of the several 25 GHz methanol maser spots mapped by (Johnston et al. 1992). In addition to the maser feature quasi-thermal methanol emission is observed. Plambeck & Menten (1990) mapped the 95 GHz emission toward DR 21 and DR 21(OH) and found in the case of DR21(OH) a strong agreement between the 95 GHz maser positions and the positions of the 84 GHz methanol masers determined by Batrla & Menten (1988). For DR21, their observations show that the 95 GHz masers are clustered along an interface between an outflow (traced by shock-excited H_2 emission) and a dense clump of ambient gas, traced by CS emission. It is clear that high resolution observations of other Class I maser regions are necessary to elucidate the relationship between these masers and their environment.

We have studied the 95 GHz maser line in W51 and W33, two star-forming regions toward which also 25 and 44 GHz methanol masers have been observed (Menten et al. 1986, Haschick et al. 1990). The observations were done with the BIMA array and simultaneous maps of the CS $J = 2 \rightarrow 1$ line were obtained to map the dense molecular cores.

2. OBSERVATIONS AND RESULTS

The observations were taken with eight (nine in the case of W51) configurations of the Berkeley-Illinois-Maryland Array at Hat Creek between 1991 May and December. The projected antenna spacings ranged from 1 kλ to 27 kλ. The phase and amplitude calibration were done using the quasars NRAO 530 and 3C 273. The $8_0 \rightarrow 7_1 A^+$ transition of CH_3OH at a rest frequency of 95.169489 GHz (De Lucia et al. 1989) and the $J = 2 \rightarrow 1$ transition of CS at 97.98097 GHz were observed simultaneously in the lower and upper receiver sidebands. Each sideband was further split into two bands of 128 channels each with bandwidths of 5 MHz and 40 MHz. This allowed us to observe both lines with velocity resolutions of 0.12 km s^{-1} and 0.98 km s^{-1} . The data were mapped and CLEANed using the RALINT data reduction software developed at the University of California at Berkeley.

3. DISCUSSION

a. W33

Figure 1 is a composite image of the continuum map, emission from two velocity channels of CS and the positions of the CH_3OH and H_2O masers. The solid and dashed contours correspond to CS emission at velocities of 35 km s^{-1} and 31 km s^{-1} , respectively. The open stars indicate the positions of the 95 GHz methanol maser components, all of which are at velocities between 32.5 and 33 km s^{-1} . The solid triangle marks the 25 GHz CH_3OH maser position (Menten et al. 1986). The H_2O maser is located between the two peaks of the continuum emission, which is shown as the short-dashed contours. The 95 GHz masers are situated on the edge of the 31 km s^{-1} CS clump and in good agreement with the position of the 25 GHz maser. Haschick & Ho (1983) have mapped the cloud in the emission from the (1,1) line of NH_3 and conclude that the velocity of the ambient molecular material is 36 km s^{-1} . Therefore the velocities of the methanol masers and the CS clump next to them are blue-shifted by $4 - 5$ km s^{-1} with respect to the ambient cloud velocity. In analogy to the case of DR21, one might expect that the masers are situated at the interface between high velocity gas and the ambient molecular material probed by the CS emission.

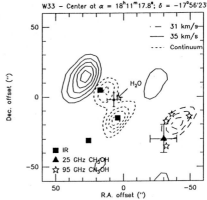

Fig. 1. Positions of the 95 GHz methanol masers (open stars), the 25 GHz methanol maser (solid triangle) and the H$_2$O maser (solid circle) superposed on a 95 GHz continuum map (dot-dashed contours) of W33-Main. The solid contours indicate CS $J = 2 \rightarrow 1$ emission at 35 km s^{-1} and the dashed contours indicate CS emission at 31 km s^{-1} . The filled squares are the positions of the infrared sources.

Unfortunately, at present no observations that could directly identify the outflowing material have been reported for W33 and measurements of CO lines to search for high velocity line wings at relatively high spatial resolution or 2.2 μm H$_2$ seem highly desirable.

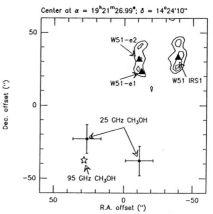

Fig. 2. Positions of the 95 (open star) and 25 GHz (solid circles) methanol masers toward W51. The solid contours outline the 95 GHz continuum emission. The three triangles indicate the positions of the continuum sources.

b. W51

Figure 2 shows the positions of the 25 GHz (errorbars) and 95 GHz (open star) CH$_3$OH masers relative to the continuum sources. The contours correspond to the 95 GHz continuum. The figure clearly indicates the offset between the continuum and the methanol masers. Within the errors, there is good agreement between the positions of the 25 GHz and the 95 GHz masers. The velocity of the 95 GHz feature (56.5 km s^{-1}) also agrees with that of the 25 GHz feature (56.8 km s^{-1}). The other velocity features detected at 25 GHz have not been detected here, possibly because they fall below our sensitivity limit. Again, as in the case of W33, high resolution observations of molecular outflows are toward the W51 region have to be made in order to to check on a possible association of the methanol masers with high velocity gas.

Figure 3a is a map of the (quasi-)thermal methanol emission toward the continuum sources W51 e1/e2. W51 IRS 1 is also shown on the map, which represents the integrated emission between 53 and 64 km s^{-1} . There are two peaks in the map, one toward W51-e2 and the other to the south of W51-e1 and coincident with some H_2O maser emission. The two peaks are at different velocities and indicate the presence of a north-south velocity gradient. Most of the H_2O, OH and the two NH_3 masers (Genzel et al. 1981; Gaume & Mutel 1989; Pratap et al. 1991) are situated in two regions, one toward W51-e2 and and another slightly north of W51-e1. Figure 3b shows an average from 49 to 60 km s^{-1} of CS $J = 2 \to 1$ emission. The CS emission is much more extended and is very similar to the HCO^+ emission mapped by Rudolph et al. (1990).

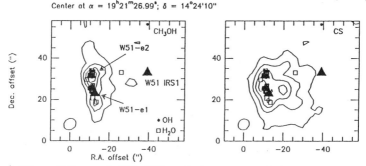

Fig. 3: a) Map of thermal CH_3OH emission averaged from 53 to 64 km s^{-1} . The triangles are the same continuum sources indicated in Fig. 2, the open squares are the positions of the H_2O masers and the dots are the positions of the OH masers. b) Map of CS $J = 2 \to 1$ emission averaged from 49 to 60 km s^{-1} . The symbols are the same as in Fig. 3a.

REFERENCES

Batrla, W., Matthews, H. E, Menten, K. M., & Walmsley, C. M. 1988, Nature, 326, 49

Batrla, W., & Menten, K. M. 1988, ApJ, 329, L117

De Lucia, F. C., Herbst, E., Anderson, T., & Helminger, P. 1989, J. Molec. Spectrosc., 134, 395

Gaume, R. A., & Mutel, R. L. 1987, ApJS, 65, 193

Genzel, R. et al. 1981, ApJ, 247, 1039

Haschick, A. D., & Ho, P. T. P. 1983, ApJ, 267, 638

Haschick, A. D., Menten, K. M., & Baan, W. A. 1990, ApJ, 354, 556

Johnston, K. J., Gaume, R. A., Stolovy, S., Wilson, T. L., Walmsley, C. M., & Menten, K. M. 1992, ApJ, 385, 232

Menten, K. M. 1991, in Skylines (Proc. Third Haystack Observatory Meeting), ed. A. D. Haschick & P. T. P. Ho (San Francisco: ASP), 119

Menten, K. M., Walmsley, C. M., Henkel, C., & Wilson, T. L. 1986, A&A, 157, 318

Plambeck, R. L., & Menten, K. M. 1990, ApJ, 364, 555

Plambeck, R. L., & Wright, M. C. H. 1988, ApJ, 330, L61

Pratap, P., Menten, K. M., Reid, M. J., Moran, J. M., & Walmsley, C. M. 1991, ApJ, 373, L13

Rudolph, A., Welch, W. J., Palmer, P., & Dubrulle, B. 1990, ApJ 363, 528

ON THE CLASSIFICATION AND LIST OF TRANSITIONS OF METHANOL MASERS

A.M. Sobolev

Astronomical Observatory, Urals State University,
Lenin ave. 51, 620083 Ekaterinburgh, Russia

ABSTRACT

A classification of methanol masers according to the type of maser sources is suggested. The list of candidates to be methanol masering transitions is compiled on the basis of general regularities analysis combined with results of statistical equilibrium calculations.

1. ON THE CLASSIFICATION OF METHANOL MASERS

According to their phenomenological characteristics methanol masers were divided by Batrla et.al.(1987) into I and II classes. Methanol masering transitions (hereafter MMT) of one series belong to the same class. Almost all of MMT belong to only one of the classes. The exclusion are 38 GHz $7_{-2}-8_{-1}E$, $6_2-5_3A^+$ and $6_2-5_3A^-$ MMT which were observed in the cites of maser sources of both classes (Haschick et.al., 1989).

The later fact shows that the existing classification of methanol masers by *transitions* is not certain. That is why we suggest to classify methanol masers according to the *type of maser sources*.

We suggest to denote the sources connected with the boundaries of HII regions by the roman figure II and the sources connected with the hot cores and giant molecular clouds as a whole by the roman figure I.

According to their phenomenological characteristics Class I masers are naturally divided into 2 subclasses.

Class Ia masers appear to originate in the isolated clumps on the earliest stages of star formation (Strel'nitskij,1981). J_2-J_1E series, $7_{-2}-8_{-1}E$, $6_2-5_3A^-$, $6_2-5_3A^+$ and some other observed methanol masering lines are the attributes of this type of maser sources.

Class Ib masers represent the whole bulk of the cloud (Sobolev, 1990). Note, that some of corresponding transitions are also masering in isolated clumps, manifesting spike features in the line profiles (see,e.g., Haschick et.al., 1990, Berulis et.al.,1991).

As for Class II masers, profiles of 38 GHz lines are similar to those of 2_1-3_0E and $9_2-10_1A^+$ lines and differ much from those of $2_0-3_{-1}E$ and $5_1-6_0A^+$ masers (see,e.g.,observations of Haschick et.al., 1989, Batrla et.al.,1987 and Menten,1991). Corresponding sources are probably different also (this supposition can be examined by the observations with high spatial resolution).

According to the above supposition we distinguish Class IIa and Class IIb masers. Note, that classification of the sources in some lines is not definite because of the lack of observational data.

2. GENERAL REGULARITIES IN THE METHANOL MASERING TRANSITIONS

The number of observed up-to-date MMT exceeds 15. The discoveries of new such transitions are still in progress (Menten,1991). That is why here we consider the possible candidates to be MMT.

In search of MMT candidates we can not confine ourselves only to the results of statistical equilibrium calculations (hereafter SEC) because the details of pumping mechanisms and rates of elementary processes (e.g., collisional cross-sections) are not clear yet. Some regularities of the observed MMT characteristics in addition to SEC results will help us to complete the list of possible MMT.

Firstly, most of the observed MMT are arranged in $(J+\alpha)_{k+\beta} J_k$ series. Here $\alpha= 0,\pm1$, $\beta=\pm1$ and quantum number k specify the series. This statement can be illustrated by the observed $J_2-J_1 E$ (J=2-10), $(J+1)_{-1}-J_0 E$ (J=3,4) and $(J+1)_0-J_1 A^+$ (J=6,7) series of MMT.

Secondly, SEC results indicate that $J_k A^+$ levels and their $J_k A^-$ counterparts have the close values of population numbers and show the similar behaviour with variation of physical conditions. That is why A^- counterparts of the observed A^+ MMT should be masering if they satisfy constraints on the frequency. One pair of such transitions ($6_2-5_3 A^+$ and $6_2-5_3 A^-$) was observed already (Haschick et. al.,1989).

Noted frequency constraint is the consequence of the general tendency of the observed methanol masers to increase their strength with decreasing frequency ν. This tendency can be easily understood from consideration of the maser energetics.

With this tendency in mind in our search for MMT candidates from the general viewpoint we confined ourselves to the transitions with ν < 100 GHz. Candidates to be MMT which were found as a result of SEC were considered free from frequency restriction. The energy of the levels of considered transitions was less than 300 cm^{-1}.

3. ON THE LIST OF CLASS II METHANOL MASERING TRANSITIONS

Because of the noted above uncertainty here we discuss MMT in maser sources of the both Class II subclasses simultaneously.

Pumping mechanisms of Class II MMT were discussed,e.g.,in Menten et.al.(1986) and Zeng & Lou (1990). Though still unclear, they seem to have IR radiation as a dominating source. With this hypothesis in mind Zeng & Lou (1990) on the basis of SEC have shown that the current list of Class II transitions is not complete and suggested $8_2-9_1A^+$ and $7_2-8_1A^+$ transitions as possible candidates for its extention.

Examination of E-type methanol transitions with given above regularities suggests that $1_0-2_{-1}E$ and 1_1-2_0E transitions are the next candidates to be the MMT of Class II. Among A-type methanol transitions such candidates are $7_2-6_3A^+$ and $7_2-6_3A^-$ transitions. Corresponding lines appear to be the attributes of Class IIb masers. 1_1-2_0E, $8_2-9_1A^+$ and $7_2-8_1A^+$ lines possibly represent these sources also.

4. ON THE LIST OF INVERTED TRANSITIONS IN CLASS Ia SOURCES

Kinetics of collisional-radiative pumping with the sink of photons on relatively cold dust which appears to operate in Class Ia masers was studied by Sobolev & Strel'nitskij (1983). Reexamination of SEC results obtained in that study shows that there exist a number of inverted transitions in our model of maser in OMC-1. They are listed in the table 1.

General considerations show that to the list of candidates must be added $7_2-6_3A^-$ and $7_2-6_3A^+$ transitions.

Table 1. Transitions, inverted in Class Ia maser model (only E-methanol)

transition	ν, GHz
J_2-J_1E	25
$4_{-1}-3_0E$	36,2
$5_{-1}-4_0E$	84,5
$6_{-1}-5_0E$	132,9
$7_{-1}-6_0E$	181,3
4_0-3_1E	28,3
5_0-4_1E	76,5
$7_{-2}-8_{-1}E$	37,7
$6_{-2}-7_{-1}E$	85,6
$5_{-2}-6_{-1}E$	133,6
$9_{-1}-8_{-2}E$	9,9
$10_{-1}-9_{-2}E$	57,3
$J_2-(J-1)_2E$	>190

Table 2. Methanol transitions, inverted in Class Ib maser model (T_b corresponds to parameter values, given in the text)

transit.	ν, GHz	T_b, K	transit.	ν, GHz	T_b, K
$4_{-1}-3_0E$	36,2	32	$7_0-6_1A^+$	44,1	1,27
$5_{-1}-4_0E$	84,5	18	$8_0-7_1A^+$	95,2	0,6
$6_{-1}-5_0E$	132,9	7,1	$1_1-1_1A^{-+}$	0,837	0,4
$7_{-1}-6_0E$	181,3	2,1	$2_1-2_1A^{-+}$	2,50	0,2
4_0-3_1E	28,3	1,1	$9_0-8_1A^+$	146,6	0,2
5_0-4_1E	76,5	0,8	$3_1-3_1A^{-+}$	5,01	0,1
$8_{-1}-7_0E$	229,8	0,5	$6_3-6_2A^{-+}$	251,7	0,1
6_0-7_1E	124,6	0,4	$6_3-6_2A^{+-}$	251.9	0,1
7_0-6_1E	172,5	0,1	$4_1-4_1A^{-+}$	8,34	0,1
$9_{-1}-8_{-2}E$	9,9	0,1	$10_0-9_1A^+$	198,4	0,1

5. ON THE LIST OF INVERTED TRANSITIONS IN CLASS Ib SOURCES

Pumping of Class Ib masers has definitely collisional-radiative mechanism. Here we present SEC results for E-methanol transitions, inverted in molecular cloud with temperature $T_K= 50$ K, hydrogen density $n_H= 10^4 cm^{-3}$, specific methanol column density $N_M/V_d= 3x10^9 cm^{-3}$s, optical depth of dust on 200μ $\tau_{200}=0,01$ and dust temperature $T_d= 40$ K. Here N_M is the methanol column density and V_d is the doppler width. The transitions with model brightness temperature $T_B> 0,1$ K are listed in the table 2.

Table 3. List of probably masering methanol transitions.

Class Ia		Class Ib		Class IIa		Class IIb	
trans.	ν,GHz	trans.	ν,GHz	trans.	ν,GHz	trans.	ν,GHz
$7_2-6_3A^-$	86,6	table 2		$1_0-2_{-1}E$	60,5	$6_{-2}-7_{-1}E$	85,6
$7_2-6_3A^+$	86,9			1_1-2_0E	68,3	$7_2-6_3A^-$	86,6
&				$8_2-9_1A^-$	29,0	$7_2-6_3A^+$	86,9
table 1				$8_2-9_1A^+$	66,9	1_1-2_0E	68,3
				$7_2-8_1A^+$	111,3	$8_2-9_1A^-$	29,0
						$8_2-9_1A^+$	66,9
						$7_2-8_1A^+$	111,3

Notation of lines for the objects which are not definitely classified is given in italics.

REFERENCES

Batrla,W., Matthews,H.E., Menten,K.M.,& Walmsley,C.M. 1987,Nature 326, N6108, 49.

Berulis,I.I., Kalenskii,S.V., Sobolev A.M. & Strel'nitskij, V.S. 1991, AATr 1, 231.

Haschick,A.D., Baan,W., & Menten,K.M. 1989, ApJ 346, 330.

Haschick,A.D., Menten,K.M., & Baan,W. 1990, ApJ 354, 556.

Menten,K.M. 1991, Center for Astrophysics, Preprint No.3284.

Menten,K.M., Walmsley,C.M., Henkel,C., Wilson,T.L., Snyder,L.E., Hollis,J.M., & Lovas,F.J. 1986, A&A 169, 271.

Sobolev,A.M. 1990, Astron.Tsirk. N1543, 7.

Sobolev,A.M. & Strel'nitskij,V.S. 1983, Pis'ma v Astron.Zhurn. 9, 26.

Strel'nitskij,V.S. 1981, Pis'ma v Astron.Zhurn. 7, 406.

Zeng,Q., & Lou,G.F. 1990, A&A 228, 480.

PHYSICAL AND EVOLUTIONARY STATUS OF G1.6-0.025 GIANT MOLECULAR CLOUD AS REVEALED BY $4_{-1}-3_0E$ METHANOL WEAK MASER

A.M. Sobolev

Astronomical Observatory, Urals State University,
Lenin ave. 51, 620083 Ekaterinburgh, Russia

ABSTRACT

G1.6 giant molecular cloud consists of the clumps with sizes varying from 3" (0,1 pc) to 5' (10 pc). Column densities of hydrogen and methanol in the large clumps are about $(1 \div 4) \times 10^{23}$ and $(1 \div 2) \times 10^{15} cm^{-2}$, correspondingly. Hydrogen densities in the large clumps vary from 10^2 to $5 \times 10^4 cm^{-3}$ and in the small clumps reach the values $> 10^6 cm^{-3}$. High kinetic temperature in the cloud $T_K > 50 \div 100$ K partially is a consequence of collisions with individual clumps. Such collision probably occurs nowadays toward ($\alpha = 17^h 46^m 12^s$, $\delta = -27^o 33'15"$), where the 165 km/s clump is observed.

Shocks connected with this collision can lead to enhancement of methanol abundance in some clumps with masses $70 \div 300$ $M_\odot$ and sizes about 3" ($4 \times 10^{17} cm$), which therefore manifest spike features in $4_{-1}-3_0E$ line. This enhancement is the result of chemical reactions initiated by the methanol evaporation from grain mantles.

A probable protostar cluster is present in the same direction.

1. INTRODUCTION

The Galactic Center GMC (giant molecular cloud) G1.6-0.025 can be found in surveys of H_2CO $1_{11}-1_{10}$ line absorption (Whiteoak & Gardner, 1979) and $^{13}CO(1-0)$, CS(2-1) lines emission (Bally et.al., 1987). Special investigations were done in a set of NH_3 emission lines (Gardner et.al., 1985, hereafter GWFPK, Gardner & Boes, 1987, hereafter GB) and CH_3OH absorption line $2_0-3_{-1}E$ (Whiteoak & Peng, 1989, hereafter WP).

Statistical equilibrium calculations (SEC) showed that $4_{-1}-3_0E$ methanol (CH_3OH) transition should be inverted in the whole range of GMC physical parameters (Sobolev, 1990). Corresponding line, consequently, must be sensitive to the column density variations and, hence, give the unique opportunity of investigation of the cloud's structure. The brief report on the observations of $4_{-1}-3_0E$ line towards G1.6 was given in Berulis et.al. (1991, hereafter BKSS).

Dimensions of G1.6 main source in $4_{-1}-3_0E$ line at the 10% level was found to be about 10'. Lesser dimensions in $4_{-1}-3_0E$ line relative to those in $2_0-3_{-1}E$ line are caused by the greater sensitivity of $4_{-1}-3_0E$ line intensity to the variations of methanol column density and hydrogen volume density n_H.

2. $4_{-1}-3_0E$ SPECTRA IN THE VICINITY OF $(17^h46^m12^s, -27^\circ33'15'')$

$4_{-1}-3_0E$ profiles in the vicinity of $(17^h46^m12^s, -27^\circ33'15'')$ differ from profiles in other directions by the presence of 2 spikes with $V_{lsr}= 54$ and 58 km/s. Below this direction will be regarded as a center with departure coordinates (0,0).

The spikes were observed towards (0,1'), (-1',0), (0,-1'), (1',-1') and (1',0). Spectra analysis states that line widths and central velocities of spikes remain the same within the observation accuracy while their intensities show substantial variations from point to point. The character of these variations says that we deal with the compact sources situated to south-east from (0,0).

To separate spectra of compact spike sources from the spectra of the surrounding cloud we have fitted the profiles by the sum of gaussian components and linear continuum.

The shapes and widths of the profiles obtained after the exclusion of the spikes change substantially for directions separated by 1' and show cardinal variations when departure exceeds 2'(HPBW). Thus, the observed emission is a complex blend of the lines originating in the sources with sizes about 1'.

SEC combined with $4_{-1}-3_0E$ (BKSS), $1_0-0_0A^+$(Kalenskij et.al., in preparation) and $2_0-3_{-1}E$ (WP) data allow to estimate methanol column densities and hydrogen densities in these sources to be $N_M \approx (1\div2)\times10^{15}$ cm^{-2} and $n_H \approx 10^3\div10^5$cm^{-3}.

3. COMPACT REGION WITH $V_{lsr}= 45$ km/s : A PROBABLE PROTOSTAR CLUSTER?

$4_{-1}-3_0E$ line toward (0,0) in the range of $V_{lsr}= 40\div50$ km/s is weak, while the corresponding components dominate in NH_3 and $2_0-3_{-1}E$ spectra.

SEC show that strong 2_0-3_{-1} absorption with $T_B<-1$ K and weak $4_{-1}-3_0$ emission with $T_B< 1$ K can occur only if n_H is less than $10^{3.5}$cm^{-3}.

The ratio of observed toward (0,0) at 45 km/s NH_3 line intensities according to SEC of Stutzki & Winnewisser (1985) and Sobolev (1989) witness of much higher values of hydrogen volume density $n_H> 10^6$cm^{-3} and kinetic temperatures $T_K> 50$ K.

Further, GWFPK outlined that high rotational and low antenna temperatures of NH_3 witness of the low value of the filling factor $\varphi_f< 0,1$.

Thus, toward (0,0) there probably exists a region with $V_{lsr} \approx 45$ km/s, containing visible in NH_3 relatively hot dense clumps ($n_H \geq 10^6 cm^{-3}$, $T_K > 50$ K) surrounded by absorbing in $2_0-3_{-1}E$ matter with substantially lower density $n_H < 10^{3,5}$ cm^{-3}.

It can be easily shown that gas pressure and magnetic fields can't provide the dynamical equilibrium of the noted components. Thus, NH_3 clumps are most probably gravitationally bound objects of the possible protostellar nature.

It can be noted, that GWFPK, GB and WP observational data, combined with SEC allow to estimate source sizes :

$$d(NH_3) \approx 60" \text{ and } d(CH_3OH) \geq 80".$$

Thus, the conclusions of this section can be examined by the observations with an angular resolution better than 30-40".

4. ESTIMATES FOR THE SIZES AND MASSES OF THE SPIKE SOURCES

Since IRAS sources and other strong factors which lead to the level population redistribution seem to be absent in the region under study, collision-radiative pumping in rotational transitions (CRr) appears to be the most probable mechanism which determines $4_{-1}-3_0E$ excitation in the sources of spikes.

Brightness temperature of $4_{-1}-3_0$ under CRr pumping has the absolute maximum $T_{s,max} = 7\ 000$ K. Hence, the lower limits of the sizes are

$$d > D_L \left(T_a / T_{s,max} \right)^{1/2} = 3,2" \text{ and } 3,1"$$

for 54 and 58 km/s spikes correspondingly.

The clumps with such sizes are typical for GMC cores. Most of them seem to be quasi-stable features. Under the assumption of virial equilibrium combined with the estimates of the sizes and the gaussian fitting results one can estimate masses of the clumps to be $\geq 70\ M_\odot$ and $\geq 300\ M_\odot$ for 54 and 58 km/s spike sources, correspondingly.

5. ON THE ORIGIN OF SPIKE SOURCES

Comparison of the obtained results with the data of Stutzki & Güsten (1990) shows that a large number of clumps in GMC core can have the parameters in the marked range. At the same time the actual number of spikes is usually small and rarely exceeds 2 - 3 in the whole bulk of the cloud.

The noted contradiction disappears only under the hypothesis that methanol column densities in the usual clumps are not high enough and the spike sources are distinguished by the high abundance of methanol relative to hydrogen $[CH_3OH/H_2] \approx 10^{-7} \div 10^{-6}$.

Recent investigations allow to suggest the following scenario of the spike origin similar to proposed by Plambeck & Menten (1990) on the basis of the fact of the close connection between spike sources and high-speed outflows.

Shock wave heats the gas of the clump and evaporates grain mantles containing considerable amount of methanol (Grim et.al., 1991). Evaporation of methanol initiates chemical reactions which result in essential enhancement of methanol abundance (Millar et.al., 1991).

Necessary shocks can appear in (0,0) direction as a consequence of collision between 165 km/s clump and G1.6 main cloud.

Note, that the same shocks can stimulate star formation in 45 km/s NH_3 emission region.

ACKNOWLEDGEMENTS

The author is grateful to V.S.Strel'nitskij, S.V.Kalenskij, I.I.Berulis, R.L.Sorochenko and A.M.Tolmachov for the fruitful critical discussions and friendly assistance.

REFERENCES

Bally,J. et.al. 1987, Ap.J.Suppl.Ser. 65, 13.

Berulis,I.I.,Kalenskii,S.V.,Sobolev A.M. and Strelnitskij, V.S. 1991, As.Ap.Trans. 1, 231 (BKSS).

Gardner,F.F. and Boes,F. 1987, Proc.astr.Soc.Austr. 7, 185 (GB).

Gardner,F.F. et.al. 1985, Proc.astr.Soc.Austr. 6, 176 (GWFPK).

Grim,R.Z.A., et.al. 1991, As.Ap. 243, 473.

Millar,T.J., Herbst,E. and Charnley,S.B. 1991, Ap.J. 369, 147.

Plambeck,R.L. and Menten,K.M. 1990, Ap.J. 364, 555.

Sobolev,A.M. 1990, Astron.Tsirk. N1543, 7.

Sobolev,A.M. et.al. 1989, Astron.-geod.issled., Sverdlovsk, 150.

Stutzki,J. and Gtsten,R. 1990, Ap.J. 356, 513.

Stutzki,J. and Winnewisser,G. 1985, As.Ap. 144, 13.

Whiteoak,J.B. and Gardner,F.F. 1979, MNRAS 188, 445.

Whiteoak,J.B. and Peng,R.-S. 1989, MNRAS 239, 677 (WP).

9. PROPER MOTION

THE PROPER MOTIONS OF THE H₂O MASERS NEAR W3(OH)

Javier Alcolea
Harvard-Smithsonian Center for Astrophysics
60 Garden St., Cambridge MA 02138, USA
and Centro Astronómico de Yebes
Apartado 148, E-19080 Guadalajara, Spain

and

Karl M. Menten, James M. Moran and Mark J. Reid
Harvard-Smithsonian Center for Astrophysics

ABSTRACT

We describe the results of three-epoch intercontinental VLBI observations of the 22 GHz H₂O masers near the compact HII region W3(OH). We have determined the relative proper motions of individual maser components. Assuming a distance to the source of 2.2 kpc, we are able to obtain a full 3-dimensional description of the kinematics of the emitting region, which indicates that the H₂O masers are part of bipolar outflow.

1. INTRODUCTION

Interstellar 22 GHz water (H₂O) masers are usually associated with mass outflows from very young stellar objects. In the case of the well studied W3(OH) star-forming region, the H₂O maser emission arises from a source that is located about 6″ (0.06 pc) east of the well-studied ultracompact HII region and OH maser source.

Because of their high brightness temperatures, H₂O masers can be observed with VLBI techniques at (sub-)milliarcsecond angular resolution. From repeated VLBI measurements of a maser region it is possible to measure proper motions of the different maser spots relative to each other. The location of the maser components in the plane of the sky and their proper motions, together with their radial motions derived from the observed Doppler shifts, give an almost complete description of the kinematics of the emitting region; only the information about the positions along the line of sight is missing.

2. OBSERVATIONS AND DATA REDUCTION

In order to measure the relative proper motions of the 22 GHz H₂O masers near W3(OH), three VLBI observations (of a 5 epoch sequence) were processed from December 1981, and March and June 1982, spanning a total time of ~ 190 days. Five telescopes were used: the Haystack 37 m, the NRAO 43 m, one VLA 25 m, the OVRO

40 m and the Onsala 20 m. The resulting synthesized HPBW was about 0.26 mas. The data were recorded with the MarkII system with a V_{lsr} coverage from −63.5 to −36.5 to km s^{-1}, and processed with the NRAO MarkII correlator, which provided a spectral resolution of 0.28 km s^{-1}.

The data were then converted into the NRAO/SAO spectral line VLBI format and calibrated using this software package. The amplitude calibration was done by comparing the intensities of the total-power spectra. The data were phase calibrated by using a channel containing strong emission as a phase reference. The calibrated data were shifted to the different centers of maser activity (found in crude large area fringe rate maps), averaged to 20 sec, and transformed into the AIPS format for further analysis. A self calibration procedure was then performed on the reference channel for the three epochs, in order to take into account phase errors due to possible structure of the reference maser feature. The visibility corrections were applied to all other channels in each of the corresponding data sets. Then synthesis maps were produced, CLEANed, and maser positions were obtained by fitting a two-dimensional gaussian model.

The positions of the maser spots have been corrected for gravitational deflection by the sun and converted into B1950 coordinates, to take into account the effects of differential precession and nutation. In this way we can estimate relative distances between different spots or between the same spot in different epochs with a precision approaching about 10 μas. The proper motions in the plane of the sky have been computed by fitting constant velocity trajectories to the locations of those maser spots found in the same channel at several epochs; we have obtained a total of 21 maser features detected in all three epochs and 27 features for which two-epoch proper motions could be obtained.

3. THE PROPER MOTIONS OF THE H$_2$O MASERS

The results of the observations are summarized in Fig. 1 where we plot the positions of the masers and their corresponding proper motions with respect to the reference feature. The H$_2$O masers are located in a $2\rlap{.}''0 \times 0\rlap{.}''5$ area centered about $6''$ east of the compact HII region. All the strong maser components are found within $0\rlap{.}''2$ of the reference feature. The general distribution of the maser emission is similiar to that found in earlier high spatial resolution studies of this source (see e.g. Genzel et al. 1978), although we note that the flux density of individual maser components can undergo dramatic changes on time-scales of months.

In order to study the kinematics of the region, we modelled the maser proper motions, adopting a distance to the source of 2.2 kpc (Humphreys 1978) to convert the measured angular proper motions into linear velocities. Assuming that the maser emission arises from a spherically symmetric expanding outflow with an expansion velocity V_{exp} at $1''$ from the center of expansion, the velocity $\vec{v} = (v_x, v_y, v_z)$ of a maser spot with respect to this center of expansion is given by the equation: $\vec{v}(\vec{r}) = V_{exp} R^\alpha \vec{r} / R$; where R is $|\vec{r}|$, $\vec{r} = (x - X_0, y - Y_0, z)$ is the position of the maser spot with respect to the center of expansion, (x, y) and (X_0, Y_0) are the offsets of the maser spot and center of expansion from the reference feature, respectively, and z is the position of the maser along the line of sight relative to the center of expansion.

Let (u_x, u_y) and (W_x, W_y) be the proper motions of the maser spots and the velocity of the center of expansion with respect to the reference feature, respectively, and u_z and W_z their corresponding radial (LSR) velocities. We then can obtain the parameters of the outflow by minimizing the expression:

$$\sum \left[\left(\frac{u_x - W_x - v_x}{\sigma_x} \right)^2 + \left(\frac{u_y - W_y - v_y}{\sigma_y} \right)^2 + \left(\frac{u_z - W_z - v_z}{\sigma_z} \right)^2 \right]$$

where the sum runs over all maser spots. $\vec{\sigma}$ accounts for the errors in the determination of the proper motions and a turbulent component of the velocity field for which we adopt a value of 10 km s^{-1}. This way we can also retrieve information about the positions of the masers along the line of sight z, leading to a full 3-dimensional picture of the emitting region.

We found that this procedure does not converge into a solution unless W_z is fixed, so we must assume a value for this parameter. We have adopted $W_z = -50\pm5$ km s^{-1}, which is the radial velocity centroid of the H_2O maser emission. Similar LSR velocities are found from observations of other molecules toward this region (Mauersberger et al. 1986)

The results of our fit are shown in Figs. 2a and 2b and Table 1 (where the uncertainty in the determination of W_z has been taken into account). The proper motions indicate that the W3(OH) H_2O masers form a bipolar outflow (Fig. 2) whose symmetry axis is nearly parallel to the X-axis (R.A.). The center of expansion is located in the center of the emitting region, about $0\overset{''}{.}8$ west of the strongest emission features. This position is coincident, within the uncertainties, with weak 6 cm radio continuum emission found by Guilloteau et al. (1985) and with a compact hot molecular condensation detected in HCN (Turner & Welch 1984) and NH_3 (Mauersberger et al. 1986) emission. The molecular gas is probably heated by an embedded young stellar object, which may also be responsible for the bipolar outflow traced by the H_2O masers. The outflow expansion velocity is moderate, 20 km s^{-1}, slightly decreasing with distance to the center of expansion ($\alpha = -\frac{1}{3}$).

Table 1. Model parameters for $W_z = -50\pm5$ km s^{-1}

W_x	-20 ± 2 km s^{-1}	X_0	$-0\overset{''}{.}83 \pm0\overset{''}{.}05$
W_y	11 ± 2 km s^{-1}	Y_0	$0\overset{''}{.}05 \pm0\overset{''}{.}05$
α	$-\frac{1}{3}$	V_{exp}	20 ± 2 km s^{-1}

Genzel R., Downes D., Moran J.M., et al., 1978, A&A 66, 13
Guilloteau S., Baudry A., Walmsley C.M., 1985, A&A 153, 179
Humphreys R.M., 1978, ApJS 38, 309
Mauersberger R., Wilson T.L., Walmsley C.M., 1986, A&A 166, L26
Turner J.L., Welch W.J., 1984, ApJ 287, L81

J.A. is partially supported by the Spanish Ministerio de Educación y Ciencia.

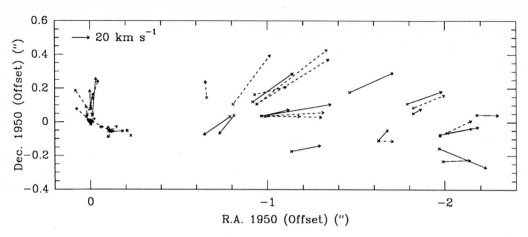

Figure 1. Proper motions of the H$_2$O masers near W3(OH). The positions and motions are with respect to the reference feature. A distance of 2.2 kpc has been assumed. The absolute B1950 coordinates and V_{lsr} velocity of the reference feature are: $\alpha=02^h23^m17\overset{s}{.}45 \pm 0\overset{s}{.}015$; $\delta=61°38'57\overset{''}{.}2 \pm 0\overset{''}{.}1$; $V_{lsr}=-46.5$ km s^{-1}. The crosses mark the positions of the maser spots and the arrows show the direction and magnitude of the proper motions. Solid arrows: 3-epoch motions; doted arrows: 2-epoch motions.

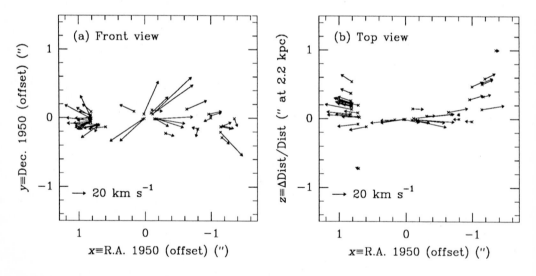

Figure 2a and b. Same as Fig. 1 but with respect to the center of expansion determined by the model fitting. Both 2- and 3-epoch proper motions are displayed with the same symbols. In Fig. 2a we represent the motions and positions of the masers in the plane of the sky (XY-plane), while in Fig. 2b we have plotted a top view (XZ-plane) of the masers. The units of the Z-axis (projected length) are in arcsec assuming a distance of 2.2 kpc. The location and proper motions of the masers clearly suggest that they arise from a bipolar outflow.

KINEMATIC STUDIES OF ULTRACOMPACT
HII REGION/MOLECULAR CLOUD INTERFACES

E. E. Bloemhof

Harvard-Smithsonian Center for Astrophysics
60 Garden Street
Cambridge, MA 02138

ABSTRACT

An exciting application of masers as an astrophysical tool is the study of proper motions through multi-epoch VLBI, which has the potential to reveal details of source kinematics on an unprecedented spatial scale. OH masers are particularly interesting, as they are typically situated in the high-density region of molecular gas lying at the interface between ultracompact HII regions and molecular clouds. We discuss OH-maser proper motion measurements we have made of W3(OH), and show that modelling methods developed in the past for H_2O maser motions may not be appropriate to this source. We then present new and fairly general analytic techniques that may be used to detect and interpret asymmetric proper motion patterns.

1. INTRODUCTION

Ultracompact HII regions are believed to be the sites of recent massive star formation. In the standard picture, a central O star, not yet visible due to obscuration by intervening dust, drives a shock wave into the ambient molecular cloud and sets up an ionized region seen in radio continuum emission. The high density and temperature and the small spatial extent of the ultracompact HII region imply a dynamical timescale perhaps as short as a few thousand years before the observed compact radio structure would dissipate.

Interstellar OH masers are found in the interfaces between ultracompact HII regions and the dense molecular clouds with which they are usually associated. Their occurrence is understood in general terms to be a result of OH density enhancement due to high temperature chemistry associated with a shock front. The OH masers are ideally located for probing the kinematics in the immediate vicinity of very young massive stars. W3(OH) is the prototype ultracompact HII region, as well as the prototype interstellar OH maser source; it is nearby (2.2 kpc), and thus not seriously affected by interstellar scattering, and both the radio continuum and OH maser morphologies exhibit a high degree of symmetry. Most compellingly, the OH maser spots appear to cover the HII region/molecular cloud interface reasonably well.

We have recently determined the OH maser proper motions in W3(OH) with a two-epoch spectral-line VLBI study (Bloemhof, Reid, and Moran 1992). At each of the epochs, which were separated by 7.5 years, we measured the positions of individual maser spots in both circular polarizations at 1665 MHz with a spatial resolution of roughly 5 milliarc seconds (mas). Positional accuracies were sufficient to detect motions as small as roughly 1 km s^{-1}, and significant ordered motions much larger than this were found.

These measurements suffer from the key observational uncertainty that affects all determinations of maser proper motions: since interferometric maps are referenced to a single bright, spatially unresolved "reference" maser feature, positions (and hence motions) are only known with respect to those of the reference feature. The motion of the reference feature with respect to other kinematic entities, such as the underlying ultracompact HII region, is not directly measured.

The motions of the OH maser spots in W3(OH) are shown in Figure 1 with the assumption that the reference feature is at rest with respect to the ultracompact HII region. There is a general appearance of expansion, at least in two dimensions, definitively ruling out the molecular infall model of Reid *et al.* (1980). With some ad hoc constraints regarding maser spot placement along the line of sight, to account for the small maser spot velocity dispersion in that direction, a spherical expansion model can be made to work in three dimensions. An alternative kinematic interpretation discussed in some detail by Bloemhof, Reid, and Moran (1992) is the cometary bow-shock model. If the reference maser feature is assumed to be moving at $(v_\alpha, v_\delta) = (6, 4)$ km s^{-1} with respect to the ultracompact HII region, a velocity field is obtained that qualitatively resembles the theoretical predictions by Van Buren *et al.* (1990) for the prototype cometary HII region G34.3+0.2 (Fig. 2).

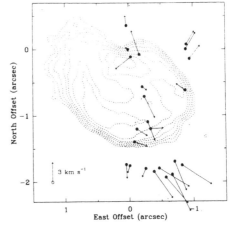

Fig. 1. Proper motions of the 1665 MHz OH masers in W3(OH) (arrows) assuming a reference feature stationary with respect to the ultracompact HII region (15 GHz radio continuum, dashed contours). The general impression of expansion is apparent, at least in two dimension.

2. A NEW, OBJECTIVE TECHNIQUE FOR DIAGNOSING MASER MOTIONS: DIAGONALIZATION OF THE VELOCITY VARIANCE/COVARIANCE MATRIX

The choice between a spherical expansion model and a cometary bow-shock model is rather subjective. A best-fit version of any model can always be found, but comparison of residuals from different models is difficult. To partially overcome these uncertainties, we developed a new technique for extracting objective information about the maser velocity field without making model assumptions. This technique is an extension and generalization of examining internal velocity dispersions of maser spot motions; these dispersions (standard deviations) have the virtue that they are insensitive to additive offsets (i.e., to motion of the reference feature). In the case of a spherical expansion, one would expect equal velocity dispersions along all three spatial axes. In the case of the cometary bow-shock model, one would expect a maximum velocity dispersion along the "cometary" axis, or axis of symmetry.

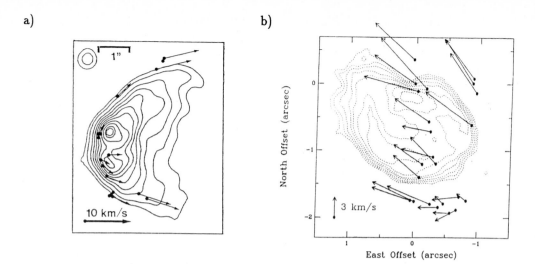

Fig. 2. a) Theoretical prediction by Van Buren *et al.* (1990) for the proper motions of OH masers around the cometary ultracompact HII region G34.3+0.2. b) Candidate velocity field for the measured proper motions of OH masers in W3(OH), assuming the reference feature moves at (6,4) km s^{-1} with respect to the ultracompact HII region.

One may thus seek a cometary axis by constructing and diagonalizing the velocity variance/covariance matrix (VVCM), σ, whose elements are given by

$$\sigma_{i,j} = \frac{1}{N-1} \sum_{n=1}^{N} (v_{i,n} - \overline{v_i})(v_{j,n} - \overline{v_j}),$$

where i, j denote spatial axes and the bar denotes averaging over N maser spots. If a marked inequality among the resulting eigenvalues were found, an asymmetric velocity field would be indicated. In the case of a cometary flow, the eigenvector corresponding to the largest eigenvalue would define the cometary axis, and the equality of the two smaller eigenvalues would describe the axial symmetry of the flow. (For a full development of the technique, see Bloemhof 1992).

3. EXAMPLE: TWO-DIMENSIONAL TREATMENT OF W3(OH) OH MASERS

A complication in the case of W3(OH) is the HII region, opaque at the frequency of the OH maser transition, that may be hiding from view some maser spots on the far side of the source. It is apparent that the VVCM diagonalization technique will work best when there is good spatial coverage, in three dimensions, of the region whose kinematics are being studied; this is true for any specific source model as well. To give a simple illustration of the VVCM diagonalization technique, we avoid the complication of the opaque HII region by treating the OH masers of W3(OH) in two dimensions: i.e., in the plane of the sky. The result is shown in Figure 3, where the derived cometary axis is drawn as a dashed line at position angle ∼30° east of north, superimposed on a radio continuum map of the general region. The eigenvalue corresponding to this cometary axis is 7.2 $(\text{km s}^{-1})^2$, significantly larger than the transverse eigenvalue 2.1 $(\text{km s}^{-1})^2$. In a spherical expansion model, one would expect the non-uniform coverage of maser spots, which extend further in declination than in right ascension, to give a maximum velocity dispersion along an axis at position angle $\sim 0°$. The low-lying radio continuum contours in Figure 3 might suggest a cometary axis at position angle $\sim 60°$; however, if the velocity field were indeed cometary, the non-uniform spatial coverage apparent in Figure 2b would have added a clockwise bias into the $\sim 30°$ cometary axis we have derived. There is thus some evidence that the internal velocity dispersions of the OH maser motions favor a cometary bow-shock model.

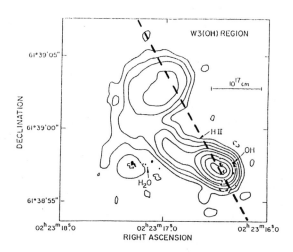

Fig. 3. Cometary axis in W3(OH) predicted by diagonalization of the OH maser velocity variance/covariance matrix (dashed line), superimposed on a map of the general region from Reid and Moran (1988). VVCM diagonalization predicts the angle of the dashed line; its position has been chosen arbitrarily.

Perhaps the most powerful application of the VVCM diagonalization technique will be in sources with more numerous maser spots, for better spatial coverage, and with maser transitions at higher frequencies, to avoid the problem of obscured maser spots. The excellent coverage of G34.3 seen in Figure 2a should permit a clear diagnosis when proper motions are measured for the OH masers: the resulting cometary axis should lie at position angle $\sim 90°$ if the flow is indeed cometary. The technique does not obviate the need for detailed, specific modelling; such models can derive important information from the proper motion data. It does, however, provide an objective and easily computed diagnostic that may be helpful in judging which model is appropriate.

REFERENCES

Bloemhof, E. E. 1992, *in preparation*.

Bloemhof, E. E., Reid, M. J., and Moran, J. M. 1992, *Ap. J.*, *in press*.

Reid, M. J., Haschick, A. D., Burke, B. F., Moran, J. M., Johnston, K. J., and Swenson, G. W. 1980, *Ap. J.*, **239**, 89.

Reid, M. J., and Moran, J. M. 1988, in *Galactic and Extragalactic Radio Astronomy*, 2nd edition, eds. G. L. Verschuur and K. I. Kellermann (New York: Springer-Verlag), 255.

Van Buren, D., Mac Low, M., Wood, D. O. S., and Churchwell, E. 1990, *Ap. J.*, **353**, 570.

SPACE VLBI AND THE MASERS

R.S. Booth
Onsala Space Observatory, Sweden

ABSTRACT

Cosmic masers are among the highest brightness temperature sources in the sky and so are natural targets for space VLBI - or are they? The limitations of space VLBI for maser research, both natural and technical are discussed. Finally some possible maser projects for the two ongoing space VLBI projects will be discused.

INTRODUCTION

The possibility of launching a radio telescope into space to increase the available VLBI baselines has been discussed for more than a decade. Significant increases in resolution are possible and the scheduled missions will give resolutions finer than 50 microarcseconds. In addition to the resolution improvements, the presence of an orbiting telescope will improve the possibilities for high resolution studies of the southern sky where many interesting sources are located.

Because of their extremely small angular diameters and correspondingly high brightness temperatures, the masers have always been considered as important targets for the space missions. There are still many unsolved problems in maser research which may benefit from increased resolution e.g. the nature of the maser - its size, shape, etc, the source of pump energy, the role of magnetic fields, the nature of extragalactic masers and the effects of interstellar scattering.

In addition, the usefulness of masers as distance indicators, through the measurement their proper motion, should be improved through increased resolution and the possibility of using the technique for extragalactic distance measurement, and therefore of an independent measurement of the Hubble constant has been discussed (Reid, 1984).

SPACE VLBI

The elements of space VLBI have been described e.g. by Schilizzi *et al*, 1984. They involve an orbiting radio telescope, ideally with multi-frequency and polarisation capability, a network of ground radio telescopes to observe in conjunction with the space borne antenna, and a network of telemetry stations to relay stable local oscillator signals to the satellite and to receive and record the down-linked signals from the orbiting telescope. The recorded data must, as is usual in VLBI, be correlated at a special purpose processing facility.

Four major space VLBI missions have been proposed: QUASAT, RADIOASTRON, VSOP, and IVS (International VLBI Satellite, a mission intended as a second generation space VLBI mission). Of these, only the Russian Radioastron mission and the Japanese VSOP mission have been selected by their respective space agencies. These missions are conceptually different: Radioastron has a large apogee height (about 80,000 km) and is intended primarily to give an order of magnitude resolution improvement over ground-based VLBI, whereas VSOP with a lower orbit, apogee height 20,000 km, is conceived as an imaging mission, i.e. in combination with the ground VLBI networks, it will enhance their imaging properties but with only a relatively modest (x2) increase in resolution. The full orbital parameters of the two missions are discussed later in these proceedings. Radioastron is equipped with a 10 m diameter antenna and VSOP will have an 8 m telescope. Both telescopes are to be equipped with receivers for the standard VLBI frequencies at 1.66 GHz, 5 GHz and 22 GHz, while Radioastron will also have a low frequency receiver at 327 MHz.

OBSERVATIONS OF MASERS

The space missions are equipped to observe main-line OH masers as well as the water masers at 22 GHz. However, there are some limitations to their usefulness, both natural and technical, and we will consider them here.

Natural limitations - interstellar scattering

It may be expected that interstellar scattering will impose natural limitations on the angular sizes of the masers and therefore limit the usefulness of space VLBI for their study. This is certainly the case for some sources but as pointed out by Booth (1984), the data on sources whose angular sizes had been measured indicated that the scattering medium was patchy giving very little apparent broadening along some lines of sight. This was confirmed by a study of low latitude extragalactic continuum sources by Dennison et al (1984).

With space VLBI in mind, we have conducted a survey of OH sources at 18 cm using a 7 station transatlantic VLBI network (Hansen et al, these proceedings). Measured sizes of 20 OH sources in the Galactic plane show that there is a trend for increased angular size as a function of increasing distance (see Fig 1 in the paper by Hansen et al, these proc.) but about a third of the sample observed have angular diameters less than 5 mas. Thus, even in the Galactic plane, regions of only small scattering are present, especially away from the inner Galaxy. (see also Gwinn et al, 1988).

Dennison et al, 1991 have described the situation at high galactic latitudes. They show that even at the maximum resolution of Radioastron, the minimum scattering angle is about the same as the resolution at 18 cm wavelength.

These studies indicate that although maser sources will be scattered, observations with space VLBI resolutions are important and will give the structure of the masers at the maximum possible resolution. This applies particularly to studies of extragalactic masers at high galactic latitudes. Observations of Galactic OH and water masers will contribute to our understanding of interstellar scattering through measurement of apparent source sizes, and even shapes.

Technical limitations

Sensitivity

The signal to noise ratio of an interferometer between a ground element, (subscript g) and a space element, (subscript s) for a correlated flux S_c, is given by

$$SNR = K.Sc \, (g_g g_s)^{1/2} \cdot (e_g e_s)^{1/2} D_g D_s (B.t)^{1/2} / (T_g T_s)^{1/2} \qquad (1)$$

where $K = 1 \cdot 8 \cdot 10^{-4}$, g_g, g_s are mispointing factors, e_g, e_s are antenna efficiencies, T_g, T_s are system temperatures, D_g, D_s are antenna diameters in m, B is the bandwidth in Hz, and t is the integration time in secs
Inserting expected values for 22 GHz for the missions which will fly, i.e.D_g=100 m, e_g= 50%, D_s=10 m, e_s=20%, T_g = 50K, T_s = 400K, B=1 Km/s (=75 kHz), t = 100 s we find that a 5 Jy source will give an SNR of about 5.
Equivalent values for 18 cm (1.6 GHz) give an SNR of 5 for a 1 Jy source in an integration time of 5 minutes.

Thus, many galactic OH and water masers will be observable with these limits. However, all known extragalactic OH mega masers are too weak to be sensible targets. In the case of extragalactic water masers, several targets are possible. These I have listed below, based on a compilation by Greenhill *et al*, 1990. These are important targets since we know so little about them; the one source observed with VLBI, NGC 3079, is unresolved with a resolution of 0.3 mas, (Haschick *et al*, 1990). Another important point about the list below is that all but one of the sources are in the southern sky, where no VLBI observations have so far been possible. A map of the component distribution in the Magellanic clouds masers would be of great interest.

Table 1. Extragalactic water maser targets for space VLBI.

Source	Luminosity (L_o)	Flux (Jy)	Distance (Mpc)	Velocity (Km/s)
LMC				
A0540-697	0.003	7.5	0.05	200
A0513-694	0.0008	15	0.05	200
SMC(S7)	0.002	7.4	0.06	200
Circinus	37	12	4.0	550
NGC 4595	85	12	4.0	650
NGC 3079	520	11	16.5	950

MASER PROPER MOTIONS

One of the most exciting proposals for space VLBI is the measurement of proper motions in extragalactic masers and hence the possible direct determination of distances to galaxies. From the list of galaxies in Table 1, we see an important target in the LMC maser, A0513-694 which has about a dozen features with flux densities in the range 2 to 15 Jy, (Whiteoak and Gardner, 1986). Let us examine the possibility of proper motion measurement in this source.
If, based on the measurements of the galactic centre maser (Reid *et al*, 1988), we assume a motion of about 1 mas per year for a galactic source at say 10 kpc, then we may expect for the Large Magellanic Cloud at approximately 50 kpc, a motion of 200 μas per year. Hence we require an accuracy of measurement of order 10 μas.
Now Greenhill*et al* (1990) have discussed the sources of error in the measurement of proper motions. We find that the accuracy of measurement, $\delta\theta$, is related to SNR as:

$$\delta\theta \sim 0.5 \, \theta_r / SNR \qquad (2)$$

where θ_r is the interferometer resolution which for a baseline of 10000 km is ~200 μas. Thus, for the required measurement accuracy of 10 μas, we require the SNR to be ~10. Using the equ 1 above, we find that for the Parkes 64m telescope and a 10 m space antenna, with parameters as listed, the SNR an the strongest (15 Jy) feature in A0513-694 is ~10 in a 100 sec integration. Once the reference feature has been detected, longer integrations can boost the SNR on the other features. Thus, it should be possible with, say Radioastron, to measure the proper motions of the LMC source over a time span of one year.

Moran *et al*, point out a second order systematic error on the measurement accuracy concerning the transfer of phase over an angle, $\Delta\theta$ between features separated in frequency by Δv, which is related to the error, ΔB, in the baseline, B and the delay error, $\Delta\tau$ as:

$$\delta\theta = (\Delta B/B). \Delta\theta + (\Delta v/v).(c\Delta\tau/B) \qquad (3)$$

where v is the observing frequency (22 GHz). Although the baseline and delay errors will be relatively large for space-ground baselines due to the uncertainty in orbit determination, if we assume that features separated in frequency be 2 MHz are separated by 1 arcsec, the terms in equ 2 amount to ~ 1 μas and 2 μas respectively, for expected errors ΔB=10 m and $\Delta\tau$=10 nanosec. i.e. they are well within the required accuracy.

CONCLUSIONS

Space VLBI observations of the OH and water masers will allow the properties of the interstellar scattering medium to be investigated. It should also be possible to investigate the size and feature distribution of Galactic masers within the limits imposed by scattering. For extragalactic water masers it will also be possible to measure limits on the size and feature distribution of a limited number of sources, including some mega masers and for at least one maser in the LMC, proper motion determinations will be possible, leading to an estimate of the distance to our nearest neighbour galaxy.

REFERENCES

Booth, R.S., 1984, in Proc of Workshop on Quasat, ESA SP-213, p 171.
Dennison, B.,Thomas, M., Booth, R.S., Brown, R.L., Broderick, J.J. & Condon, J.J. 1984, Astron. Astrophys., **135**, 199.
Dennison, B, Fiedler, R.L., Johnston, K.J., & Simon, R.L.,1991, in Proc. of Leningrad Workshop on Propagation Effects in Space VLBI.
Greenhill, L.J., Moran, J.M.,Reid, M.J., Gwinn, C.R., Menten, K.M., Eckart, A.& Hirabayashi, H. 1990, Ap. J., **364**, 513.
Gwinn, C.R., Moran, J.M., Reid, M.J.& Schneps, M.H. 1988, Ap. J., **330**, 817.
Haschick, A.D., Baan, W.A., Schneps, M.H., Reid, M.J., Moran, J.M., & Güsten, R. 1990, Ap. J., **356**, 149.
Reid, M.J., 1984, in Proc of Workshop on Quasat, ESA SP-213, p 181.
Reid, M.J., Schneps, M.H., Moran, J.M., Gwinn, C.P., Genzel. R., Downes, D. & Rönnäng, B., 1988, Ap.J., **330**, 809.
Schilizzi, R.T., Burke, B.F., Jordan, J.F. & Hawkyard, A. 1984, in Proc of Workshop on Quasat, ESA SP-213, p 13
Whiteoak, J.B. & Gardner, F.F., 1986, M.N.R.A.S., **222**, 513.

PRELIMINARY RESULTS FROM A STUDY OF THE PROPER MOTION OF WATER VAPOR MASERS IN CEPHEUS A.

T. V. Cawthorne, J. M. Moran and M. J. Reid.
Harvard-Smithsonian Center for Astrophysics
60 Garden Street
Cambridge, MA 02138

ABSTRACT

Preliminary results from a study of the proper motions of water vapor masers associated with HII regions in Cepheus A are presented. The masers lie in two distinct clumps near the peaks in intensity of the HII regions. Some 50 maser spots were detected over 3 epochs, but only 11 maser spots were identified at two or more epochs. A statistical parallax analysis from these preliminary results gives a distance between 0.4 and 0.6kpc with a statistical uncertainty of about 0.2kpc.

1. INTRODUCTION

VLBI observations of water vapor masers in Cepheus A were made at 3 epochs in 1981-82 separated by two equal intervals of 100 days using 5 antennas. Water vapor masers in Cepheus A have been mapped previously by Lada *et al.* and Cohen *et al.* (1984). Migenes *et al.* (1992) have recently published the results from a proper motion study of OH masers in Cepheus A from which they derive a distance of 320^{+140}_{-80}pc.

The radio source Cepheus A consists of a number of ultra-compact HII regions and the masers are associated with the brightest of these. There are 3 distinct peaks in continuum emission, and VLBI observations by Lada *et al.* (1981) first showed that the maser spots are associated with two of these (the Northern and Southern Peaks). Maser emission was detected also at a third (South-Western) peak by Cohen *et al.* (1984). In the observations presented here only the masers in the Northern and Southern clumps will be discussed.

2. RESULTS.

Fig. 1 shows the spatial distribution of maser spots from all three epochs. Each spot is represented by a symbol which indicates a range of radial velocities. The 1.8 mJy/beam contour from the continuum map of Hughes & Wouterloot (1984) is superimposed (heavy lines) to show approximate registration of the maser spots with the HII regions. The uncertainty in registration is at least 0.1 arcsecond.

The maser spots associated with the Northern HII region are more widely distributed than those associated with that in the South. They are also short lived compared with the Southern features: out of eleven maser spots detected at two or three epochs, only three were from the Northern clump. They also span a wider range in radial velocity than those in the South by a factor of roughly two.

237

A first attempt to obtain a distance for the source from these data makes use of the method of statistical parallax. The distance obtained from an O star determination quoted by Blaauw (1964) is 0.7kpc. Using measurements of all 11 maser spots detected at two or three epochs, we find the standard deviation of the angular speed $\sigma_\omega = 13.5 \pm 4\mu$as/day. The standard deviation of the radial velocity for all spots detected is $\sigma_{vz} = 9.4 \pm 1.4$km/s and the resulting distance is $D = 0.85\sigma_{vz}/\sigma_\omega = 0.6 \pm 0.2$kpc. in reasonable agreement with the value quoted by Blaauw. However, most of the range of radial velocity is due to components in the Northern clump, while 8 out of 11 maser spots for which proper motions were measured lie in the Southern cluster. If the calculation is repeated using only data from the southern cluster the distance obtained is 0.4 ± 0.15kpc.

Fig. 1. Distribution of maser spots from 3 epochs of observation. The 1.8mJy/ beam contour of continuum emission from Hughes and Wouterloot (1984) is shown.

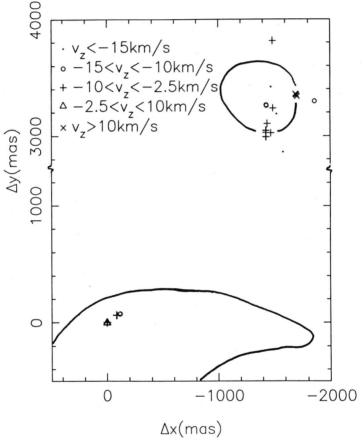

Blaauw, A. 1964, *Ann. Rev. Astr. Ap.*, **2**, 261.
Cohen, R.J., Rowland, P.R. and Blair, M.M. 1984, *M.N.R.A.S.*, **210**, 425.
Hughes, V.A. and Wouterloot, J.G.A. 1984. *Ap. J.*, **276**, 204.
Lada, C.J., Blitz, L., Reid, M.J. and Moran, M.J. 1981 *Ap. J.*, **243**, 769.
Migenes, V., Cohen, R.J. and Brebner, G.C. 1991, *M.N.R.A.S.*, **254,**, 501.

HIGH- AND LOW-VELOCITY MASER FEATURES

C.R. Gwinn
Physics Department, Broida Hall
University of California, Santa Barbara
Santa Barbara, CA 93106

ABSTRACT

Observations of 105 H_2O maser features in W49N indicate that high-velocity features lie at the periphery of the maser outflow. Model fits indicate that the high-velocity features lie at either end of a bipolar outflow. A high-velocity stellar wind may accelerate ambient material to produce the observed velocity field.

1. INTRODUCTION

The presence of large ranges of Doppler shift, particularly in H_2O masers, has puzzled investigators since the discovery of astrophysical masers. Proposed explanations have included:

- A single explosion produced fragments of different velocities (Baudry, Forster, & Welch 1974).
- The frequency shift is not due to Doppler velocity, but to Raman scattering (Radhakrishnan, Goss, & Bhandari 1975).
- Two distinct outflows with different outflow velocities are superimposed (Genzel et al. 1981).
- Outflow velocity depends on stellar latitude (Shu et al. 1991).
- All features have about the same velocity; features with low Doppler velocities are moving across the line of sight (Elitzur 1992).
- The outflow accelerates or decelerates with distance from the central star (Strelnitskii & Sunyaev 1973).

As discussed here, all but the last of these explanations can be ruled out.

The H_2O masers in W49N comprise the most populous and luminous such cluster in the Milky Way. W49N lies near the solar circle, $\frac{1}{4}$ of the way around the galaxy, in an active region of high-mass star formation. Emission from the $\lambda = 1.3$ cm transition of H_2O spans > 500 km s^{-1} in Doppler velocity, or about 300 times the speed of sound in the masing material (Elitzur, Hollenbach, & McKee 1989). The greatest concentration of features lie at low Doppler velocities, $-15 < v_z < +25$ km s^{-1} (Liljeström et al. 1989), or a range of 20 times sound speed.

2. OBSERVED STRUCTURE OF THE MASER OUTFLOW

The study of H_2O masers in W49N by Gwinn, Moran, & Reid (1992a,b,c) yields insight into the nature of high- and low-velocity maser features. Over 5 epochs, the authors measured 2423 positions of maser spots in the W49N H_2O maser cluster, from which they obtained proper motions of 105 maser features. The range of Doppler velocity v_z included the entire range of low-velocity emission and samples of red- and blue-shifted high-velocity emission, near $v_z \approx \pm 200$ km s^{-1}. Spot positions were determined to ≈ 10 μas, and proper motions were obtained from series of spot positions at 3 or more epochs. The observations yield 5 of the 6 degrees of freedom of the outflow: 2 components of position in the plane of the sky, (x, y), 2 components of proper motion (v_x, v_y), and the Doppler velocity, v_z. (See Gwinn et al. 1992a for a description of the coordinate system.)

The masers are seen to expand from a single common center, near the center of the maser cluster. The velocity distribution is bipolar, along an axis inclined 55° to the line of sight. This bipolarity appears in both velocity space and position, and corresponds to the tendency of red-shifted features to lie to the east, and blue-shifted to the west, noted by Walker, Matsakis, & Garcia-Barreto (1982). Maser features with high Doppler velocities tend to have high proper motions as well, while features with low Doppler velocities tend to have low proper motions.

The high-velocity features are spread over the ≈ 2" region of the maser cluster, indicating that they must lie throughout the distribution or at its periphery. Auto-correlation of positions of features shows that they tend to cluster together, in a scale-free or "fractal" fashion, over scales from ≈ 300 AU to < 1 AU (Gwinn et al. 1992b). Cross-correlation of positions of high- with low-Doppler velocity features shows that features from these different velocity ranges are associated on scales of ≈ 300 AU. This scale of clustering is much smaller than the $\approx 4 \times 10^4$ AU extent of the maser cluster.

3. MODEL-FITTING AND CLUSTER KINEMATICS

Models for the motions of the maser features indicate that the high-velocity features are on the outside of the cluster. Such models restore the unmeasured degree of freedom, the position along the line of sight, z, from the velocity and the observed components of position. The simplest models assume that the expansion is purely radial, and positions each maser along the line of sight so that its velocity points as nearly as possible toward the center of expansion. The model must also include parameters for the position and velocity of the center of expansion, the point toward which the motions of the masers point most accurately. For this model the direction of the velocity of the feature determines its z; the magnitude of the velocity does not affect the result.

The model fit places high-velocity features at the outer boundary of the cluster. However, the kinematic ages of the high-velocity features, given by their distances from the center of expansion divided by their outflow velocities, are actually smaller than ages of most low-velocity features. Thus the features cannot all have originated in a single explosion.

More general models reinforce placement of the high-velocity features on the edges of the cluster (Gwinn et al. 1992a). These models suppose particular functional forms for the variation of velocity with position within the cluster. Unlike the simple model discussed in the previous paragraph, such models can include rotation, and they provide more accurate values of z for masers near the line of sight to the center of expansion. These models are similar to, but more general than, those described by Bowers (1991).

Figure 1 shows the motions of the 105 maser features, from a viewpoint perpendicular to the outflow axis. The maser outflow has been deprojected using values of z from the model fit (Model 4 of Gwinn et al. 1992a), and rotated to orient the outflow

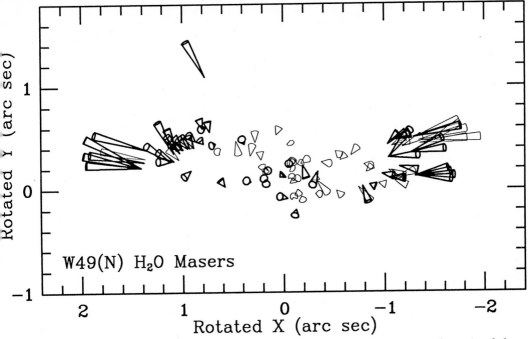

Figure 1: The W49N outflow of H_2O masers seen from perpendicular to the axis of the maser outflow. The distribution has been deprojected using a model fit, and rotated to the indicated viewpoint. Maser features at the apexes of the cones. Cones' lengths and inclinations show how far the masers would travel in 150 years at their present velocities. Heavy cones show blue-shifted features; light cones show red-shifted maser features.

axis perpendicular to the line of sight. The bipolar character of the outflow is apparent. The high-velocity features tend to lie at the ends of the outflow, at the preiphery of the maser cluster. As the model fit indicates, the outflow velocity is approximately constant at 17 ± 3 km s^{-1} from the origin to a radius of 1", where the outflow velocity increases to > 200 km s^{-1} in a distance of ≈ 0.3 arc sec. The masers appear to accelerate from $10\times$ to $100\times$ sound speed in a radial distance of 300 AU, at 10^4 AU from the central source.

4. DISCUSSION

High- and low-velocity features are physically different, although they probably represent two parts of a continuous distribution. High Doppler velocities correspond to high proper motions, indicating that Doppler velocity and proper motion have a common, physical, origin. High- and low-velocity features form in a single outflow, and are associated on scales much smaller than that of the outflow. These observations rule out all the explanations for high-velocity features mentioned in Section 1 except that of acceleration.

Strelnitskii & Sunyaev (1973) discussed mechanisms for accelerating H_2O masers in W49N, not long after the discovery of the broad range in Doppler velocity of maser emission. They consider acceleration by a stellar wind, by the "rocket" effect, and by radiation pressure from the central object. They conclude that acceleration by a stellar wind is the most likely cause of the high-velocity emission.

More recently MacLow and Elitzur (1992) have modeled our data as the interaction of a high-velocity stellar wind with ambient material. Where the ambient material is tenuous, along the axis of the outflow, the wind can accelerate it to high velocity, producing high-velocity maser features. Perpendicular to that axis, the ambient material is more dense, and a wind with the same momentum flux can accelerate it only to much lower velocity, producing low-velocity maser features. They achieve excellent agreement of the calculated velocity field with our observed velocity field. They suggest that the age of the outflow is only a few hundred years, the period in which the tenuous ambient material responsible for the high-velocity features would be swept away. If this is the case, we should be able to detect the resulting global changes of the outflow in a few more years of VLBI observations.

5. CONCLUSIONS

The high Doppler shifts observed for some maser features represent high physical velocities. High-velocity masers appear at the periphery of the maser cluster, but in association with low-velocity features. These observations strengthen theoretical pictures in which energetic mass outflow from the young star near the center of the maser outflow accelerates the high-velocity masing gas in situ.

6. ACKNOWLEDGEMENTS

The VLBI study of H_2O masers in W49N was completed through the cooperative efforts of many individuals over many years, including J.M. Moran, M.J. Reid, M.H. Schneps, and associates of the US VLBI Network. I thank M. Elitzur and M.-M. MacLow for many useful conversations. This work was supported by the National Science Foundation (AST-90-01426).

7. REFERENCES

Baudry, A., Forster, J.R., & Welch, W.J. 1974, A&A, 34, 217
Bowers, P.F. 1991, ApJS, 76, 1099
Elitzur, M. 1992, Astrophysical Masers, (Dordrecht: Kluwer), 269
Elitzur, M., Hollenbach, D.J., & McKee, C.F. 1989, ApJ, 246, 983
Genzel, R., Reid, M.J., Moran, J.M., & Downes, D. 1981, ApJ, 244, 884
Gwinn, C.R., Moran, J.M., & Reid, M.J. 1992a, ApJ, in press
Gwinn, C.R., Moran, J.M., & Reid, M.J. 1992b,c, in preparation
Liljeström, T., Mattila, K., Toriseva, M., & Anttila, R. 1989, A&AS 79, 19
MacLow, M.M., & Elitzur, M. 1992, ApJL, in press
Radhakrishnan, V., Goss, W.M., & Bhandari, R. 1975, Pramãna, 5, 51
Shu, F.H., Ruden, S.P., Lada, C.J., Lizano, S. 1991, ApJ, 370, L31
Strelnitskii, V.S., & Sunyaev, R.A. 1973, Sov Ast, 16, 579
Walker, R.C., Matsakis, D.N., & Garcia-Barreto, J.A. 1982, ApJ, 255, 128

PROPER MOTIONS OF OH MASERS IN W75N

V. Migenes and R. J. Cohen
University of Manchester
Jodrell Bank
Macclesfield, Cheshire SK11 9DL, U.K.

ABSTRACT

MERLIN measurements have been made between 1983 and 1989 to search for proper motions of OH masers in the source W75N. The 1665 MHz line was observed in left and right hand circular polarizations, with a velocity resolution of 0.3 km s^{-1}, and an angular resolution of 0.28 arcsec. After three epochs of measurements we have detected proper motions of typically a few milliarcseconds per year in about a third of the 1665 MHz masers. These proper motions are substantially larger than we would expect from the usually assumed distance of 2 kpc, and the observed spread in 1665 MHz radial velocities. There is no clear systematic pattern to the motions. However the distribution of the 1665 MHz masers down the western side of the HII region, together with the absence of absolute position information, means that the measurements do not rule out a bulk motion of the masers towards the west, such as might arise from simple expansion of the HII region. Further measurements are planned with the improved angular resolution of 0.16 arcsec provided by the new MERLIN 32-m telescope at Cambridge (UK). It is also planned to include the 1667 MHz masers in future measurements, as they have a more widespread distribution (G. Brebner 1988, PhD thesis, University of Manchester).

PROPER MOTION STUDIES OF H_2O MASERS

J.M. Moran, M.J. Reid
Harvard-Smithsonian Center for Astrophysics
60 Garden Street, Cambridge MA 02138, USA

and

C.R. Gwinn
University of California at Santa Barbara
Department of Physics, Broida Hall
Santa Barbara, CA 93106

ABSTRACT

The work of the CfA group over the past decade to measure the proper motions of H_2O masers in our Galaxy is described. The distances to 7 masers have been measured and the distance to the Galactic Center, R_o, is estimated to be 7.6 ± 0.6 kpc. Various technical aspects of the measurements are discussed here.

1. INTRODUCTION

The spectrum of a water vapor maser usually consists of many features spread over a velocity range of 10–300 km s^{-1}. By the mid-1970s, it was generally accepted that these velocities were kinematic in nature, due to physical motion of cloudlets of masing gas. However various non-kinematic mechanisms had been proposed to explain for the high velocity features such as Raman scattering (see Genzel et al 1981a). Strelnitskii and Syunyaev (1973), in a seminal paper, described the acceleration of maser clouds in the stellar wind of a newly formed massive star.

In the late 1970s, it became clear that the motions of masers should be measurable by VLBI techniques. Most images in that era were made from analyses of the fringe frequency spectrum of each velocity component. Positions of maser features could be measured to an accuracy of a few milliarcseconds, indicating that motions of ~ 30 km s^{-1} could be measured over a few years. However it was by no means obvious that attempts to measure proper motion would be successful, since masers were known to be rapidly time variable and it was not known whether masers would persist for a long enough time with sufficiently stable angular structure to allow for accurate proper motion measurements. The work of the past decade shows that most water masers do trace the motion of radially expanding flows. There is still no evidence for non-kinematic Doppler shifts.

2. ACCURACY OF MEASUREMENTS

The angular velocity of a clump of masing gas moving with transverse velocity V and distance D is

$$\theta = 206 \left(\frac{V}{\text{km s}^{-1}} \right) \left(\frac{D}{\text{kpc}} \right)^{-1} \mu\text{as yr}^{-1} \tag{1}$$

Hence a velocity of 35 km s^{-1} results in an angular velocity of 1 mas and 10 μas at distance of 7 kpc ($\sim$Galactic Center) and 700 kpc (nearby galaxy) respectively.

The synthesized beam of an intercontinental array has a width θ_r that is limited to about 300 μas. However, in the absence of systematic errors, the relative positions of masers can be estimated to an accuracy of about $\delta\theta \sim 0.5\theta_r/\text{SNR}$ where SNR is the signal-to-noise ratio. Only masers with reference features strong enough to be detectable within the interferometer's coherence time can be readily studied. With current instrumentation the coherence time is about 2 minutes and limiting sensitivity is about 1 Jy. There are about 300 known interstellar masers with flux densities exceeding this level (Cesaroni et al 1988). After a reference feature is detected, the SNR improves as the square root of the integration time and features as weak as $\sim$ 20 mJy can be detected. A prime source of systematic error is due to structural changes in the maser as a function of time. That is, two features with velocities separated by less than the linewidth and positions separated by less than the interferometric beamwidth cannot be readily distinguished and any change in their relative amplitudes will give rise to an apparent shift in position. In addition, although phase referencing removes geometric errors to first orders, second order systematic errors remain, as given by the equation

$$\delta\theta \sim \frac{\Delta B}{B} \Delta\theta + \frac{\Delta\nu}{\nu} \frac{c\Delta\tau}{B} \tag{2}$$

where $\Delta\theta$ and $\Delta\nu$ are that offsets from the reference feature in angle and frequency, respectively, B is the baseline length, ν is the frequency, c is the speed of light and ΔB and $\Delta\tau$ are the errors in the baseline length and instrumental delay. For ΔB and $c\Delta\tau \sim 10$ cm, $\Delta\theta = 2''$ and $\Delta\nu = 2$ MHz, $\delta\theta$ is below 1 μas. Hence geometric errors are significant primarily for the study of extragalactic masers. Note that the astrometric accuracy of space VLBI will depend critically on the accuracy of baseline determinations (e.g., Reid 1984).

3. STATISTICAL PARALLAX

If the motions of the maser cloudlets are assumed to be random and isotropic then the distance to the source can be estimated by the method of statistical parallax. If σ_μ is the dispersion in transverse angular velocities in one coordinate and σ_z is the the disperson in radial velocity then the distance estimate is given by

$$D = \frac{\sigma_z}{\sigma_\mu} . \tag{3}$$

Note that the observed dispersion in transverse angular velocity, $\sigma_{\mu_{\text{obs}}}$, will be larger than σ_μ because of the effects of error in the position measurements, such that

$\sigma_\mu^2 = \sigma_{\mu_{obs}}^2 - \sigma_m^2$, where σ_m is the measurement error in angular velocity. Hence an underestimate of σ_m leads to an underestimate in the distance. This dispersion correction is important primarily for extragalactic work. If σ_m is negligible then the accuracy of the statistical error in the distance estimate is given by the relation

$$\frac{\sigma_D}{D} \sim \frac{1}{\sqrt{2N}} \ , \tag{4}$$

where N is the number of maser spots. If the flow is not isotropic and random then this method can be biased. For example, if the maser cloudlets were flowing from a central point with constant flow velocity, V_o, but were confined to the hemisphere facing us, then the spread in angular velocity would be $2V_o/D$ and the spread in radial velocity would be V_o. Hence the distance estimate would be one-half the true distance. Comparison of the ambient velocity of the molecular cloud and the mean velocity of the maser features is an important diagnostic test for assessing the isotropy and randomness of the maser emission and the applicability of the statistical parallax method.

4. MODEL FITTING

When the masers show organized motion such as uniform outflow, their velocities can be fit to a parameterized model where a distance parameter relates angular and linear velocities. Such an analysis is not entirely straigtforward because it is non-linear and the line-of-sight positions for the features are unknown parameters. Hence an observation of 25 masers features (and a reference feature) yields 75 velocity-component measurements (and 50 coordinate measurements). A model of simple outflow at constant velocity has 7 global parameters and 25 line-of-sight position parameters. A key element in the solution is the estimation of initial parameters. For example, the systemic velocity and position can be estimated by plotting transverse velocity versus position (see Reid *et al* 1988a). The least mean square analysis involves both radial and transverse velocities, which are experimentally quite distinct quantities, and the solution depends on the relative weights assigned to them. The quantity to be minimized can be written

$$\chi^2 = \frac{1}{3N_m - N_p} \sum \left[\frac{\left(\dot{\theta}_i^2 - \frac{V_{mi}}{D} \right)^2}{\sigma_i^2} + \frac{(V_{z_i} - V_{zm_i})^2}{\sigma_{z_i}^2} \right] \tag{5}$$

where $\dot{\theta}_i$ and σ_i are the measured angular velocities and their errors, V_{zm_i} and σ_{z_i} are the measured radial velocities and their errors, and V_{m_i} and V_{zm_i} are the model velocities in the transverse and radial directions respectively. The weights are given by

$$\sigma_i^2 = \frac{\sigma_V^2}{D^2} + \sigma_{n_i}^2 \quad \text{and} \quad \sigma_{z_i}^2 = \sigma_V^2 \tag{6}$$

where σ_V is the rms turbulent velocity of the flow and σ_{n_i} is the measurement error. We assume that σ_V is greater than measurement errors in radial velocities. Note that we have an unusual situation where the weighting depends on the distance parameter. If the σ_{n_i}s can be calculated from the residuals in the position versus time analyses, then σ_V can be found by requiring that $\chi^2 = 1$. If the measurement noise in the proper motions is negligible with respect to the σ_V/D then the expression for χ^2 is substantially simplified.

5. RESULTS

The proper motions of eight Galactic masers have been analyzed. The results are summarized in Table 1. Statistical parallax measurements have also been made for the maser in IC 133 in the nearby galaxy M33 (see Greenhill 1992). It is interesting to note that the flow velocities fall in the narrow range of 18–45 km s^{-1}. The flow in W3(OH) is well collimated and the distance cannot be estimated reliably. The distance estimates for W51M, W51N, Sgr B$_2$N, Sgr B$_2$M, and W49N can be used to derive values for R$_o$, the galactic center distance, based on the model of circular rotation of the Galaxy at 220 km s^{-1}. The statistical average of these values is R$_o$ = 7.6 $\pm$ 0.6 kpc.

Table 1: Parameters of Galactic H$_2$O Masers Derived from Proper Motions

Source	Method[a]	Number of Maser Features	Flow[b] Velocity (km s^{-1})	D^c (kpc)	$R_o{}^d$ (kpc)	Ref.
Orion	M	26	18	0.5 ± 0.1	–	Genzel et al (1981a)
W51M	SP	27	–	7 ± 2	10 ± 4	Genzel et al (1981b)
W51N	SP	10	–	7 ± 2	10 ± 4	Schneps et al (1981)
Sgr B$_2$N	M	24	45	7.1 ± 1.5	7.1 ± 1.5	Reid et al (1988a)
Sgr B$_2$M	M	27	35	6.5 ± 1.5^e	6.5 ± 1.5	Reid et al (1988b)
W49N	M	105	18	11.4 ± 1.2	8.1 ± 1.1	Gwinn et al (1992)
W3(OH)	M	42	18^e		–	Alcolea et al (1992)
Cep A	SP	11	–	0.5 ± 0.2^e	–	Cawthorne et al (1992)

^{a}SP = statistical parallax; M = model fit. ^{b}Based on isotropic radial outflow. ^{c}Distance. ^{d}Distance between Sun and Galactic center. ^{e}Preliminary result.

REFERENCES

Alcolea, J., Menten, K.M., Moran, J.M., and Reid, M.J., 1992, these proceedings.
Cawthorne, T.V., Moran, J.M., Reid, M.J., 1992, these proceedings.
Cesaroni, R. et al 1988, Astr. Ap. Suppl., 76, 445
Genzel, R., et al 1981b, ApJ, 247, 1039.
Genzel, R., Reid, M.J., Moran, J.M., and Downes, D. 1981a, ApJ, 244, 884.
Greenhill, L.J. 1992, these proceedings.
Gwinn, C.R., Moran, J.M., and Reid, M.J. 1992, ApJ, in press.
Gwinn, C.R., Moran, J.M., Reid, M.J., and Schneps, M.H. 1988a, ApJ, 330, 817.
Reid, M.J. 1984 in *QUASAT–A VLBI Observatory in Space Proceeding of a Workshop at Gross Enzersdorf*, Austria, European Space Agency, p. 181.
Reid, M.J., Gwinn, C.R., Moran, J.M., and Matthews, A.H. 1988b, Bull. AAS, 20, 1017.
Reid, M.J., Schneps, M.H., Moran, J.M., Gwinn, C.R., Genzel, R., Downes, D., and Rönnäng, B. 1988a, ApJ, 330, 809.
Schneps, M., Lane, A.P., Downes, D., Moran, J.M., Genzel, R., and Reid, M.J. 1981, ApJ, 249, 124.
Strelnitskii, V.S. and Syunyaev, R.A. 1973, Soviet Astr.-AJ, 16, 579.

10. SCATTERING

INTERSTELLAR SCATTERING OF MASER RADIATION

James M. Cordes
Astronomy Department
Cornell University
Ithaca, NY 14853

ABSTRACT

Interstellar scattering of pulsars, extragalactic sources, and galactic masers is used to discuss the overall galactic distribution of scattering material and the implied angular scattering vs galactic coordinates. I discuss the prospects for observing diffractive and refractive scintillations of maser signals.

1. BASICS

Free electrons in the interstellar medium show a number density δn_e that varies by large amounts over a wide range of length scales, from $\sim 10^2$ km to 1 kpc. The smallest length scales are probed through diffractive interstellar scintillations and angular broadening (~ 100 km to 10^7 km); intermediate length scales (10^7 km to 10 AU) are probed by refractive variations, such as angular wandering and refractive scintillations; while large scale variations ($\gtrsim 0.01$ pc) are probed directly by telescopes with the requisite resolving power. Remarkably, the density variations are consistent with a Kolmogorov wavenumber spectrum (eg. Rickett 1990 & references therein).

Variations in refractive index $\delta n_r \approx -(r_e \lambda^2/2\pi)\delta n_e$ yield a phase perturbation $\phi(x_\perp) = (\omega/c)\int_0^D dz\, \delta n_r(x_\perp, z)$ that underlies a host of observable phenomena, including (1) 'seeing' effects, such as angular broadening (eg. Fey $et\ al.$ 1991) and angular wander (Gwinn, Moran, Reid, & Schneps 1988); (2) pulsar pulse broadening and pulse timing fluctuations (Cordes $et\ al.$ 1990); (3) intensity scintillations (diffractive and refractive; Rickett 1990). Measurements of many of these effects has led to a good first cut at understanding the galactic distribution of free electrons, the wavenumber spectrum of δn_e, and the role of the interstellar plasma in altering signals from radio sources of various kinds.

The wavenumber spectrum (q) is often modeled as a power-law with cutoffs,

$$P_{\delta n_e}(q) = C_n^2 q^{-\alpha}, \quad \frac{2\pi}{\ell_0} \le q \le \frac{2\pi}{\ell_1} \tag{1}$$

where α is empirically determined to be $11/3 \pm 0.3$, The spectral coefficient, C_n^2, varies by > 4 orders of magnitude. More closely related to observable quantities is the path-length integrated C_n^2, the $scattering\ measure$, $SM \equiv \int_0^D ds\, C_n^2(s)$. For example, measurement of the scattering diameter of an extragalactic source, θ_{FWHM}, yields $SM = (\theta_{FWHM}/128\,m.a.s.)^{5/3}\nu_{GHz}^{11/3}$ for SM in the 'user-friendly' units of kpc m$^{-20/3}$. In these relations, the value of SM is actually the integral of C_n^2 times a weighting function that depends on the kind of measurement. For angular broadening, $w = 1$ for extrgalactic sources, $w \approx x^2$ for galactic sources, where $x \equiv s/D$. For scintillation or temporal broadening measurements, $w = x(1-x)$.

2. GALACTIC DISTRIBUTION OF SCATTERING REGIONS

Study of SM for ~ 250 lines of sight indicates that the electron density consists of (1) an inner Galaxy component; (2) an outer Galaxy component; (3) and clumps of large n_e also having large fluctuations δn_e. Modeling of SM (Cordes et $al.$ 1991) shows that the outer Galaxy component has large scale height ~ 1 kpc and relatively low density, compared to the high density inner Galaxy component with scale height ~ 0.1 kpc. The 'fluctuation' parameter $\zeta = (\delta n_e/n_e)^2/f\ell_0^{2/3}$ (with $f = $ filling factor and $\ell_0 = $ 'outer scale' ≈ 1 pc) is 20 to 50 times larger in the inner Galaxy, compared to the outer Galaxy. The larger ζ in the inner Galaxy implies a much larger amount of angular broadening of sources viewed through it and sufficiently large amounts of pulse broadening that most radio pulsars on the far side of the Galaxy are unobservable as pulsating sources.

The axisymmetric model (in coordinates with respect to the galactic center) of Cordes et $al.$ (1991) may be used to predict the angular broadening for any direction. Figure 1 shows predicted broadening for $extragalactic$ $sources$ at $\nu = 1$ GHz. Angular broadening scales as $\nu^{-2.2}$. For galactic sources, the broadening is smaller because of the difference in path length but also because a correction for spherical waves must be made.

The scattering data, along with data on pulsar dispersion measures and distances (eg. Frail & Weisberg 1990), indicate that spiral arm structure is discernable. At present, the author and Joe Taylor are using all data to construct a new galactic model for free electrons that incorporates spiral arms. This model should improve both the pulsar distance scale and predictions of interstellar scattering.

3. SPECIAL LINES OF SIGHT

Several galactic sources are scattered by amounts much greater than predicted from the large scale galactic model, suggesting that there are small regions where scattering is enormously enhanced.

Galactic center sources, including Sgr A* (Lo et $al.$ 1985) and OH/IR masers (van Langevelde et $al.$ 1992), all show enhanced scattering, with angular diameters $\theta_{FWHM} \approx$ 0.5 to 3 arc sec at 1.7 GHz compared to modeled diameters ≈ 0.2 arc sec. The location of the region that causes the enhanced scattering toward these sources is controversial. Sufficient ionized material exists within the region to produce the needed SM, but this also requires a large fluctuation parameter, ζ. If the outer scale $\ell_0 \approx 1$ pc, however, the emission measure implied by the estimated SM is large enough to make the region optically thick at > 1 GHz. This conclusion holds if the region is near the galactic center such that geometrical deleveraging of the scattering is significant (eg. van Langevelde et $al.$ 1992). If, on the other hand, the scattering region is in the foreground, more modest SM and EM are implied, thus avoiding the 'absorption crisis.' Another way out would involve a much smaller outer scale ℓ_0 than is suggested for other regions.

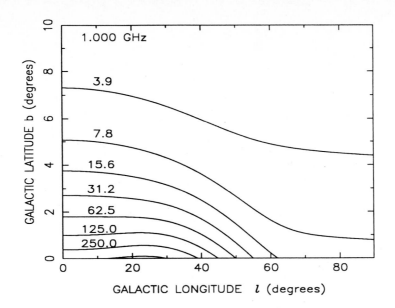

Figure 1. Predicted angular broadening for extragalactic sources in the plane-wave approximation and using the axisymmetric model of Cordes *et al.* (1991). Contours are labelled in milli-arc sec and scale as $\nu^{2.2}$.

Scattering data on other objects indicates that 'clumps' of intense scattering material are found in many regions of the Galaxy, including along the line of sight toward the Vela pulsar ($\ell = 264°$, $b = -2°.8$) and towards Cygnus X-3 at $\ell = 79°.8$, $b = 0°.7$ (Wilkinson *et al.* 1988). The HII complex NGC6334, broadens both an extragalactic source behind it, with $\theta_{FWHM} \approx 2.2$ arc sec at 1.7 GHz, and the OH masers within it, with $\theta_{FWHM} \approx 10^{-2}$ arc sec at 1.7 GHz (Moran *et al.* 1989). The ratio of these angular diameters is caused by geometrical leveraging if the scattering comes from, effectively, about 6.5 pc from the masers, about the size of the complex. NGC6334 is only about 1.7 kpc from the Sun, indicating that intense scattering regions need not require the special conditions that are encountered in the Galactic center. Indeed, a region like NGC6334 placed midway along the line of sight to the galactic center could fully account for the scattering—and its spatial variation— across the galactic center region.

4. MASER SCINTILLATIONS?

Radio pulsars show diffractive (DISS) and refractive scintillations (RISS). DISS has characteristic time and frequency scales at 0.4 to 2 GHz, $\Delta t_d \sim$ seconds to hours, $\Delta \nu_d \sim$ Hz to MHz, and is saturated with modulation index (rms/mean) $m_d = 1$ and a critical or isoplanatic angular size

$$\theta_{crit} \sim \frac{\ell_d}{D} \sim \frac{V_\perp \Delta t_d}{D} \lesssim 10^{-6} \text{arc sec.}$$

Extended sources, such as AGNs and masers will show quenched scintillations with modulation index

$$m_d \sim \frac{\theta_{crit}}{\theta_{source}} \lesssim \frac{0.1\%}{\theta_{source}(m.a.s.)}$$

assuming that the frequency resolution is $\lesssim \Delta \nu_d$. Such scintillations have been looked for and not found from AGNs and are probably not a viable explanation for the short term variability of OH masers seen by Clegg & Cordes (1991).

Refractive scintillations are more difficult to quench with finite source-size effects and, indeed, have been identified in AGNs as well as pulsars. The scales for pulsars are $\Delta t_r \sim$ days to years, $\Delta \nu_r \sim \nu$, and the modulation index and isoplanatic angle are

$$m_r \sim 10 - 20\%$$

$$\theta_{crit} \sim \theta_d \sim 1\,\text{m.a.s. to } 1 \text{ arc sec.}$$

The time scale of refractive scintillations is modified by the intrinsic source size for extended sources. There is all the reason to believe that masers should show refractive scintillations, though they have not been identified definitively.

REFERENCES

Armstrong, J. W., Cordes, J. M., & Rickett, B. J. 1981, *Nature*, **291**, 561.

Clegg, A.W. & Cordes, J.M. 1991, *ApJ*, **374**, 150.

Cordes, J.M., Wolszczan, A., Dewey, R. J., Blaskiewicz, M., & Stinebring, D. R. 1990; *ApJ*, **349**, 245.

Cordes, J.M., Weisberg, J.M., Frail, D.A., Spangler, S.R., & Ryan, M. 1991, *Nature*, **354**, 121.

Fey, A. L., Spangler, S.R., & Cordes, J.M. 1991, *ApJ*, **372**, 132.

Frail, D.A. & Weisberg, J.M. 1990, *AJ*, **100**, 743.

Gwinn, C.R., Moran, J. M., Reid, M. J., Schneps, M. H. 1988, *ApJ*, **330**, 817.

Lee, L.C. & Jokipii, J. R. 1976, *ApJ*, **206**, 735.

Lo, K.Y. *et al.* 1985, *Nature*, **315**, 124.

Moran, J.M., Rodriquez, L. F., Greene, B. & Backer, D.C. 1989, *ApJ*, **348**, 147.

Rickett, B. J. 1990, *Ann. Rev. Ast. Ap.*, **28**, 561.

Van Langevelde, H.J., Frail, D.A., Cordes, J.M., & Diamond, P.J. 1992, *ApJ*, in press.

Wilkinson, P.N., Spencer, R.E., & Nelson, R.R. 1988, in *Proc. IAU Symposium 129*, ed. J. Moran & M. Reid (Dordrecht: Reidel), 305.

A VLBI SURVEY OF INTERSTELLAR BROADENING OF GALACTIC OH MASERS

J. Hansen and R.S. Booth (Onsala Space Observatory)
B. Dennison (VPI) and P.J. Diamond (NRAO)

ABSTRACT

Results of a 7-station VLBI experiment to measure the apparent sizes of OH maser sources are presented. It is assumed that the sources are broadened by interstellar scattering, and that the sizes have implications for the distribution of the scattering medium.

INTRODUCTION

The effects of interstellar scattering on the sizes of interstellar maser spot components have been discussed by several authors (see e.g. Booth, 1984 and Gwinn et al, 1988), but there has never been a systematic attempt to measure their angular sizes as a function of Galactic coordinates. We have attempted to rectify the situation by observing a relatively large sample (39) of OH masers distributed in Galactic coordinates and over a range of distances. These measurements are useful in defining target sources for space VLBI observations, as well as for studying interstellar scattering.

THE OBSERVATIONAL RESULTS

The observations were carried out using a 7-station VLBI array consisting of the following telescopes in Europe: Effelsberg 100 m, Jodrell Bank 76m, Onsala 26 m, Westerbork 25 m, and in the USA: Green Bank 40 m, Haystack 37 m and Owens Valley 37 m. The source list contained both OH/IR masers which were observed at 1612 MHz and masers associated with HII regions which were observed at 1665 MHz. The Mk II VLBI system was employed and the bandwidth used was either 250 kHz or 62.5 kHz, the choice depending on the velocity extent of the individual spectra. Each source was observed for at least two periods of 15 minutes duration to obtain a reasonable spread of baseline lengths. In addition, regular observations of compact continuum sources were made for delay calibration. Single dish spectra were also measured with the Effelsberg telescope.

After correlation at the MPI correlator in Bonn, the visibility amplitudes of the strongest spectral features were fitted with a single Gaussian function of baseline length on the assumption that the point maser features were simply broadened by scattering. This naive procedure produced reasonable fits on 20 of the observed sources but failed in several cases, presumably due to blends of features in individual spectral channels with insufficient resolution. Such blending occurred for some of the OH/IR sources at 1612 MHz and in at least two cases, OH 104.9+2.4 and OH 25.6+0.6, where strong features exhibited complex visibility curves on the shorter baselines, there was still correlated flux on the very longest baselines indicating component sizes of order 5 mas. Thus we must have reservations about the meaning of at least some of the 1612 MHz data in terms of a scattering interpretation. This will be discussed further in a forthcoming paper (Hansen *et al*, in prep).

TABLE 1

Source	Type*	Galactic longitude	Galactic latitude	Velocity (km s^{-1})	Angular size (mas)	Distance (Kpc)
OH12.2-0.1	Type I	12.22	-0.13	29.0	25.6±2.6	16.1 a
W 33a	Type I	12.91	-0.26	39.6	3.8±0.5	4.4 a
				37.4	3.5±0.5	
OH20.1-0.1	Type I	20.08	-0.14	46.5	41.1±0.4	4.1 a
				45.3	40.7±0.5	
OH25.1-0.4	Type II	25.06	-0.36	154.8	61.0±6.3	9.1 b
				129.7	74.2±7.0	
OH28.5-0.0	Type II	28.52	-0.02	120.6	38.6±2.5	8.8 b
				93.5	43.2±3.4	
OH32.8-0.3	Type II	32.83	-0.32	74.0	31.9±3.0	5.0 c
				45.0	36.3±2.6	
OH35.2-1.7	Type I	35.19	-1.74	41.7	2.4±0.3	2.9 a
				43.5	0.8±0.5	
OH39.7+1.5	Type II	39.72	1.48	34.8	25.4±2.5	1.5 c
				2.9	30.1±3.0	
OH40.6-0.2	Type I	40.62	-0.15	32.4	21.0±1.0	2.3 a
OH43.8-0.1	Type I	43.80	-0.14	41.4	8.0±1.0	2.7 a
OH44.8-2.3	Type II	44.79	-2.32	-88.2	36.6±5.0	6.0 d
OH45.5	Type I	45.47	0.05	59.0	3.0±1.0	9.7
W 75S	Type I	81.72	0.57	0.5	4.0±0.3	2.5
Cep A	Type I	109.87	2.11	-13.8	3.0±0.7	0.73
N 7538	Type I	111.54	0.78	-59.5	3.7±1.0	3.5 e
OH127.8-0.0	Type II	127.82	-0.02	-66.4	9.7±0.3	2.9 c
W 3	Type I	133.95	1.07	-44.3	3.5±0.5	2.2
				-46.4	4.4±0.4	
OH138.0+7.3	Type II	137.97	7.26	-47.4	7.9±0.2	2.4 b
				-28.9	8.4±0.2	
OH141.7+3.5	Type II	141.73	3.52	-69.8	10.1±0.4	3.7 b
Mon R2	Type I	213.74	-12.59	11.2	3.4±0.3	0.8 e

*)   Type I sources are associated with HII regions.
       Type II sources are associated with OH/IR stars.

Distances taken from a: Caswell and Haynes (1983), b: Bowers et al (1980), c: Langvelde et al. (1990), d: Fix and Mutel (1984), e: Genzel and Downes (1977).

ANGULAR SIZE - DISTANCE RELATION

The 19 sources for which we judge the Gaussian fit to be good are listed in Table 1, together with their Galactic coordinates, distance and the velocity and angular size of the spectral feature(s) which were used for the fit. Type I sources are main line masers associated with HII regions and type II are OH/IR star masers observed at 1612 MHz.

A plot of the angular size v. distance is shown in Fig. 1. The plot shows considerable scatter although there is certainly a trend for the more distant sources to have the larger angular sizes.The solid line shows the best fit to the data. It is of the form

$$\theta \propto D^{0.65}$$

close to the expected scaling with distance for homogeneous turbulence with a Komolgarov power law spectrum, but this is probably fortuitous since the scatter in the data is so great. Perhaps a better indication of the meaning of the data can be derived from the size in relation to Galactic coordinates seen in Table 1. Here we see several sources in the distance range 2 to 5 kpc whose angular sizes range over a factor of more than 10 (from 2 to 40 mas). Even sources close in galactic longitude and at similar distances have rather different sizes. Thus we may conclude that the scattering material is unevenly distributed.

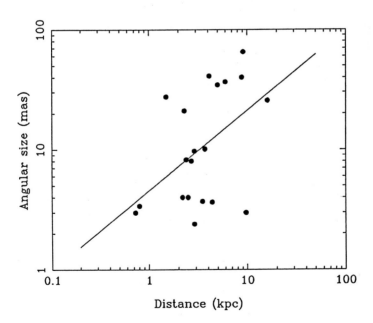

Fig. 1.   A plot of the angular size v. distance for the OH masers observed. The solid line is the best fit to the data on the assumption that there is a simple power law relationship, and has a slope of 0.65. i.e. $\theta \propto D^{0.65}$.

Dennison *et al* (1984) from VLBI studies of low latitude extragalactic radio sources suggested that heavy scattering is confined to the inner Galaxy, being caused by a clumpy medium with the same scale height as the population I objects (z = 80 pc). Similarly, from observations of pulsars, Cordes *et al* (1985), confirm that the interstellar medium has a highly inhomogeneous component of low scale height (z < 100 pc), and another, more homogeneous component of larger scale height, but found it difficult to reach a conclusion on whether or not the low scale height component was confined to the inner Galaxy.

Our results imply that the largest scattering takes place in the inner Galaxy. All sources with l > 45º have angular sizes less than about 10 mas, with a mean of ~ 5 mas. However, more than half of this group of the masers are situated ± 1º from the galactic plane and so we are unable to be more precise than CordeS at al.

CONCLUSIONS

A systematic study of OH masers close to the galactic plane shows, in common with previous observations, that there is a trend for increased size with distance. This is assumed to be the result of interstellar scattering. However, the scattering medium is inhomogeneous and strong scattering seems to be confined to the inner Galaxy. These results show that higher resolution observations, e.g with space VLBI, could be useful for studies of high latitude masers in the Galaxy, and for extragalactic masers, where scattering is not too great.

REFERENCES

Booth, R.S., 1984, in Proc of Workshop on Quasat, ESA SP-213, p 171.
Bowers, P.F., Reid, M.J., Johnston, K.J., Spencer, J.H.& Moran, J.M. 1980, Ap.J., **242**, 1088.
Caswell, J.L.. & Haynes, R.F., 1983, Aus. J. Phys. **36**, 417.
Cordes, J.M., Weisberg, J.M. & Boriakoff, V. 1985, Ap.J., **288**, 221.
Dennison, B., Thomas, M., Booth, R.S., Brown, R.L., Broderick, J.J. & Condon, J.J. 1984, Astron. Astrophys., **135**, 199.
Fix, J.D.. & Mutel, R.L., 1984, Astron. J., **89**, 406.
Genzel, R. & Downes, D., 1977, Astron. Astrophys. Suppl., **30**, 145.
Gwinn, C.R., Moran, J.M., Reid, M.J. & Scneps, M.H., 1988, Ap.J., **330**, 817.
van Langevelde, H.J., van der Heiden, R. & van Schooneveld, C., 1990, Astron. Astrophys. **239**, 193.

ANGULAR WANDER MEASUREMENTS OF MASER CLUSTERS

Robert L. Mutel

Department of Physics and Astronomy
University of Iowa
Iowa city, Iowa 52242

ABSTRACT

Angular wander measurements of the relative positions of closely spaced maser features provides a powerful probe of interstellar turbulence associated with regions of star formation. Differential angular wander is easily measured in a maser complex and can strongly distinguish between shallow and steep power-law turbulence. The best candidates for such measurements appear to be the 6 and 12 GHz type II methanol masers.

1. INTRODUCTION

Interstellar plasma turbulence along the line of sight to compact radio sources causes a variety effects on the received radiation, including intensity fluctuations, angular broadening, and angular wander of the centroid of the broadened image. These fluctuations in the phase and amplitude of the received wave are related to the path integral of the strength and spatial spectrum of turbulence along the line of sight. Based both on theoretical considerations and *in situ* measurements of atmospheric and interplanetary turbulence, a power-law is often assumed. The usual form is

$$P_n(q) = C_n^2 q^{-\alpha} \quad q_0 < q < q_1 \tag{1}$$

where q is the spatial wavenumber, C_n^2 is the 'strength of turbulence', and q_0, q_1 are the wavenumbers corresponding to the largest and smallest turbulent eddies (outer and inner scale) respectively.

Historically, the study of interstellar scattering of maser sources has focussed on angular broadening observations (see papers by Cordes and Booth in these proceedings). These observations have been used primarily to determine the line integral of the strength of scattering, called the scattering measure

$$SM = \int_0^D ds \, C_n^2(s) \tag{2}$$

For angular broadening dominated by diffractive scattering ($2 < \alpha < 4$, 'shallow' turbulence), the FWHM broadening angle is related to the scattering measure by

$$\theta_{FWHM} = 10^{\frac{6(\alpha-4)}{(\alpha-2)}} \left[f(\alpha) \lambda_{cm}^\alpha SM \right]^{1/(\alpha-2)} \quad \text{mas} \tag{3}$$

where the normalization function is

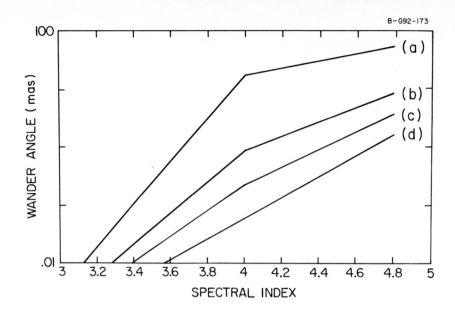

Figure 1 Differential angular wander as a function of spectral index α for a pair of closely spaced compact sources at (a) 1.66, (b) 6.6, (c) 12.2, and (d) 22.3 GHz, where we have assumed $SM = 1$ kpc-cm$^{-(20/3)}$ and a screen distance $z = 5$ kpc and component separation 1 arcsec at the screen.

$$f(\alpha) = \frac{2^{(1-\alpha)}\Gamma(\frac{4-\alpha}{2})\Gamma(\frac{\alpha-2}{2})}{\Gamma(\frac{\alpha}{2})^2} \tag{4}$$

and SM is in units of kpc-cm$^{-(3+\alpha)}$. A recent tabulation by Cordes *et al.* (1992) of 14 hydroxyl maser sources at 1.6 GHz shows values $10^{-1.5} < SM < 10^1$ kpc-m$^{-20/3}$ using an assumed spectral index $\alpha = 11/3$.

2. DIFFERENTIAL ANGULAR WANDER

Angular wander of the centroid of the diffraction disk occurs on a refractive timescale $t \sim 2z\theta_d/v$ where z is the screen distance, v is the bulk velocity across the line of sight, and θ_d is the scattering disk size[a]. This is typically weeks to years at centimeter wavelengths. Several recent papers have developed models for angular wander (e.g. Cordes *et al.* 1986; Romani *et al.* 1986; Rickett and Coles 1988). Fey and Mutel (1992) compare these models and find that they all give similar results aside from a normalization function.

---

[a] The diffractive scattering size θ_d is related to the measured FWHM scattering size by $\theta_d = \theta_{FWHM}/1.67$

For shallow turbulence the RMS *differential*[b] wander angle between two closely spaced components can be written (Cordes *et al.* 1986; Fey and Mutel 1992)

$$\Delta_{24}\theta = g(\alpha)\left(\frac{z}{\lambda}\right)^{\frac{\alpha-4}{2}}\theta_d{}^{\alpha-3} \tag{5}$$

where

$$g(\alpha) = \frac{2^{(\alpha-1/2)}\Gamma(\alpha/2)(2\pi)^{(\alpha/2-2)}}{[(4-\alpha)\Gamma(\alpha/2-1)\Gamma(2-\alpha/2)]^{\frac{1}{2}}} \tag{6}$$

For the steep spectrum case $(4 < \alpha < 6)$, the scattering disk is not well defined since it has a fractal geometry (e.g. Narayan and Goodman 1989). However, for any given observation, the centroid is well defined, as is the diffraction scale r_d, *viz* the $1/e$ halfwidth of the visibility modulus. The equivalent diffraction disk is then $\theta_d = 1/kr_d$. The RMS wander will be extemely large for long time separations (up to $t \sim L_0/v$) but the differential angular wander can be shown to be (Fey and Mutel 1992)

$$\Delta_{46}\theta = h(\alpha)\left(\frac{r_{comp}}{r_{mp}}\right)^{\frac{\alpha-4}{2}}\theta_d \tag{7}$$

where r_{comp} is the linear separation between lines of sight to each component at the effective screen, $r_{mp} = 2z\theta_d$ is the multipath scale , and the normalization is

$$h(\alpha) = \frac{\left[2^{(7-2\alpha)}(\alpha-4)\Gamma(\frac{6-\alpha}{2})\Gamma(\frac{\alpha-4}{2})\right]^{\frac{1}{2}}}{\Gamma(\frac{\alpha-2}{2})} \tag{8}$$

The differential angular wander saturates after $t = r_{mp}/v$ because for larger turbulent scales both components move in unison.

3. OBSERVATIONS OF ANGULAR WANDER

There have been few attempts to measure angular wander to date, probably because of observational difficulties: sources need to have closely spaced compact components and must be observed several times over a period of months to years. Maser clusters satisfy the first criterion, so long as intrinsic motion of the maser spots can be accounted for.

---

[b] The differential angular wander is simply $\sqrt{2}$ times the angular wander of a single component if the linear separation between lines of sight at the screen exceeds the multipath scale $r_{mp} = 2z\theta_d$.

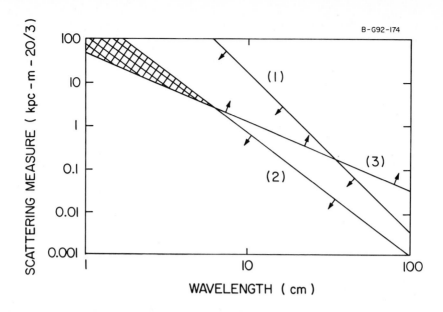

Figure 2 Scattering Measure versus observing wavelength for $\alpha = 11/3$, $z = 1$ kpc and observing constraints (1) Refractive timescale $t < 10^8$ sec, (2) $\Delta\theta/\theta_d > 0.2$, and (3) Wander resolvable on Earth baseline, $(\Delta\theta > 0.1\lambda_{cm}$ mas). Note that (2) is always more restrictive than (1). The optimal area for SM as a function of λ is the shaded area. *N.B.* This area will vary significantly with both α and z.

Observations of angular wander at radio wavelengths have typically resulted in upper limits only. Backer (1988) looked for angular wander in a series of 5 GHz VLA maps of the heavily scattered source Sgr-A. He found no wander at a level < 1 mas. Likewise, Gwinn *et al.* (1988) found a upper limit of 18 μarcsec to deviations from rectilinear motion of water maser spots in Sgr-B2. Finally, Fey and Mutel (1992) report an upper limit of 4 mas for differential angular wander between components of the compact double source 2050+364 at 0.61 GHz over 4 years. These upper limits are generally consistent with shallow power-law turbulence with $\alpha \lesssim 11/3$ and appear to rule out steep power-law spectra.

4. OPTIMIZING MEASUREMENTS OF ANGULAR WANDER

In order to design an optimal observational test of angular wander, several constraints must be met. For specificity, we will choose $\alpha = 11/3$ and $v = 100$ km-s^{-1} in the following discussion. The dimensions of SM, λ, and z are kpc-m$^{-20/3}$, cm, and kpc respectively.

1. The refractive timescale cannot be so long that the observer loses interest in repeated observations, say less than 10^8 seconds ($\sim$3 years). This implies

$$\theta_{FWHM} \lesssim 50z^{-1} \text{ mas} \quad \text{or} \quad SM^{0.6}\lambda^{2.2} \lesssim 850z^{-1} \tag{9}$$

2. The RMS differential wander angle must be a reasonable fraction of the diffractive scattering disk, e.g. $\Delta\theta \gtrsim 0.2\theta_d$. This implies

$$\theta_{FWHM} \lesssim 0.5(\lambda/z)^{0.5} \text{ mas} \quad \text{or} \quad SM^{0.6}\lambda^{1.7} \lesssim 40z^{-0.5} \qquad (10)$$

This is almost always more restrictive than the timescale requirement (1).

3. The wander angle must be resolvable using terrestrial baselines, say $\Delta\theta > 0.1\lambda_{cm}$ mas. This implies

$$\theta_{FWHM} \gtrsim 0.7z^{0.25} \text{ mas} \quad \text{or} \quad \lambda^{1.6}SM > 50z^{0.42} \qquad (11)$$

These constraints are plotted on figure 2. It is clear that in order to observe a wander angle large enough to be observable with terrestrial baselines, but avoid $\Delta\theta << \theta_{FWHM}$, the observing wavelength should be shorter than 10 cm and the scattering measure $SM \gtrsim 1$ kpc cm$^{-20/3}$. While some H_2O masers at 22 GHz certainly satisfy this requirement, a better choice might be the Type II methanol masers at 6.6 and 12.2 GHz. Recent high resolution observations suggest that both transitions are compact and nearly spatially coincident, have relatively simple spatial structure, and arise from the same physical region (Norris, this conference). They also appear to be less variable than H_2O masers, perhaps indicating fewer intrinsic kinematic motions. If both transitions are scattered by the same turbulent screen, they allow a stringent test of the wavelength dependence of both the scattering disk and angular wander, since both quantities have different $\lambda(\alpha)$ dependencies.

REFERENCES

Backer, D. C. 1988, in *Radio Wave Scattering in the Interstellar Medium*, Ed. J.M. Cordes, B.J. Rickett, and D.C. Backer, A.I.P. Conference Proceedings 174 (New York:AIP)

Cordes, J. M., Pidwerbetsky, A., and Lovelace, R. V. E. 1986, *Astrophys. J.*, **310**, 737.

Cordes, J.M., Spangler, S.R., Weisberg, J.M., and Ryan, M. 1992, *preprint*

Fey, A. and Mutel, R.L. 1992, *submitted to Ap.J.*

Gwinn, C., Moran, J., Reid, M., and Schnepps, M. 1988, *Astrophys. J.*, **330**, 817.

Narayan, R. and Goodman, J. 1989 *Monthly Notices Roy. Astron. Soc.*, **238**, 963.

Rickett, B. and Coles, W. 1988 in *IAU Symposium 129, The Impact of VLBI on Astrophysics and Geophysics*, ed. J. Moran and M. Reid (Dordrecht: Reidel), p.287.

Romani, R. W., Narayan, R., and Blandford, R. 1986, *Monthly Notices Roy. Astron. Soc.*, **220**, 19.

11. VARIABILITY

THE POLARIZED WATER MASER SOURCE IN ORION

Zulema Abraham
Instituto Astronômico e Geofísico
Universidade de São Paulo
CP 9638, 04301 São Paulo, SP, Brazil

J.W.S.Vilas Boas
Centro de Radioastronomia e Aplicações Espaciais
CP 8174, 05558 São Paulo, SP, Brazil
and
Harvard-Smithsonian Center for Astrophysics
Cambridge, MASS 02138, USA.

ABSTRACT

We report the time behaviour of the highly polarized 8 km s^{-1} water maser source in the Orion nebula. During the last few years the flux density was low and it presented three small outbursts. In each case, although the total intensity was of the same order of magnitude, the degree of polarization and polarization angle were different, as well as their variation across the line profile. In November of 1991, twelve years after its first detection, the source disappeared or became too weak to be detected.

1. INTRODUCTION

By the end of 1979 a new and very strong water maser source, characterized by a very high degree of linear polarization was discovered in the Orion nebula. During a long time this was the strongest feature in the cloud, presenting four well defined flares in the intensity and variability in the degree of polarization and position angle of the polarization plane. The line width was 40-60 kHz, that is, less than 1 km s^{-1} (Matveenko et al. 1980, Abraham et al. 1981). VLBI observations showed, however, that the source was formed by several components, with slightly different velocities, each with well defined degree of polarization and polarization angle but variable intensity

(Matveenko et al. 1988). The behaviour of the source up to 1988, obtained from single dish observations was reported by Abraham et al. 1981, 1986, Vilas Boas and Abraham 1988 and Garay et al. 1989. Due to the high degree of linear polarization, single dish observations allowed the study of the source even when its intensity became lower than some of the other features in the spectrum, as was the case in the last few years. In this paper we report observations during the low intensity phase, since 1988 until its complete disappearence in November 1991.

2. OBSERVATIONS

The observations were made with the 13.7 m radome enclosed Itapetinga radiotelescope. The feed consisted of two rectangular horns, separated by 18' in azimuth and sensitive to the vertical and horizontal components of the E vector respectively. The receiver was a K-band room temperature mixer. The signal passed through either a 100 kHz or a 15 kHz filter, in the last case the central frequency was changed several times to cover the whole line.

3. RESULTS AND DISCUSSION

The time behaviour of the 8 km s^{-1} water maser source in Orion is shown in Fig. 1, where (a) represents the total antenna temperature, (b) the degree of polarization and (c) the position angle of the polarization plane. Since 1987 the total intensity remained very low, but presented at least three flares, as can be seen in the insert of Fig.1(a). The variation of the degree of polarization and polarization angle across the line was different during the three flares. In November 1991, the polarized source disappeared, or it become weaker than the other sources with the same velocity. The antenna temperature measured with the 100 kHz filter centered at 7.8 km s^{-1} was 40 K, with no polarization, the nearby non-polarized line centered at 10 km s^{-1} had an antenna temperature of 250 K.

The interpretation of this unusually polarized source has changed along the time as new data become available and refined theoretical calculations of the polarization of astronomical masers were developed.

Abraham et al. (1986) proposed a model in which the quasi-periodic variation in the polarization angle was explained by the beamend radiation of a saturated maser precessing along with the magnetic field within a small angle. Matveenko et al. (1988) based in VLBI observations interpreted the spectra as due to the directional radiation of a family of concentric rotating rings. The variation of the polarization angle with the velocity would be a consequence of the correlation between the radial velocity and the distance from the source to the center of the ring.

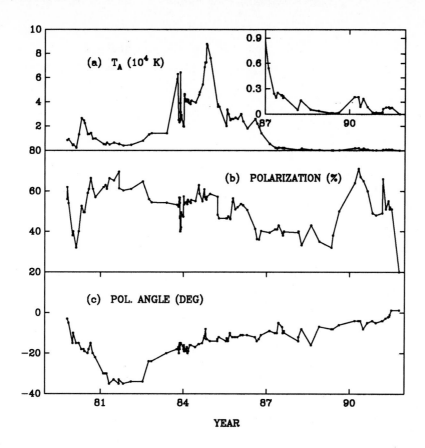

Fig.1 Antenna temperature, degree of polarization and polarization angle of
the 8 km s^{-1} water maser line in Orion.

In Fig.2 we plotted our observations of the polarization angle along the line profile
for the three small outbursts seen in the insert of Fig. 1. We found that, except for
the wing of the maser line at 6.9 km s^{-1} in 5 Apr. 1988, all the points lie at higher
polarization angles, well above the expected line. It is possible then, that the main
central component of the source had become very weak, or had already disappeared in
1988, and what was seen during the last few years were other components, at a position
where the polarization angle was about 0-5°.

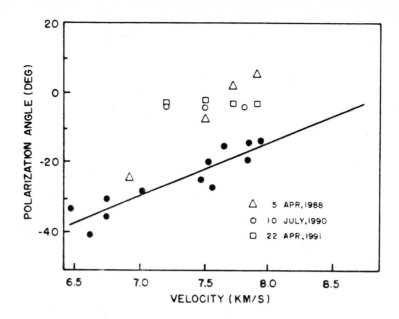

Fig. 2 Position angle of the polarization plane as a function of velocity in the line profile. Full circles are from Matveenko et al. (1988).

Aknowledgments: This work was partially supported by the brazilian agencies CNPq and FAPESP.

REFERENCES

Abrahamn Z., Cohen, N.L., Raffaelli, J.C., Zisk, S.H. 1981, Astr.Ap.Lett. **100**, L10.
Abraham, Z., Vilas Boas, J.W.S., del Ciampo, L.F. 1986, Astr.Ap. **167**, 311.
Garay, G., Moran, J.M., Haschick, A.D., 1989, Ap.J. **338**, 244.
Maatvenko, L.I., Kogan, L.R., Kostenko, V.I. 1980, Sov. Astron. Lett. **6**, 279.
Matveenko, L.I., Graham, D.A., Diamond, P.J. 1988, Sov. Astron. Lett. **14**, 468.
Vilas Boas, J.W.S., Abraham, Z. 1988, Astr.Ap. **204**, 239.

VARIATIONS OF OPTICAL AND WATER MASER EMISSION
FROM LONG PERIOD VARIABLES

Priscilla J. Benson[2] and Irene R. Little-Marenin[1,2]
[1]CASA, U. of Colorado, Boulder, CO 80309-0391
[2]Whitin Observatory, Wellesley College, Wellesley, MA   02181

Robert R. Cadmus, Jr.
Department of Physics, Grinnell College, Grinnell, IA   50112

ABSTRACT
We have monitored the 22 GHz H_2O maser line from 22 LPVs since 1987. All stars show highly variable water maser emission with most Miras having a phase lag of 0.3 phase between the water maser and optical emission. VX UMa appears to show variations associated with a rotating circumstellar disk.

1. INTRODUCTION

In order to provide observational constrains for the pumping mechanism for the 22 GHz water maser emission in long-period variable stars (LPVs), we have monitoring the 6_{16}-5_{23} transition from 17 Miras and 5 semi-regular (SR) variables (see Table 1). Over half of the stars have been monitored since 1987 with the 37m telescope at Haystack Observatory at approximately monthly intervals (1K=10 Jy before 1989 and 8 Jy after that). The photometric light curves of RX Boo and V CVn were obtained with the 0.61 m telescope of the Grant O. Gale Observatory of Grinnell College.

2. RESULTS and CONCLUSIONS
All stars show highly variable water maser emission. For Mira variables we typically see a phase lag of 0.3 phase between maximum water maser emission and the optical maximum, independent of the period of the star (Table 1). This phase lag corresponds to a time lag of about 50 to 100 days.

The water maser emission of R Ceti (Mira, P=166d) clearly tracks the visual light curve with a phase lag of 0.30 (50d) (Fig 1a). The LSR velocity of the feature has remained fairly constant ($\Delta v < 0.1$ km s^{-1}) except when very weak. The 22 GHz emission of T UMa (Mira, P=257d) has a phase lag of 0.30 (80d)

TABLE 1
STARS IN THE OBSERVING PROGRAM

Name	RA (1950)	Dec (1950)	Period	SiO	OH	observe	Phase lag
o CET	02 16 49	-03 12 13	M-332	Y	N	01/90-pre.	0.30 (100^d)
S PER	02 19 15	+58 21 36	SRc-822	Y	Y	04/87-pre.	....
R CET	02 23 29	-00 24 12	M-166	Y	Y	11/86-pre.	0.30 (50^d)
R TRI	02 34 00	+34 02 54	M-267	Y	N	11/88-pre.	no H_2O
U ORI	05 52 51	+20 10 07	M-368	Y	Y	05/90-pre.	0.27 (100^d)
LO AUR	05 53 35	+48 22 23	M- ?	..	..	08/89-pre.	....
R LEO	09 44 52	+11 39 42	M-310	Y	Y	02/90-11/90	no H_2O
VX UMA	10 52 08	+72 08 12	M-215	..	..	11/88-pre.	....
T VIR	12 12 02	-05 45 29	M-339	Y	Y	11/86-pre.	0.16 (55^d)
T UMA	12 34 07	+59 45 44	M-257	Y	N	12/86-pre.	0.30 (80^d)
RS UMA	12 35 42	+58 45 30	M-259	..	N	12/88-pre.	0.05: (15^d)
V CVN	13 17 17	+45 47 24	SRa-192:	N	Y	04/87-pre.	0.83 (160^d)
W HYA	13 46 12	-28 07 06	M-361	Y	Y	06/90-pre.	0.0:
R BOO	14 34 59	+26 57 12	M-223	N	N	11/88-pre.	....
RX BOO	14 21 57	+25 55 48	SRb-165:	Y	N	04/87-pre.	0.16:($\sim$$30^d$)
R SER	15 48 23	+15 17 02	M-357	Y	N	04/87-07/89	no H_2O
RZ SCO	16 01 36	-23 57 54	M-157	Y	N	04/87-11/90	0.37: (60^d)
R DRA	16 32 31	+66 51 30	M-246	N	N	11/88-pre.	0.20 (50^d)
Z CYG	20 00 02	+49 54 12	M-264	Y	Y	11/88-pre.	0.30 (80^d)
AC CYG	20 11 21	+49 17 54	SRb-142	..	N	04/87-pre.	....
Z AQL	20 12 31	-06 18 18	M-129	..	N	12/87-pre.	0.50: (65^d)
SV PEG	22 03 17	+46 30 06	SRb-145	Y	N	11/86-pre.	....

(Fig 1b). The LSR velocity of the -88 km s^{-1} feature of T UMa shows only small, random variations ($\Delta v < 0.2$ km s^{-1}). The 22 GHz spectra of LO Aur (listed as a Mira in the GCVS) show features that extend over 15 km s^{-1}, typical of supergiants or semi-regular variables.

The three 22 GHz features of VX UMa, extending over 4 km s^{-1}, show periodic variations with P=380d-400d, distinctly different from the 215d period listed in the GCVS (Fig 1d). The velocity of the -52 and -50 km s^{-1} features also shows 380-440 day periodic variations with an amplitude of about 0.25 km s^{-1}. We suggest that these variations are due to the rotation of a circumstellar disk (Cesaroni 1990).

The response of the water maser emission in semi-regular variables (SR) is more complex than for the Miras. The visual magnitude of RX Boo shows semi-regular small magnitude variations (0.2 mag) with P=165±30d. Its water maser spectrum shows several components extending over about 15 km s^{-1}. After 1988, the general decline in the optical light curve is followed by a decline of the water maser emission with a possible phaselag of 27 days (0.16 phase). The visual light curve of V CVn shows semi-periodic variations with P=192±15 days. Its water maser spectrum has several features extending over 7 km s^{-1}. Water maser emission is not observed during all cycles. The integrated area appears to have a phase lag of 0.83. Martinez, et al. (1988) find that SiO maser emission from about half of the Mira variables show a secondary maximum at 0.8 phase. The H$_2$O maser spectrum of S Per is typical for supergiants showing over a dozen features extending over about 25 km s^{-1} (for more details see Little-Marenin, et al. 1991). The water maser emission of SV Peg (SRb-145d) is semi-periodic with a period similar to that of the star. AC Cyg has not been detected since 1988, even though its emission was strong during the previous year.

The maximum energy output for the Long-Period Variables occurs between 1-3 μm and the IR maximum lags the optical maximum typically by about 0.1 phase. Hence, the length of the time lag between maximum luminosity and maximum maser intensity of 1-3 months (0.2 phase), indicates that the water maser emission is not pumped directly by the optical or IR radiation of the stars, since the light travel time through the water masing region is only on the order of days. Neither is the maser line pumped directly by material from the star moving to the masing region, since the travel time (at typical outflow velocities of 5-15 km s^{-1}) is on the order of months to years. The water maser line appears to be collisionally pumped (see Cooke and Elitzur, 1985) with the radiation heating the gas in the circumstellar shell which then collisionally pumps the water maser.

Most stars we monitored are also observed as SiO masers. The only exceptions are V CVn, R Boo and R Dra. Since the SiO emission, as is the case of the water maser emission, is highly variable, we suggest that these three stars be searched again. Similarly the stars not yet searched for SiO emission, LO Aur, VX UMa, RS UMa, AC Cyg and Z Aql (indicated by .... in Table 1), are promising candidates for an SiO search. Less than half of the stars (8 out of 20) have been detected as OH masers. In general the longer period Miras appear to have a greater probability of being detected as OH masers. As suggested by the chronology of B.M. Lewis (1989), Miras and SRs have evolved circumstellar shells that can support SiO and H$_2$O masers but many are not yet evolved to the stage of supporting OH emission.

We would like to thank the members of the AAVSO, especially J. Mattei, for providing the visual light curves. Haystack Observatory of the Northeast Radio Observatory Corporation is supported by a grant from the National Science Foundation. This research is in part supported by NSF grant RII-9002960, AST-8913001, AST-8914917 and NASA grant NAG-1667.

Cesaroni, R. 1990, Astron.Ap., 233, 513.
Cooke, B. and Elitzur, M. 1985, Ap.J., 295, 175.
Lewis, B.M. 1990, Ap.J. 338, 234.
Little-Marenin, I.R., Benson, P.J., McConahay, M.M., and
Cadmus, R.R.,Jr. 1991, Astron.Ap., 249, 465.
Martinez, A., Bujarrabal, V. and Alcolea, J. 1988, Astron.Ap.
    (Suppl), 74, 273.

Fig 1. (a) The light curve and H_2O maser intensity of R Cet
(1987-1992); (b) the light curve and H_2O maser intensity of T
UMa (1987-1992); (c) the light curve and H_2O maser intensity of
RX Boo; (d) VX UMa: velocity and intensity variations of the 22
GHz lines (1989-1992)

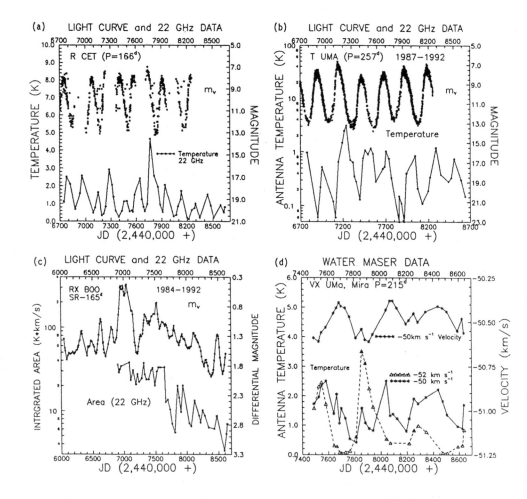

AN H_2O FLARE IN W49N

David A. Boboltz, John H. Simonetti
Virginia Polytechnic Institute and State University
Blacksburg VA 24061

P. J. Diamond
National Radio Astronomy Observatory
Soccoro NM 87801

Jeffrey A. Uphoff, and Brian Dennison
Virginia Polytechnic Institute and State University
Blacksburg VA 24061

ABSTRACT

We monitored the 22-GHz water vapor emission from W49N using a single VLBA antenna on intervals of a few days to about a week from December 1989 through May 1990. During this period, a strong flare occurred in one of the spectral features. This feature increased in flux density by a factor of $\gtrsim 20$ over a period of approximately 30 days. Line narrowing accompanied the rise in flux density; the decrease in linewidth with flux density follows the power law $\Delta V \propto F^{-0.22}$.

1. INTRODUCTION

Interstellar H_2O masers are known to flare on time scales as short as a week (Rowland and Cohen 1986). The increase in the flux density of a spectral feature is sometimes accompanied by a decrease in the linewidth and a variation in the center velocity of the feature (Sullivan 1971). In many cases, the narrowing of the linewidth (ΔV) has been observed to follow a power law relationship with flux density (F) given by $\Delta V \propto F^{-\alpha}$, where $\alpha \approx -1/2$. In this paper we discuss the flux density, linewidth, and center velocity of a flaring feature in the spectrum of W49N.

2. OBSERVATIONS

We monitored the 22-GHz line for W49N as part of a larger monitoring program which included 10 OH and 4 H_2O masers (Simonetti et al. 1992). Observations were made with a single VLBA antenna (25m) in left circular polarization with on-source integration times of 20 minutes. The signal was correlated using the NRAO Mk II correlator in autocorrelation mode. NRAO's Astronomical Image Processing System (AIPS) was used to produce 288-channel spectra from these autocorrelations. We removed a flat baseline from each spectrum, then computed the integrated flux density over the region not containing the flare. A nominal day was chosen, and the flux density scale of each spectrum was adjusted relative to the nominal day. We computed cross correlation functions between spectra and used them to correct any velocity shifts in the spectra relative to a nominal day. Finally, we fit a Gaussian to the flaring line which provided the flux density at the peak (F), the full width at half maximum value of the line width (ΔV_{FWHM}), and the velocity of the peak (V_{center}).

3. RESULTS

Figure 1 shows the spectra for five monitoring days at various stages of the flare. These spectra have had a baseline removed, but have not been adjusted for integrated flux density or velocity shifts.

Figure 2a is the light curve for the flaring line at -66.3 km s^{-1} from day 92 on. Monitoring days prior to day 92 are not shown because the flux densities found for these days only represent upper limits of their actual values. These upper limits for the 8 monitoring days prior to day 92 were $\lesssim 200$ Jy. Figure 2a shows the peak flux density rising to approximately 4100 Jy over a period of about 30 days, then falling off to a final value of approximately 1400 Jy on the last day of monitoring. Figure 2b shows the narrowing of the line as a function of time to a minimum value of about 0.8 km s^{-1}. Figures 2a and 2b

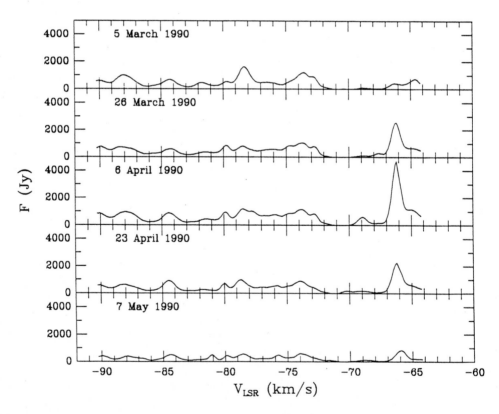

Fig. 1.— Spectra for five monitoring days during the flare in W49N. The flare occurs at a velocity of about -66.3 km s^{-1}.

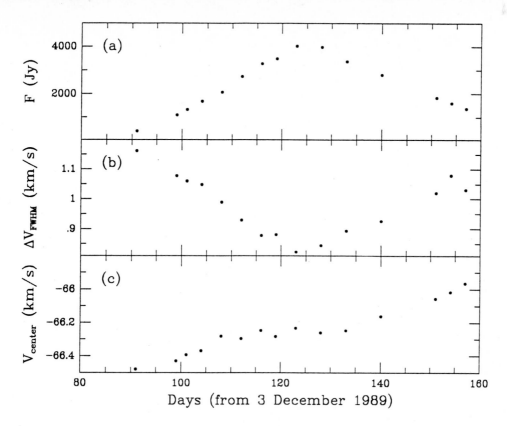

Fig. 2.— Results for the flaring feature in W49N. (a) Light curve. (b) Time variation of linewidth. (c) Time variation of V_{center}.

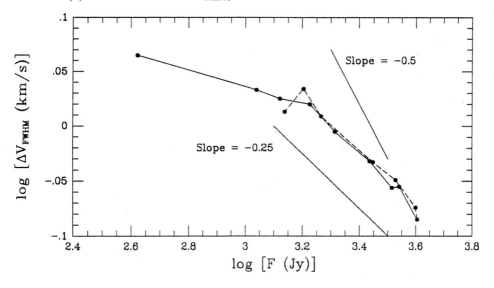

Fig. 3.— Relationship between flux density and linewidth for the flaring feature in W49N.

clearly show that the flux density and linewidth are anticorrelated. Figure 2c shows the shift in the center velocity as a function of time. The total shift is approximately 0.6 km s^{-1} over the flaring period.

Figure 3 shows the dependence of line width on the flux density. The points connected by the solid line are data taken during the rise of the flare, while those connected by the dashed line are data taken during the fall. The slope of a line through the data in Figure 3 yields the exponent α in the relationship $\Delta V \propto F^{-\alpha}$. Two slopes are shown in Figure 3 for reference. One is the slope of -0.25 which is close to the value found by performing a least squares fit on the data with $\log F \gtrsim 3.2$; both the rise and fall of the flare in W49N follow this slope. The second is a slope of -0.5 which has been reported for past observations.

4. DISCUSSION

There are three strong hyperfine components of the $6_{16} - 5_{23}$ H$_2$O transition. Adjacent components lie at 33 kHz (0.5 km s^{-1}) below and 43 kHz (0.6 km s^{-1}) above the central component at 22235.08 MHz (Bluyssen et al. 1967). Variations in V_{center} of $\lesssim 0.5$ km s^{-1} for for isolated spectral features may indicate a variation in the relative gains of the hyperfine components. For the flaring feature observed in W49N the change in V_{center} was about 0.6 km s^{-1}.

For unsaturated masers, standard theory predicts the line narrowing to be relatively independent of flux density with $\Delta V \propto (\log F)^{-1/2}$. For saturated masers, Goldreich and Kwan (1974) showed that when the cross relaxation rate is greater than the stimulated emission rate, line narrowing will continue with the same dependence. Once the stimulated emission rate exceeds the cross relaxation rate, the line rebroadens to its original Doppler width. A least squares fit to those data in Figure 3 with $\log F \gtrsim 3.2$, indicates a relationship between the flux density and line width of the form $\Delta V \propto F^{-0.22}$.

In conclusion, the flaring feature in W49N showed a shift in V_{center} of ~ 0.6 km s^{-1} which could indicate a variation in the relative gains of the hyperfine components of the $6_{16} - 5_{23}$ transition. Also, the line narrowing is inversely proportional to the flux density and follows a power law given by $\Delta V \propto F^{-0.22}$.

We would like to thank Gerald E. Nedoluha for helpful discussions.

REFERENCES

Bluyssen, H., Dynamus, A., and Verhoeven, J. 1967, Phys. Lett., 24A, 482
Goldreich, P. and Kwan, J. 1974, ApJ, 190, 27
Rowland, P.R., and Cohen R.J. 1986, MNRAS, 220, 233
Simonetti, J.H., Diamond, P.J., Uphoff, J.A., Boboltz, D., and Dennison, B. 1992, these proceedings
Sullivan, W.T. III 1971, ApJ, 166, 321

TIME-VARIABLE OH EMISSION FROM S269

ANDREW W. CLEGG
NRC/NRL Cooperative Research Associate
Naval Research Laboratory, Code 7210, Washington, D.C. 20375-5000

ABSTRACT

The 1665 MHz OH maser emission from S269 (a.k.a. PP59) is variable on year-like time scales by factors as great as 26,000%. There is no obvious correlation between the light curves of individual spectral features and their relative positions in the maser region. These results suggest that some OH masers have short lifetimes (<2 yr), and that caution must be exercised when using OH maser variability to study interstellar refraction effects over long time scales.

1. INTRODUCTION

S269 is an optical bipolar nebula and HII region approximately 2 kpc distant towards Galactic coordinates l = 196 deg, b = -2 deg. Optical, infrared, and radio continuum observations of S269 are presented by Israel (1976), Wynn-Williams, Becklin, & Neugebauer (1974), and Turner & Terzian (1985).

The 1665 MHz OH maser spectrum of S269 has been monitored over the period 1988.8 - 1992.1, and the projected spatial distribution of the 1665 MHz masers has been mapped out with the VLA. These are among the first detailed observations of the OH maser emission from this region. The results of this investigation are reported here, with particular emphasis on the extreme long time scale variability of the maser spectrum.

2. OBSERVATIONS

S269 was monitored with the 305 m Arecibo telescope and the 43 m Green Bank telescope on six occasions between 1988.8 and 1992.1 (Figure 1). Details of the Arecibo observing procedure can be found in Clegg & Cordes (1991); the Green Bank observing procedure was similar. To obtain observations of the angular distribution of maser features, S269 was observed with the A-configuration VLA in 1991.7 using observing techniques as outlined in Clegg & Cordes. The VLA data are presented in Figure 2.

3. RESULTS AND DISCUSSION

The 1665 MHz OH maser spectrum of S269 is highly variable over long time scales. This source is significantly more variable than nine other maser regions that were monitored over the same period. The variability of the 17.9 km/s feature is best described as a flaring event. In 1988.8 the feature was at or below the radiometer noise level of 60 mJy, while in 1991.5 its LHCP flux density was ~16 Jy, a brightness increase of greater than 26,000%. The source had diminished to ~1 Jy (LHCP) by 1992.1.

The VLA data show that all of the maser features are contained within approximately 2 arc sec of each other, which is much smaller than the ~2.5 arc min and ~18 arc min beams of the Arecibo and Green Bank telescopes, respectively, so the spectral changes are not due to differences in response between the two instruments. Also, prior to the flare observed with the Green Bank telescope in 1991.5, the feature had been growing steadily in brightness as observed with Arecibo.

There is no clear relationship between the observed light curves for individual features and their relative positions within the maser complex. In the RHCP, where there are more spectral features visible, the VLA data suggest a bipolar structure to the maser emission, with blueshifted masers lying to the west and red-shifted masers to the east. Roughly speaking, the position angle of this bipolar structure is the same as that of the optical bipolar nebula.

There is some hint that the polarization properties of one or more spectral features are also variable. One example is the 14.4 km/s feature which has shown opposite trends in the brightness of its LHCP and RHCP emission. Errors in the absolute flux density calibration for each polarization confound the analysis of polarization variability, although differences in the relative trends of LHCP/RHCP light curves between various features are significant since the same calibrations are applied to each feature.

S269 is also variable on time scales as short as minutes (Clegg & Cordes 1991). The modulation index for individual spectral features over short time scales is <5-10%. The relationship (if any) between long and short time scale variability is unknown.

4. CONCLUSIONS

The 1665 MHz OH maser emission from S269 is highly variable. One feature in particular flared by a factor of >26,000% over a 1.8 yr time scale. This region is more active than nine other regions monitored over the same time period, with the other regions showing typical long time scale variability of ~0% to ~50%.

This extreme variability suggests that some OH masers may have relatively short life times (<~2 yr) like their water maser counterparts. The extreme variability, which is likely intrinsic to the masers, also suggests that caution must be exercised in using OH maser variability to study interstellar refraction effects over long time scales.

REFERENCES

Clegg, A. W., & Cordes, J. M. 1991, ApJ, 374, 150

Israel, F. P. 1976, A&A, 52, 175

Turner, K. C., & Terzian, Y. 1985, AJ, 90, 59

Wynn-Williams, C. G., Becklin, E. E., & Neugebauer, G. 1974, ApJ, 187, 473

FIGURE CAPTIONS

Figure 1. The 1665 MHz OH spectrum of S269 at six epochs. The solid and dashed lines are RHCP and LHCP respectively.

Figure 2. The projected spatial distribution of the OH maser emission. The RHCP and LHCP emission are shown in panels *a* and *b*, respectively. The inserts are spectra obtained at Green Bank in 1991.5; the VLA data were obtained in 1991.7. The error bars on the relative coordinates of individual features are conservative: +/- 0.1 of the synthesized beam (FWHM). The absolute positional uncertainty is ~1 synthesized beam, or ~1 arc sec.

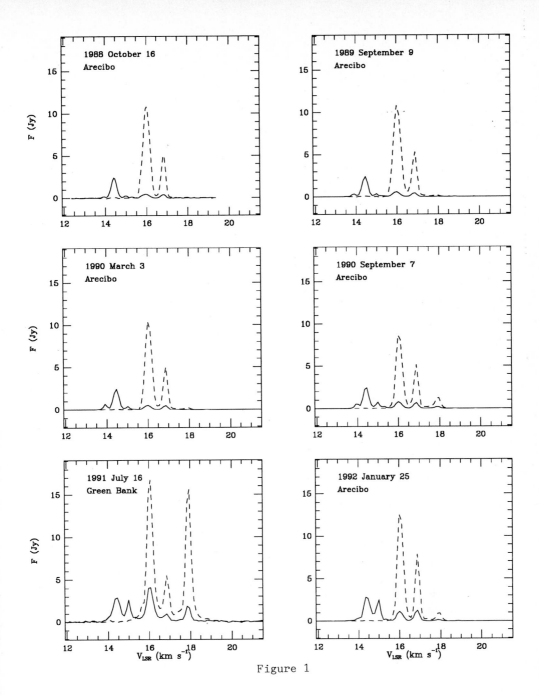

Figure 1

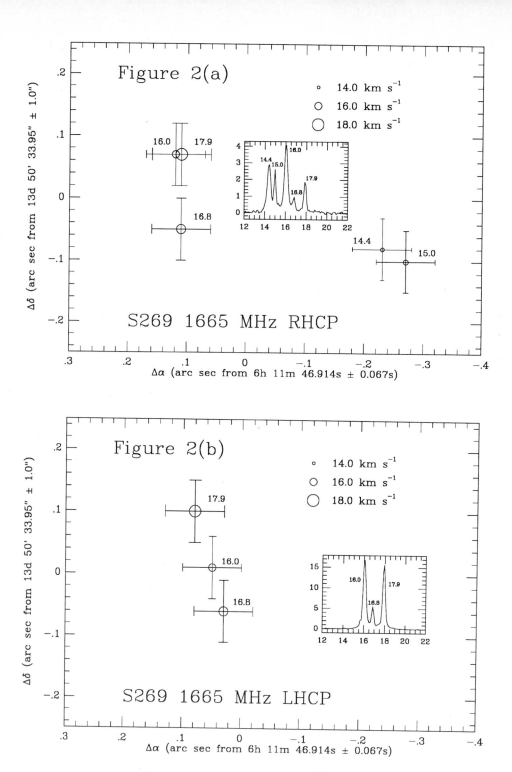

VARIATIONS IN POLARIZATION OF IRAS 19566+3423 AT THE 1612 MHZ OH LINE FROM 1987 TO 1992

John Galt

Dominion Radio Astrophysical Observatory, Herzberg Institute of Astrophysics,
National Research Council, Box 248, Penticton B.C. Canada V2A 6K3.

Unusual behaviour of the 1612 MHz OH spectrum of IRAS 19566+3423 was noted during a survey of IRAS sources whose infrared spectra showed silicate features (Galt, Kwok and Frankow 1989). This high luminosity (6.8×10^4 $L_\odot$) source shows a blackbody temperature of 226 K with a deep silicate absorption at 9.7 μm. Zuckerman and Lo (1987) suggested that this object is an oxygen-rich evolved post-main-sequence star located near the tip of the asymptotic giant branch. Likkel (1989) has observed both H_2O (22 GHz) and OH (1612 MHz) emission from the source.

The Penticton 26m telescope was used to observe both circular polarizations simultaneously. A 3-level spectrometer produced spectra with a resolution of 1.2 km/s. Each spectrum required at least one day's observing (15.5 hours) but most spectra represent an average of several days. Observing efficiency was low because of interference from GLONASS satellites.

Spectra at 21 epochs over a 4.7 year period have been combined in Figure 1 to show the sums and differences of the two circular polarizations. The 1612 MHz OH velocities observed in 1986.8 by Likkel (1989) for this source are indicated by squares in Figure 1a. Her H_2O velocities are indicated by stars. To a first approximation, it appears that the line intensities rather than their velocities change with time. Assuming this to be the case, I have attempted to measure the intensities of these lines by fitting Gaussians to the spectra. To begin the process, an average spectrum was generated for each polarization and for the sum of both polarizations. Each average spectrum was then analyzed by fitting six Gaussians with no restrictions on initial parameters. Finally, velocities and line-widths determined from the average spectra were fixed while Gaussians were fitted to each spectrum. The resulting intensities, plotted in Figure 2, show a gradual increase but no convincing evidence of periodic behaviour.

Additional spectra obtained at 1665, 1667, and 1720 MHz are shown in Figure 3. Only the 1665 MHz line shows emission at velocities near those observed at 1612 MHz. The 1667 frequency was observed in 1988 and 1991 for a total of 41.6 hours but no line was seen. The line was also reported as a nondetection by Likkel (1989). It may be significant to note that the two lines detected arise from the same upper energy level.

Several spectra were observed at four times higher resolution. Two of these are shown at the bottom of Figure 3. Due to spectrometer limitations, these spectra show almost no baseline but they do indicate that most spectral features are unresolved at the lower resolution.

Percentage polarization for a typical observation is shown in Figure 4. The polarization is strongest at the edges of the spectrum, probably because the other lines are blended. The isolated feature at -37.7 km/s shows a small velocity shift between the two polarizations which corresponds to a magnetic field of about 1.2 mG.

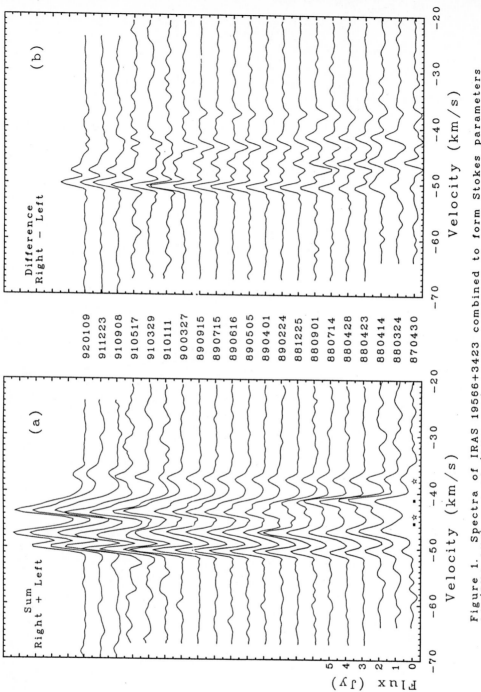

Figure 1. Spectra of IRAS 19566+3423 combined to form Stokes parameters S0 (RHC+LHC) and S3 (RHC-LHC). The squares and stars indicate velocities at which Likkel (1989) observed OH and water vapour respectively.

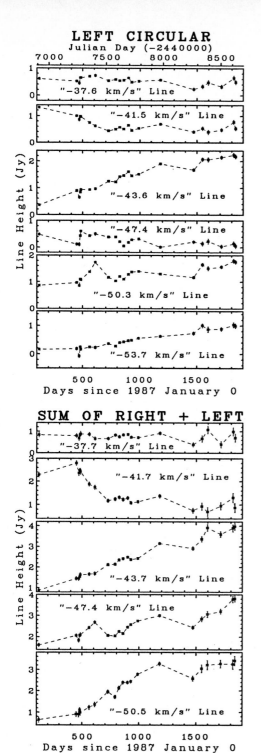

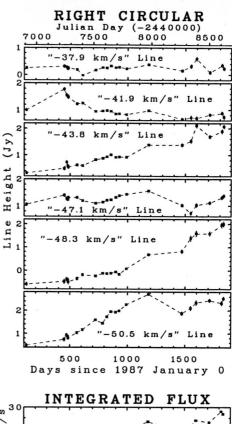

Figure 2. Intensities of lines determined by fitting Gaussians to observed spectra, plotted against time.

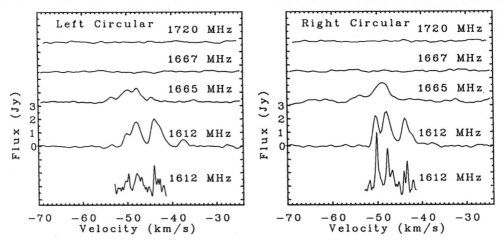

Figure 3.   Spectra of all four OH lines observed 1992.0. The bottom spectra were recorded at spectral resolution 0.3 km/s which is four times higher than the main survey (1.2 km/s).

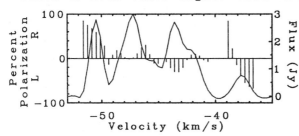

Figure 4.   Sum spectrum from Figure 1a (890915) and its percentage polarization.

The spectra suggest that the source may consist of three concentric shells. Similar, though thinner, shells have been considered by Norris *et al.*(1984) and by Rowan-Robinson (1982). If the velocities of the six lines in Figure 2L are averaged in pairs, we obtain almost the same mean velocity (-45.7 km/s) for each pair.   In many OH/IR stars intensity fluctuations arise from changes in the flux of infrared radiation from the central star. A crucial confirmation of any such shell hypothesis would be the discovery of correlations (with appropriate time lags) between intensity fluctuations in the various lines. There is no obvious evidence for this effect in Figure 2 although observations were not made often enough to show fast fluctuations or for a long enough time to show slow periodic variations.   The observed spectra resemble some of the models for ellipsoidal shells calculated by Bowers (1991).

REFERENCES

Bowers, P.F., 1991. Astrophys. J. Suppl. **76**, 1099.
Galt, J.A.,Kwok, S. and Frankow, J., 1989. Astron. J. **98**, 2182.
Likkel, L., 1989. Astrophys. J. **344**, 350.
Rowan-Robinson, M., 1982. Mon. Not. R. astr. Soc., **201**, 281.
Norris, R.P., Booth, R.S., Diamond, P.J., Nyman, L.-A., Graham, D.A. and Matveyenko, L.I., 1984. Mon. Not. R. astr. Soc. **208**, 435.
Zuckerman, B. and Lo, K.Y., 1987. Astron. Astrophys. **173**, 263.

THEORY OF FLUCTUATIONS IN UNSATURATED MASERS

Michael J. Kaufman and David A. Neufeld
Department of Physics and Astronomy
The Johns Hopkins University, Baltimore MD 21218

ABSTRACT

We develop a theoretical framework for the interpretation of variability in astronomical masers, assumed here to result from the passage of a disturbance through a region of maser activity. The source of the disturbance is not explicitly identified, but is assumed to give rise to a small amplitude, wave-like perturbation in some physical property of the masing region to which the opacity or emissivity is sensitive. The resulting change in the output flux is dependent upon the properties of the masing material and upon the geometry of the masing region. Fluctuations of frequency ω in the output flux must arise from Fourier components of the input perturbation at the same frequency. Spherical and cylindrical maser geometries are considered and perturbations traveling nearly parallel or nearly perpendicular to the maser axis are discussed. In particular, it is shown that fluctuation time scales may be smaller than the light crossing time for all maser dimensions if the wavevector of the perturbation is well aligned with the observer's line of sight and the perturbation wave is travelling at or near the speed of light.

1. INTRODUCTION

Interstellar masers have long been known to be variable on timescales from years to days, and recent observations by Clegg and Cordes (1991) have suggested that OH maser sources may show variability on timescales of *minutes*. (Note, however, the unsuccessful search for such rapid variability reported by Matsakis in this volume.) We develop a framework for interpreting maser variability, assumed here to be an intrinsic property of masers rather than the result of interstellar scintillation. In §2 below we derive the amplitude of fluctuations in the maser flux resulting from a small-amplitude wave-like perturbation in the physical conditions within the masing region. In §3 we consider the consequences of our results for the interpretation of observations of maser variability. We confine our attention to unsaturated masers, and thereby exclude from consideration the radiative instabilities discovered recently by Scappaticci and Watson (1992). Such instabilities may arise in partially-saturated masers when the light-crossing timescale is comparable to the lifetime of the molecular states.

Unsaturated masers respond passively to an input perturbation. They act as a low-pass filter which responds only to long-period perturbations. Elitzur (1991) has emphasized the importance of the maser level loss rate Γ in determining the cutoff frequency. In the present paper we discuss this effect quantitatively. We also consider the cutoff frequencies which result from the finite length and width of the masing region. As we shall argue in §3 below, it is the maser *size* that actually controls the minimum timescale for variability, except when perturbation waves propagate towards us along the line-of-sight at the speed of light.

2. FLUCTUATION AMPLITUDE

We consider how the output flux of an interstellar maser responds to a small wave-like perturbation in some physical property of the gas within the masing region. The perturbed property is characterized by the variable V, which might represent the gas temperature, gas density, molecular abundance, or any other property on which the opacity or source function may depend. The variation of V with position and time is given by

$$V = V_0(1 + Re[\delta_V \exp(i[\vec{k} \cdot \vec{x} - \omega t])]),\tag{1}$$

where V_0 is the unperturbed value of V, and where δ_V, $\vec{k}$, and ω are the amplitude ($\ll 1$), wavevector and angular frequency of the perturbation wave, respectively. In the regime where δ_V is small, only frequency components present in the input perturbation will be present in the output flux fluctuation, which is given by

$$F(t) = F_0(1 + Re[\delta_F \exp(-i\omega t)]).\tag{2}$$

The flux $F(t)$ may be related to $V(\vec{x}, t)$ by a complex filter function, defined by $\mathcal{R}_{FV} \equiv \delta_F/\delta_V$. The function $\mathcal{R}_{FV}$ depends both on the microphysics of the masing region and on the maser geometry.

The filter function $\mathcal{R}_{FV}$ is a product of several factors, representing (i) the response of the opacity, κ_ν, and the source function, S_ν, to a perturbation in variable V; (ii) the response of the emergent intensity, I_ν, to fluctuations in κ_ν and S_ν; and (iii) response of the total output flux, F_ν, to variations in the total intensity. The opacity, source function, emergent intensity and total output flux have average values $\kappa_{\nu0}$, $S_{\nu0}$, $I_{\nu0}$ and $F_{\nu0}$, respectively, and are assumed to show small sinusoidal variations of fractional amplitude δ_κ, δ_S, δ_I and δ_F. The factors making up $\mathcal{R}_{FV}$ may each be written in the form $\mathcal{R}_{ab} \equiv \delta_a/\delta_b$, where a is a physical quantity which depends on the physical quantity b.

In determining $\mathcal{R}_{\kappa V} \equiv \delta_\kappa/\delta_V$ and $\mathcal{R}_{SV} \equiv \delta_S/\delta_V$, we assume that κ_ν and S_ν relax exponentially following a sudden change in V, with time constants τ_κ and τ_S respectively. This assumption leads to the results

$$\mathcal{R}_{\kappa V}(\omega) = \frac{\partial \ln \kappa_{\nu0}}{\partial \ln V_0} \frac{1}{(1 - i\omega\tau_\kappa)}\tag{3}$$

and

$$\mathcal{R}_{SV}(\omega) = \frac{\partial \ln S_{\nu0}}{\partial \ln V_0} \frac{1}{(1 - i\omega\tau_S)}.\tag{4}$$

If the pump rates and the decay rates Γ for the maser levels respond instantaneously to a change in V, τ_κ and τ_S are simply given by $1/\Gamma$ (e.g. Elitzur 1991).

288

We compute the effect of the perturbation on the intensity of a single ray travel-ing along the z-axis, by integrating the standard time-dependent equation of radiative transfer (Cook 1969) along a path from $z = -L$ to $z = 0$. The emergent intensity is

$$I_\nu(0,t) = I_{\nu b} \exp[-\tau_\nu(0,t)] + \int_{-L}^{0} S_\nu(z,t') \exp[-\tau_\nu(z,t')]\kappa_\nu(z,t')dz, \tag{5}$$

where

$$\tau_\nu(Z) \equiv \int_{-L}^{Z} \kappa_\nu(z,t')dz, \tag{6}$$

$I_{b\nu} = I_\nu(-L,t)$ is the (constant) incident intensity, and $t' \equiv t - z/c$ is the retarded time. The retarded time appears within these integrals because the emergent intensity depends on the value of the opacity and source function at the instant the radiation passes a particular location in the maser. The filter function $\mathcal{R}_{IV} \equiv \delta_I/\delta_V$ may be conveniently be written in the form

$$\mathcal{R}_{IV} = \mathcal{R}_{I\kappa}\mathcal{R}_{\kappa V} + \mathcal{R}_{IS}\mathcal{R}_{SV}. \tag{7}$$

In the limit of $\delta_\kappa, \delta_S \ll 1$ and $\exp(-\kappa_{\nu 0}L) \equiv \exp(-\tau_{\nu 0}) \gg 1$, we find that

$$\mathcal{R}_{I\kappa} = \tau_{\nu 0} \left(\frac{\exp(ik'_z L) - 1}{ik'_z L} \right), \tag{8}$$

and

$$\mathcal{R}_{IS} = \left(\frac{S_{\nu 0}}{I_{\nu b} + S_{\nu 0}} \right) \left(\frac{\tau_{\nu 0}^2 - (ik'_z L)\tau_{\nu 0}}{\tau_{\nu 0}^2 + k'^2_z L^2} \right) \exp(ik'_z L), \tag{9}$$

where $k'_z \equiv k(\mu - \beta)$, $k = |\vec{k}|$ is the magnitude of the fluctuation wavevector, μ is the cosine of the angle between the fluctuation wavevector and the z-axis, and $\beta = \omega/kc$ is the speed of propagation of the perturbation divided by the speed of light.

From these results, we see that $|\mathcal{R}_{I\kappa}|$ has a maximum value of $\tau_{\nu 0}$ and begins to decline when $|k'_z L| \gtrsim 1$, while $|\mathcal{R}_{IS}|$ has a maximum value of $S_{\nu 0}/(I_{\nu b} + S_{\nu 0})$ but begins to decline when $|k'_z L| \gtrsim \tau_{\nu 0}$. The decline of these quantities at large $|k'_z L|$ represents the fact that the perturbations in the opacity and source function that are "seen" by a given photon during its passage through the maser begin to cancel.

Observations of astrophysical masers represent flux measurements over entire maser spots. To compute the observed flux fluctuation, we must average the intensity fluctua-tions over the face of the maser. We adopt a Cartesian coordinate system in which the z-axis is oriented along the line-of-sight and the x-axis placed so that $\vec{k}$ lies in the the x-z plane. The response of the output flux is characterized by the filter function

$$\mathcal{R}_{FV} = \mathcal{R}_{F\kappa}\mathcal{R}_{\kappa V} + \mathcal{R}_{FS}\mathcal{R}_{SV}. \tag{10}$$

We have computed $\mathcal{R}_{F\kappa}$ and $\mathcal{R}_{FS}$ both for prism-shaped and for spherical masers. For a prism-shaped maser (a cylinder, for example), the computation is straightforward since all rays traverse the same distance within the maser. If the maser has cross sectional area A, we find

$$\frac{\mathcal{R}_{F\kappa}}{\mathcal{R}_{I\kappa}} = \frac{\mathcal{R}_{FS}}{\mathcal{R}_{IS}} = \frac{1}{A} \int \Delta y(x) \exp(ik_x x)dx, \tag{11}$$

where $\Delta y(x)$ is the y-extent of the maser spot at projected position x, and k_x is the x-component of the perturbation wavevector. The analysis is slightly complicated for the spherical maser because rays lying parallel to the maser axis have different amplification path lengths. In the limit $|k_z' L| \ll \tau_{\nu 0}$, we find

$$\frac{\mathcal{R}_{F\kappa}}{\mathcal{R}_{I\kappa}} = \frac{\mathcal{R}_{FS}}{\mathcal{R}_{IS}} = \exp(-k_x^2 d^2/2), \tag{12}$$

where d is the diameter at which the mean intensity has fallen by a factor e from its value at the center of the maser spot.

3. OBSERVATIONAL CONSEQUENCES

The largest fluctuations in the maser output flux are possible when the quantities $\mathcal{R}_{\kappa V}$, $\mathcal{R}_{I\kappa}$ and $\mathcal{R}_{F\kappa}/\mathcal{R}_{I\kappa}$ lie close to their maximum values. For a perturbation with period $P = 2\pi/\omega$, which propagates at velocity βc and at an inclination $\theta = \cos^{-1}\mu$ to the line-of-sight, $\mathcal{R}_{\kappa V}$ lies within a factor 2 of its maximum value whenever

$$P \geq 2\pi\tau_\kappa \sim 2\pi/\Gamma. \tag{13}$$

For perturbations of shorter periods than this, the molecular level populations are unable to respond sufficiently rapidly to allow a large fluctuation in the output flux.

Perturbations propagating at the speed of light can give rise to the most rapid variability. Such perturbations may result when a varying external radiation field controls some property of the masing gas. When $\beta = 1$, the quantity $\mathcal{R}_{I\kappa}$ lies within a factor 2 of its maximum value provided

$$P \geq 3.3\,(L/c)\sin^2(\theta/2). \tag{14}$$

For perturbations of shorter periods than this, the effect of the perturbation is "smeared out" during the passage of maser radiation along the line-of-sight. The ratio $\mathcal{R}_{F\kappa}/\mathcal{R}_{I\kappa}$ lies within a factor 2 of its maximum value provided

$$P \geq 5.3\,(d/c)\sin\theta. \tag{15}$$

For perturbations of shorter periods than this, the effect of the perturbation is smeared out because of phase differences across the maser spot. It should be noted that if the perturbation is well-aligned with the line-of-sight ($\theta \ll 1$), the variability timescale may be smaller than the light crossing time for all maser dimensions.

REFERENCES

Clegg, A. W., and Cordes, J. M. 1991, ApJ, 374, 150.
Cook, A. 1969, Physica, 41, 1.
Elitzur, M. 1991, ApJL, 370, L45.
Scappaticci, G. A., and Watson, W. D. 1992, ApJL, 387, L73.

LONG-TERM OBSERVATIONS OF OUTBURSTS AND VARIABILITY OF THE W49N WATER MASERS

T. Liljeström
Helsinki University Observatory
Tähtitorninmäki, SF-00130 Helsinki, Finland

Abstract. A long-term monitoring of the 22 GHz water masers in W49N has been carried out with the 14-m Metsähovi radio telescope (operated by Helsinki University of Technology) during November 1980 - June 1983 and February - April 1985 (Liljeström et al., 1989). Here some new results are presented : (1) some basic statistics of the H_2O outbursts, (2) the locations of the correlated outbursts on the VLBI maps (Gwinn et al., 1992) obtained from the same time period as our observations, (3) the long-term total flux variability, and (4) the observed line narrowing relation $\Delta V_{maser} \propto F_0^{-0.5}$ of saturated masers has been interpreted according to the suggestion of Strelnitskij (1986) that it may result from changes in the kinetic temperature T_k. Using this approach upper limits to the kinetic temperature can be derived from 3 outbursts of the flare feature ($V_{LSR} = -0.9$), and are some 940 K at $F_0(max)$ and 1160 K at $\Delta F(outburst)/2$. If a 75% turbulence is assumed in ΔV_D the above temperatures are 408 K (at $F_0(max)$) and 509 K, respectively, yielding a cooling rate of some 4.8 K/day during a time period of 3 weeks. The big majority of the outbursts does, however, not show any line narrowing suggesting that these masers are in the constant temperature plateau of post-shocked gas.

1. Short-term flux variability

The 22 GHz H_2O (6_{16} - 5_{23}) observations of W49N show strong time variations of the individual velocity features. Thus the shape of the H_2O spectrum is highly variable. To ensure that the flux variability is not due to polarization effects, we measured the linear polarization, p, of W49N and found $p < 2$ %. This characteristic result of very strong water masers has been explained by Deguchi and Watson (1986).

From our W49N observations some 170 H_2O outbursts (in the velocity range -233 to 178 kms^{-1}) were documented by fitting gaussian profiles to the individual spectra of the variable features as function of time. The gaussian fits provided the peak antenna temperature, the radial velocity, the line width (FWHM) and the line area for each feature as function of time. Of the 170 H_2O outbursts some 150 were completely covered during the whole outburst, and form the data base for our statistics.

Of the 150 H_2O outbursts some 22% occured in the low-velocity range ($-15 < V_{LSR} < 30$ kms^{-1}) and some 78% in the high-velocity range ($V_{LSR} < -15$ kms^{-1} ; $V_{LSR} > 30$ kms^{-1}). The majority (some 78%) of the outburst features were blue-shifted with respect to the ambient cloud. The low-velocity outbursts lasted 17 to 206 days (rise time 5 to 147d; fall time 9 to 128d). The median durations of the low-velocity outbursts, rise and fall times were 63d, 31d and 31d, respectively. The high-velocity outbursts lasted 6 to 224 days (rise time 3 to 147d; fall time 2 to 131d). The median durations of the high-velocity outbursts, rise and fall times were 49d, 22d and 22d, respectively. We looked also for very rapid outbursts (< 1 day) but found none. The fastest outburst we registered was a weak feature at $V_{LSR} = -10$ kms^{-1} (February 22, 1985) which had a rise time less than 2 days and a fall time of 1 day. Some strong outburst features persisted a few years with one outburst following the other. For example the feature at -0.9 kms^{-1} (the flare feature) started in July 1981 and was still seen at the end of our monitoring period (April 1985). During these years the radial velocity of this narrow feature gradually changed from -0.8 kms^{-1} (1981) to -0.9 kms^{-1} (1982-83) and to -1.0 kms^{-1} (summer 1983 and 1985). This small shift in V_{LSR} during a few years may indicate a large scale systematic motion of this masering cloudlet.

There were also some 3 bigger events per year during which 6 to 14 outbursts had their maximum intensity at the same time. Because these correlated outbursts occured typically in the high-velocity range, the variable maser pump seems to be generated locally when cloud fragments in the high velocity outflow collide with ambient dense medium. A preliminary comparison of our correlated outbursts with the VLBI data of Gwinn et al. (1992) reveals that (1) in some events the correlated outbursts are tightly clustered together, (2) correlated outbursts at different times tend to reappear at certain "centres", and (3) about half of the centres of correlated outbursts are located in the maser spot concentrations seen in the VLBI maps of W49N.

The strongest outbursts of W49N during our monitoring period are listed in Table 1. For comparison, the big flares in Orion (some 10^6 Jy) have an isotropic H_2O luminosity of only a few percent of the isotropic L_{H_2O} of W49N. Because there are hundreds of H_2O masers in W49N the average photon rate is probably of the same order as the isotropic value (some 10^{48} photons/s) even if the masers are beamed. On the other hand, $L_{IR} \propto 10^{45}$ photons/s in W49N (Becklin et al., 1973). Thus radiative pumping seems to be ruled out.

Table 1. The strongest outbursts of W49N during our monitoring period

| date | feature (km/s) | F(peak) (Jy) | isotropic L_{H_2O} (for $d = 11.4$ kpc) | |
			(L_o)	$(10^{48}$ photons/s)
03.07.81	5.8	49200	0.14	3.7
27.07.82	-0.9	35800	0.06	1.6
10.04.83	-0.9	112000	0.19	4.9
14.02.85	-1.0	45900	0.08	2.1

2. Long-term total flux variability

If the masering cloudlets were powered independently a relatively constant total flux would result. Although this seems to be true for the short-term variations, a comparison of our W49N observations with previous observations reveal long-term variations of the integrated H_2O flux density in the low-velocity range ($-15 < V_{LSR} < 30$ kms^{-1}). If we adopt the same relative accuracy ($< 15\%$) for our observations as Sullivan (1973) and Little et al. (1977), the mean error in the flux densities of Table 2 is less than $\pm 0.3\ 10^5$ Jy kms^{-1}. Thus the changes seen in Table 2 seem to be real. According to Table 2 the time interval between $L_{tot}(max)$ and $L_{tot}(min)$ is some 7 yr. In addition, the time interval between the very big flares in W49N ($0.8\ 10^5$ Jy in March 1970 and $1.1\ 10^5$ Jy in April 1983), so far observed, is 13 yr. Thus, a rough estimate of the "period" of the long-term variability of W49N is some 13 - 14 yr. This long-term variability of the total H_2O flux probably reflects the long-term variations in the outflow activity of the embedded massive star.

Table 2. A comparison of the average integrated flux density and average total isotropic H_2O luminosity of W49N in the velocity range $-15 < V_{LSR} < 30$ kms^{-1}

year	1969-70	1975-76	1981	1982	1983	1985
$\int F_v dv$ (10^5 Jy km/s)	2.3 a)	1.3 b)	2.8	2.8	3.6	2.2
$L_{tot}(H_2O)$ (L_o)	0.60	0.34	0.72	0.70	0.91	0.58

a) data of Sullivan (1973)
b) data of Little et al. (1977)

3. Line narrowing and cooling of saturated masers

Some 11% of the well defined H_2O outbursts show line narrowing according to the relation $\Delta V_{maser} \propto F_o^{-0.5}$ (where F_o = flux density at line center). This is a much stronger dependence on F_o than predicted for an unsaturated maser ($\Delta V_{maser} \propto (\ln F_o)^{-0.5}$) or a saturated maser (see Likhachev et al., 1989). To get an explanation for the observed $\Delta V(F_o)$ relation the following approach is made.

The adopted ideas are : (1) H_2O masers are collisionally pumped filamentary or planar structures in post-shock warm gas, where the shocks result from the interactions of dense high-velocity clumps with ambient dense medium (Hollenbach and McKee, 1989; Elitzur et al., 1989), and (2) Strelnitskij's suggestion (1986) that the observed $\Delta V(F_o)$ relation may result from changes in the kinetic temperature T_k. Because for a saturated maser $F_o \propto |\tau_o| \propto \Delta P /\Delta V_D$ (where τ_o is the optical depth at line centre, ΔP the pumping rate , and $\Delta V_D \propto T_k^{0.5}$ the Doppler line width), and because for a collisional pump the relation $\Delta P \propto$

$T_k^{-0.5}$ is expected (Strelnitskij, 1986), one easily obtains $\Delta V_D \propto F_o^{-0.5}$ from the above proportionalities. But already the observed line narrowing relation is of this form. Thus $\Delta V_D = C \, \Delta V_{maser}$, where the factor C must be quite constant for saturated masers. This has also theoretically been shown by Likhachev et al. (1989) who give for saturated masers the result $C = (y + 1)^{0.5}$ where $y = |\tau_o|$. The expression $\sqrt{y}$ indicates how many times the profile has been narrowed during the phase of unsaturated maser emission (for unsaturated masers $\Delta V_D = \sqrt{-\tau_o} \, \Delta V_{maser}$).

Adopting $|\tau_o| = 9$ (saturation condition fullfilled for typical maser geometries, see Elitzur et al., 1992) the following T_k estimates are derived from the outbursts of the flare feature ($V_{LSR} = -0.9$ kms^{-1}), which probably is unblended (the line width of this gaussian profile is mostly < 0.6 kms^{-1} with $\Delta V_{min} = 0.48$ kms^{-1}). This frequently shocked feature consists of 3 "smaller" outbursts ($F_o \approx 23000$ to 36000 Jy) and the flare ($F_o \approx 10^5$ Jy) which can be interpreted as a superposition of two outbursts : a more long lived outburst 4 (rise time some 4 months) and a shorter outburst 5 (rise time 39 days) which starts just after the maximum of outburst 4. The outbursts 1, 2, 5 and the fall of outburst 3 obey the relation $\log\Delta V/\log F_o = -0.5$. During the rise of outbursts 3 and 4 the slope of $\log\Delta V/\log F_o$ is -0.15 and -0.22, respectively. The simplest explanation for these somewhat flatter slopes is an increasing turbulence during the rise of outbursts 3 and 4. Considering only those outbursts which show the relation $\Delta V_{maser} \propto F_o^{-0.5}$ the mean values of 940K (at $F_o(max)$) and 1160K (at $\Delta F(outburst)/2$) are obtained as the upper limit estimates of T_k (no turbulence included in ΔV_D). Assuming a 75% turbulence in ΔV_D (which still enables velocity coherence to occur) the above temperatures are 408K (at F_{max}) and 509K, respectively. Taking into account the mean rise time of 21 days (from $\Delta F(outburst)/2$ to F_{max}) of these 3 outbursts a mean cooling rate of 4.8 K/day results.

The majority (some 81%) of the well defined H_2O outbursts does, however, not show any line narrowing. Thus the bulk of the 22 GHz masers in W49N seems to be formed behind fast interstellar J shocks in the temperature plateau ($T_k \approx 400K$) of post-shocked gas as predicted by Hollenbach and McKee (1989; 1980).

References

Becklin, E.E., Neugebauer, G., and Wynn-Williams, C.G. 1973, Astrophys. Letters 13, 147
Deguchi, S. and Watson, D. 1986, in A.D. Haschick (ed.) , *Masers, Molecules and Mass Outflows in Star Forming Regions*, Haystack Observatory, p. 327
Elitzur, M., Hollenbach, D.J., and McKee, C.F. 1989, ApJ. 346, 983
Elitzur, M., Hollenbach, D.J., and McKee, C.F. 1992, ApJ. preprint (July 20)
Gwinn, C.R., Moran, J.M., and Reid, M.J. 1992, ApJ. preprint (June 1992)
Hollenbach, D.J. and McKee, C.F. 1980, ApJ. 241, L47 - L50
Hollenbach, D.J. and McKee, C.F. 1989, ApJ. 342, 306
Likhachev, S.F., Strelnitskij, V.S., and Sumin, A.A. 1989, Astronomicheski Tsirkular 1538, pp.25-26
Liljeström, T., Mattila, K., Toriseva, M., and Anttila, R. 1989, A&A Suppl. Ser. 79, 19
Little, L.T., White, G.J., and Riley, P.W. 1977, MNRAS 180, 639
Strelnitskij, V.S. 1986, Astronomicheski Tsirkular No. 1465, pp. 1-3
Sullivan, W.T. 1973, ApJ. Suppl. Ser. 25, 393

WATER MASER AND OPTICAL VARIABILITY FOR V778 CYG AND EU AND 1987-1991

Irene R. Little-Marenin[1,2] and Priscilla J. Benson[2]
[1]CASA, U. of Colorado, Boulder, CO 80309-0391
[2]Whitin Observatory, Wellesley College, Wellesley, MA   02181

Robert R. Cadmus, Jr.
Department of Physics, Grinnell College, Grinnell, IA   50112

ABSTRACT

The water maser spectra of V778 Cyg show periodic variations with the same period as the optical carbon star (P=302d).
However, a phaselag of 0.23 is present. The central velocity of
the features may show long term variability. The 22 GHz spectra
of EU And and the optical light curve show semi-periodic
variations with periods ranging between 140 to 175 days.

1. INTRODUCTION

In 1986, Little-Marenin and Willems and de Jong discovered
that seven optically classified carbon stars are associated
with circumstellar shells that exhibit strong silicate emission
features (the silicate-carbon stars). Since then, water maser
emission has been detected from three of these stars and OH
emission from two. Since 1987, we have monitored the 22 GHz
maser line from V778 Cyg (C5,5) and EU And (C4,4) at approximately monthly intervals with the 37m telescope at Haystack
Observatory. Photoelectric light curves were obtained with the
0.61m telescope of the Grant O. Gale Observatory at Grinnell
College.

2. RESULTS

The 22 GHz spectrum of V778 Cyg usually has three components centered at about -15, -17 and -20 km s^{-1}. The intensity
of the components has varied by over a factor of 25 during the
last four years. The component at -15 km s^{-1} was weak in 1987
but became the strongest line in 1991 whereas the -17 km s^{-1}
line has become the weaker one (Fig. 3). The -20 km s^{-1} line is
currently not visible. Three component maser spectra suggest an
expanding or rotating shell or disk. The optical light curve of

V778 Cyg (1989-1991), typical of C star LPVs (long period variables), shows broad maxima and sharp minima with a P=302 days and an amplitude of 0.4 mag. The maser flux varies in phase with the light curve; however, a phaselag of about 70 days (0.23 phase) is present (Fig. 1). It appears that the maser flux is strongly coupled to the optical variability of the C star. Both the -15 km s^{-1} (Fig 1, middle panel) and the -17 km s^{-1} line appear to show systematic velocity variations with an amplitude of about 0.7 km s^{-1} over a time span of about 4-5 years (but the two lines have different phase dependencies).

The 22 GHz spectra of EU And usually shows two components at -30 and -31 km s^{-1} as well as possibly a component at -42 km s^{-1} during the summer of 1988. The maser weakened shortly after detection and has been very weak since then until the summer of 1991 when the -30 km s^{-1} component has suddenly brightened. Its optical light curve shows semi-periodic variations with the interval between successive minima ranging from 140 to 175 days. The maser intensities also appear to show semi-period variations; however, the intensities in general are too weak to determine if the optical and maser variability correlate and to establish a phaselag (Fig.2).

3. DISCUSSION

The following information about silicate-carbon stars is available. Several of the stars were classified as C-stars more than 50 years ago. Spectra shortward of 4 μm can not be distinguished from typical of C-star spectra. All silicate -carbon stars appear to be J-type stars, enriched in ^{13}C with ^{12}C/^{13}C ratios between 10-15. The circumstellar shell at distances greater than about 10 R_* from a central star (not necessarily the C star) is dominated by an oxygen-rich chemistry (silicate dust, H_2O and OH masers). The chemistry close to a central star has not yet been established (no SiO maser emission has been detected). No HCN microwave emission, produced at large distances from a central star, has been detected (the chemistry in the shell is oxygen-rich to large distances).

No currently available model is able to explain all the observations. 1. A single AGB star seen in transition from an oxygen-rich M giant to a C star does not provide convincing reasons why these stars are ^{13}C enriched or why the oxygen-rich material has not yet passed through the H_2O maser and silicate emission region since the stars became C stars. 2. A single strongly N-enriched C star (tieing up C in HCN and C_2H_2 molecules) is unlikely to produce an O-rich chemistry in the shell with strong silicate features. 3. There is no evidence for a companion to the C star (around which the oxygen-rich material resides). However, there is a tantalizing suggestion (Fig. 1) that the -17 and -15 km s^{-1} maser lines of V778 Cyg show long-term velocity variations with an amplitude of about 0.7 km s^{-1}.

Visual light curves from members of the AAVSO are gratefully acknowledged. Haystack Observatory of the Northeast Radio Observatory Corporation is supported by a grant from the NSF. This research is in part supported by NSF grant RII-9002960, AST-8913001, AST-8914917 and NASA grant NAG-1667.

Figure 1. The optical light curve (top) from 1989-1991, the central velocity of the -15 km^{-1} feature (middle) and the intensity of the 22 GHz lines at -15 and -17 km s^{-1} (bottom) for V778 Cyg from 1986-1992 are plotted. A phase lag of 0.23 phase (70 days) between the optical light curve and the maser intensities is present.

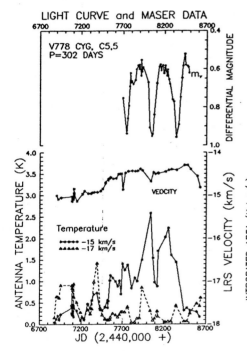

Figure 2. The optical light curve (top) and the integrated area of the water maser features (bottom) for EU And are shown. Both sets of data show semi-periodic variations.

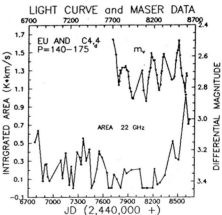

Figure 3. The water maser spectra of V778 Cyg from 1986-1991. The variation of the three major components are clearly visible.

ANTENNA TEMPERATURE (K)

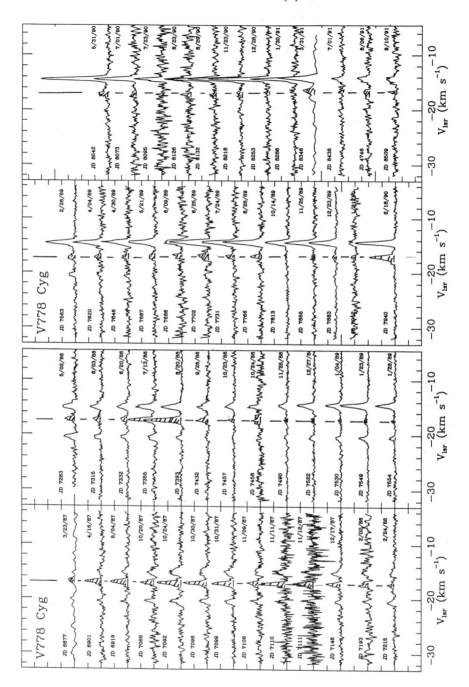

Search for Rapid Variations in OH and H2O Masers

Demetrios Matsakis

U.S. Naval Observatory, Washington, DC 20392

Abstract: Upper limits of a few percent have been found in 8 bright OH masers, on time scales of 7.5 seconds to several hours. A possible variation in an H2O maser is presented.

It has been recently reported by Clegg and Cordes, 1991, (CC hereafter) that several OH masers exhibit variations on time scales of minutes to hours. More than half the masers studied appear to exhibit these variations in some form.

As part of a larger study on the statistics of astronomical masers, data were taken using the Mark III VLBI recorders at the NRAO 140' antenna at Green Bank. Although the bit stream is filtered, digitized (1-bit samples), and contaminated by various tones, timing information, and parity bits it is possible to "detect" the bit stream using an analog system. With our in-house spectrum analyzer, we have achieved frequency resolution as low as 10 Hz, and time resolution as low as .03 seconds. The results presented herein, however, were achieved by digital autocorrelation of the data recorded on VLBI tapes. The USNO Mark III correlator has the capability of autocorrelating at lags at integer values of up to 10,000 times the Nyquist interval of .25 microseconds, which are averaged into time-bins of no less than 7.5 seconds. Since autocorrelations at negative lags need not be sampled, it is possible to Fourier transform the data so as to obtain spectra at a formal resolution of 200 Hz. After Hanning-smoothing these data, effective resolutions of .005 Km/s at K band (water) and of .072 km/s at L band (OH) are achieved over a spanned width of 2 MHz/recorded frequency channel.

The system has some drawbacks. To achieve full resolution, it takes about 30 hours to reduce one hour's data for each 2 MHz channel. Since correlator time is expensive, the autocorrelation mode is only used when it is possible, by running the correlator in split mode, to use components that would otherwise remain idle. Another problem is that observatories are not normally set up to shift the Local Oscillator for VLBI observations, or to fire the noise diode at 7.5 sec intervals (although one could in principle do both). Even at our resolution, some of the OH line components vary rapidly enough between frequency resolution elements that data taken a few hours apart can not be simply compared due to the change in the projected rotation angle. In our case, interpolation was achieved by fitting a cubic spline. At K-band, there are receiver gain variations of a few percent which occur on time scales of minutes. Special calibration techniques are called for, which will not be discussed, as the water data shown below were taken with the standard system.

Although the collecting area of the 140' antenna is considerably less than that of Arecibo or the VLA (both used by CC), in the case of the strongest OH masers there was sufficient signal-to-noise that we should have detected the variations reported in W51 and W49. Spectra of those two sources are shown in Figure 1. We note that CC considered their W49 variations tentative due to the existence of a contaminating source at the edge of their beam. In our case, that source is well within our beam.

Table 1 summarizes the data. In no cases were variations observed. Table 1a gives the peak-to-peak limits for 7.5-second integrations over a 12-minute scan. Table 1b gives the limits when the 12 minute scan-averaged data are compared between different scans separated on hourly time scales.

Although it is possible that the conclusions of CC are erroneous, due perhaps to interference (as was present in some of our Green Bank data), it is also possible that the sources simply happened to be quiescent at the precise time of the observations. It is not a subject worthy of great debate, since CC are planning to repeat their observations, using widely-separated antennas.

Figure 2a shows an average of 54 20-sec scans of the water masers in W75S using the Green Bank online autocorrelator (not the Mark III VLBI system described above). Figure 2b gives the integrated line strength of the weaker component for each 20-second scan going into the average. For calibration, the plots were normalized to the stronger component; any variation of the stronger component would cause an apparent inverse variation of the weaker. The observed variations are too large to be due to random noise alone. The overall downward slope is not statistically significant, but corresponds to a 100% variation in about 6 hours. Figures 2c and 2d show two of the

Figure 1

1A) W51   OH (RCP, 1665 MHz)

1B) W49   OH (RCP, 1665 MHz)

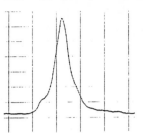

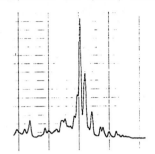

Figure 2   W75S   H2O   (RCP)

2A)   12-minute Scan Average

2B) Components Ratio vs Time

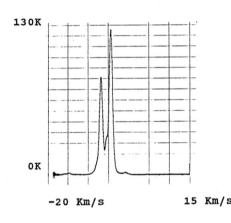

130K

0K

-20 Km/s   15 Km/s

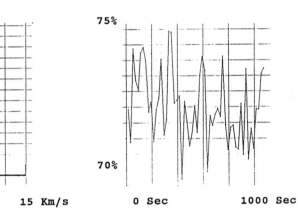

75%

70%

0 Sec   1000 Sec

2C) 20-sec subscan – Scan Average

2D) 100 sec later – Scan Average

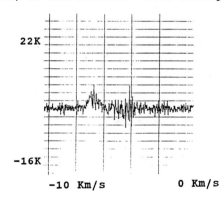

22K

-16K

-10 Km/s   0 Km/s

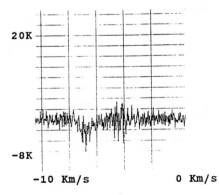

20K

-8K

-10 Km/s   0 Km/s

individual normalized 20-second scans differenced from the average. The region at the location of the stronger peak, which must average to 0 due to the normalization, is unusually noisy because the line itself increases the system temperature there. These two spectra seem to show an 8% variation over 1.5 minutes. Unfortunately, simultaneous observations at a VLBA site were unsuccessful, due to a system problem and therefore this variation should probably be considered tentative. I recommend that future monitoring of water masers (as well as OH masers, of course) be conducted in such a manner as to catch such short-term variations if they occur.

It is a pleasure to thank Valerie Bocarie, Roland Cameau, Michelle Diaz, Craig Ewan, Joe Granderson, Reginald Kerney, Mark Lyght, and Harvest Macon for their assistance at the correlator.

Reference

Clegg, A.W. and Cordes, J.M. 1991, ApJ. 374, 150.

Table 1a. Peak/peak Limits to 7.5-second OH Variations
    Limits are for strongest peak, weaker components have proportionally larger limits. CC results are the largest variation seen at the VLA over 20 minutes, extracted from their figures by dividing the largest variation by the largest peak component.

Source	Pol	Freq	Variation	Number of Scans	Clegg&Cordes
W3OH	LCP	1665	<3%	(four 12-min scans)	–
	RCP	1665	<3%	(five 12-min scans)	–
	LCP	1667	<10%	(four 12-min scans)	–
	RCP	1667	<15%	(five 12-min scans)	–
VYCaMa	RCP	1665	<12%	(one 12-min scan)	–
	RCP	1667	<9%	(one 12-min scan)	–
G231.8+4.2	RCP	1667	<15%	(one 12-min scan)	–
NGC6334F	RCP	1665	<15%	(<6% if bin by 1-min)	C&C: 5%
	RCP	1667	<15%	(one 12-min scan)	–
W49	RCP	1665	<6%	(seven 12-min scans)	C&C: 25% ?
	RCP	1667	<6%	(seven 12-min scans)	–
W51	RCP	1665	<5%	(two 12-min scans)	C&C: 12%
ON1	RCP	1665	<10%	(two 12-min scans)	C&C: 0
DR21OH	RCP	1665	<30%	(one 12-min scan)	C&C: 5%

Table 1b. Peak/peak Limits for Hourly OH Variations
           See explanations of Table 1a.

W3OH LCP 1665 <1%   (four 12-min scans spanning 60 min))
     RCP 1665 <1%   (five 12-min scans spanning 350 min)
     LCP 1667 <2%   (four 12-min scans spanning 60 min)
     RCP 1667 <1.3% (five 12-min scans spanning 350 min)
W49 RCP 1665 <1.3% (seven 12-min scans spanning 310 min)
     RCP 1667 <1.3% (seven 12-min scans spanning 310 min)
W51 RCP 1665 <2%   (two 12-min scans spanning 57 min)
ON1 RCP 1665 <3%   (two 12-min scans spanning 170 min)

WATER MASER BURSTS IN W49N (1989-92)

G.M. PACHECO, E. SCALISE JR.

INPE - Instituto Nacional de Pesquisas Espaciais (SCT) - Divisão de Astrofísica
C. P. 515, 12201 S. J. Campos, São Paulo, Brasil

Abstract

After an extensive monitoring program of the water vapour emission originated from W49N we were able to detect very strong eruptions at +28 km/s, -61 km/s and -27 km/s. Here we present the results of the monitoring of these explosions and discuss the possible correlated variability of the events recorded at +28 km/s with one of them, not so intense, that appeared at +62 km/s.

Introduction

The discovery of the 616-523 rotational transition of interstellar water in maser emission by Cheung et al. (1969), and subsequent observations revealed that several of its characteristics like intensity, polarization, width, etc. varied with time. The understanding of these changes are an important tool to help us to understand the physical conditions of the masers sites.

In order to determine these parameters several monitoring programs were carried out and the varied results gave rise to the proposition of different models. The strong Galactic maser burst in W49N cannot be satisfactorily explained by the existing collisional and radiative models, as pointed out by Kylafis at al. (1991).

To study the variability of several strong water masers we started a long basis monitoring program in 1985 (Sestokas and Scalise, 1991). Here we present the results of this program for the W49N source, covering the period of 1989-1992.

Equipment and observations

The observations were made using the Itapetinga 13.7 m radiotelescope which halfpower beamwidth at 22.2 GHz is 4.2 arc min. At this frequency the system temperature is of the order of 1000 K. An acousto-optical spectrometer with 1000 channels of 40 kHz each (0.54 km/s) was used for data processing. Observations were made using the ON-ON beam switching technique.

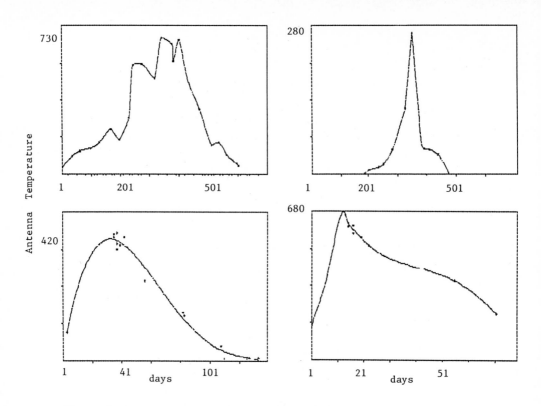

Figure 1: Antenna temperature variations of the observed features:
a) + 28 km/s; b) + 62 km/s; c) – 62 km/s; d) – 27 km/s.

Analysis of the +28 km/s and +62 km/s features

The time evolution of the strong flare that happened at +28 km/s and of the possible correlated feature at +62 km/s are shown in Figures 1a and 1b.

The +28 km/s feature was observed for the first time on October 7, 1989, reaching a relative maxima by April Scalise et al. (1991).

Superimposed to this activity a strong flare happened, causing the emission to rise, almost linearly, three folds in only 50 days. It remained at this high level for the next 200 days and presented some fluctuations, which origin could be due to several explosions, similar to those seen in Orion (Vilas Boas and Abraham, 1990), where several sources with different degrees and angles of polarization were detected within the 60 kHz line centered at about 8 km/s.

The decreasing to the pre–flare level was slower than the rise up and was reached in 90 days but the return to the October 1989 level was faster, happening in only 100 days, and presenting no fluctuations like the one observed by Mattila et al. (1985) or Vilas Boas et al. (1988).

Assuming the half power width of this feature at its maxima as 100 kHz, the temperature of 730 K and the source placed at 14 kpc, it reached an intensity of 0.17 solar luminosities.

The other feature at +62 km/s first detected on April 20, 1991 rose slowly, reaching a maximum value, 280 K, in 145 days, followed by a fast and monotonic decrease in 30 days to a quiescent level and disappearing completely after further 80 days.

Based on Gammon's (1976) results one could say that the variation of these features was correlated. Studying the spatial position of these masers (Walker et al. 1982) we found out that there was at that time one source at +62 km/s placed about 0.230 miliarcseconds (2.8E15 cm) from the +28 km/s. If the +62 km/s source was triggered by the +28 km/s, and both were placed at the same distance, its excitation front should travel at more than 2000 km/s, destroying all molecules of the medium. To avoid this, and if there was a common center of activity, it should be placed somewhere in the cloud.

Analysis of the −27 and −62 km/s features.

The −61 km/s was detected on July 28, 1991. Its time profile, as seen on Figure 1c, shows that in 40 days it rose from 95 K to maximum and returned in 80 days almost linearly to the same level. The rise up phase was faster than the decrease phase which becomes linear until the intensity reaches a value smaller than the first detected one. The maximum antenna temperature is equal to 420 K, corresponding to 0.095 solar luminosities.

The −27 km/s feature was observed since November 21, 1991. The successive observations made in the following days showed that from November 21 to 26 it rose 2.5 times and from November 26 to December 3, when it achieved its maximum value of 740 K, it rose 1.7 times. It took only 16 days to reach its maxima and another 40 days to return to the pre-burst level. The time intensity profile seen in Figure 1d shows that its rise up evolution was much faster than the corresponding decreasing phase. The highest antenna temperature achieved corresponds to 0.17 solar luminosities.

FINAL COMMENTS

As one can see, this star formation region is passing by a very active period showing several strong bursts spreaded out over a velocity range of the order of 120 km/s. The duration of the events are varied from less than 100 days up to 600 days. The rise time also varied from 11 to 150 days and the duration from 80 to 650 days. The measured width of 100 kHz of the structures remained constant during all the

the event duration. This invariance, as pointed out by Walker et al. (1982), could be due to the existence of a velocity gradient along the line of sight.

The four time intensity profiles are completely different and none can be fitted by the diffusive model (Burke et al. 1976). This is another important result for it certainly reflects different local physical conditions within the same cloud. These profiles reflect the region complexity and cannot be fitted by a simple model.

Regarding the correlation between +28 km/s and +62 km/s, the inexistance of a VLBI map taken during the event makes it impossible to determine the precise position of both spots. Without this information it is impossible to infer what happened in the region and which was the most probable mechanism responsible for the correlation.

The peak luminosities achieved in these events is a direct evidence of the activity stage under way in this region. But even being one of the strongest water vapour eruptions ever recorded in Galaxy, they are still 3 orders of magnitude weaker than an extragalactic megamaser (Henkel et al. 1984).

A possible correlation of many transitions in W49 spectrum was shown by Menten et al. (1990). If intensity variation at other transitions is detected, this information can help us to construct a more realistic energy diagram. Since the monitoring program will continue, these last two mentioned aspects will be considered for the next observation period.

Acknowledgements

The authors wish to express their gratitude to the Itapetinga Radio Observatory staff, to Avelino M. Gómez Balboa and Nori Beraldo by the reduction data software. This program was partially supported from local financial agencies, FAPESP and CNPq.

References

Burke, B.F., Giufrida, T.S. and Haschick, A.D.: 1978, Astrophys. J. **226**, L21
Cheung, A.C., Rank, D.M., Townes, C.H., Thornton, D.C. and Welch, W.J.: 1969, Nature **221**, 626.
Gammon, R.H.: 1976, Astron. Astrophys. **50**, 71.
Henkel, C., Gusten, R., Downes, D., Thum, C., Wilson, T.L. and Biermann, P. : 1984, Astron. Astrophys. Let, **141**, L1.
Kylafis, N.D., and Norman, C.A.: 1991, Astrophys. J. , 373, 525.
Mattila, K., Holsti, N., Toriseva, M., Anttila, R. and Malkamaki, L.: 1985, Astron. Astrophys. **145**, 92.
Memten, K. M., Melnick, G.J. and Phillips, T.G. : 1990, 29th Liege International Astrophysical Colloquium
Scalise Jr., E.; Pacheco, G.M.; Balboa, A.G. and Abrahamm, Z.: 1991 IAU Simposium No 150, Campos do Jordão, in press.
Sestokas Filho, B. and Scalise Jr., E.:1989, Astron. Astrophys. Suppl. **88**, 177.
Vilas Boas, J.W.S. and Abraham, Z.: 1988, Astron. Astrophys. **204**, 239.
Walker, R.C., Matsakis, D.N. and Garcia-Barreto, J.A.: 1984, Astrophys. J. **255**, 128.

RADIATIVE INSTABILITIES IN ASTROPHYSICAL MASERS

Gerardo A. Scappaticci and William D. Watson
Department of Physics, University of Illinois
Urbana, IL 61801-3080

ABSTRACT

We have previously found that the solutions to the equations of radiative transfer for astrophysical masers are unstable. Here, we provide additional information on the behavior of the interior of the masing region during the variations that result from the instability. It is plausible that the relevant parameters for the interstellar, OH 1665 MHz masers are within the regime for which instability occurs.

Although the issue of the stability of the radiative transfer in astrophysical masers has previously been investigated, no clear evidence for instability was found. Based on computations by Salem and Middleton (1978), it has been recognized that a perturbation to a maser leads to relaxation oscillations when there is radiation propagating in both directions in the maser as would be expected. Reports by Clegg and Cordes (1991) of fluctuations in the radiation from OH 1665 MHz masers on time scales as short as 1000 s has prompted us to reconsider radiative instabilities (Scappaticci and Watson 1992a, b). Based on the linear maser idealization which we believe to be satisfactory, we do find that these masers can be radiatively unstable for conditions that are plausible. After a long time (exactly how long is determined by the nature of the perturbation and other conditions), the emergent maser radiation of an unstable idealized masing tube has a periodic variation in time. In realistic astrophysical environments, a number of considerations can be expected to reduce significantly the amplitude of the variation from that of the calculations, and to mask or prevent the pure, periodic time variation. We have calculated the unstable behavior of isolated masers, as well as of interacting pairs of masers where the components are separated by some distance. We demonstrate the instability in two ways. A stability analysis for small perturbations is performed using the steady state configurations. In addition, we integrate numerically the time dependent equations of radiative transfer.

The instability depends upon four parameters that describe i) the degree of saturation, ii) the "matching" of the light travel time through the maser and the characteristic time scale of rate equations for the molecular populations, iii) the flux of incident continuum radiation, and iv) the relative importance of spontaneous emission and the incident continuum radiation as sources of the "seed" variation. From the calculations performed so far, spontaneous emission seems to quench the instability for masing transitions other than the 18 cm transitions of OH. The period for the oscillations of the emergent maser radiation is roughly the length L of the maser divided by the speed of light. For instability, it is necessary that this time is comparable with the characteristic decay time $1/\Gamma$ for the molecular populations.

The behavior of the emergent radiation and the criteria for instability have been delineated in our other publications on this topic. In Figures 1 and 2 here, we present information that describes the internal changes in the masing region during a representative period of the permanent oscillations that are generated by the instability. The information in Figure 1 demonstrates how the intensity grows as the radiation that enters at x = 0 propagates through the maser and emerges at the other end (x = L) at the times that are used as the labels for the curves in the Figure. The insert in Figure 1 relates this time to the phase of the periodic variation of the emergent maser radiation. This Figure shows how the central region amplifies by a constant factor as expected for true unsaturated amplification regardless of the phase. It is toward the ends where the differences occur during a period of the variation. The ends are thus where the instability is generated. We present in Figure 2, the normalized molecular population difference which determines the amplification at each location for the maser radiation that emerges at the indicated times. That is, in both Figures 1 and 2 the quantities given by a curve at a location x are those at a time earlier than the time with which the curve is labelled by an amount (L−x)/c. As in Figure 1, it is evident in Figure 2 that the variations in the emergent intensity are a result of changes at the ends of the maser.

References

Clegg, A.W. and Cordes, J.M. 1991, Ap.J., 374,150.
Salem, M. and Middleton, M.S. 1978, MNRAS, 183, 491.
Scappaticci, G.A. and Watson, W.D. 1992a, Ap.J., 387, L74.
Scappaticci, G.A. and Watson, W.D. 1992b, Ap.J., in press.

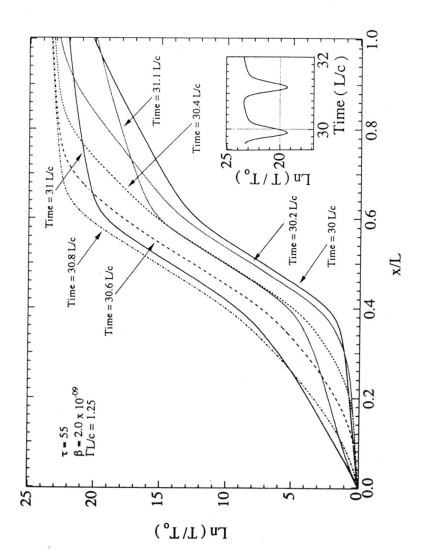

Figure 1: The natural log of the brightness temperature T for unstable maser radiation propagating in the positive x direction divided by the brightness temperature T_0 of the incident continuum radiation, as a function of the location x within a maser of length L. The curves are labelled according to the time at which the radiation emerges from the maser at x = L. The insert depicts ln (T/T_0) for the emergent radiation versus time during the relevant period. Calculations are performed with representative maser parameters using the methods described by Scappaticci and Watson (1992a, b).

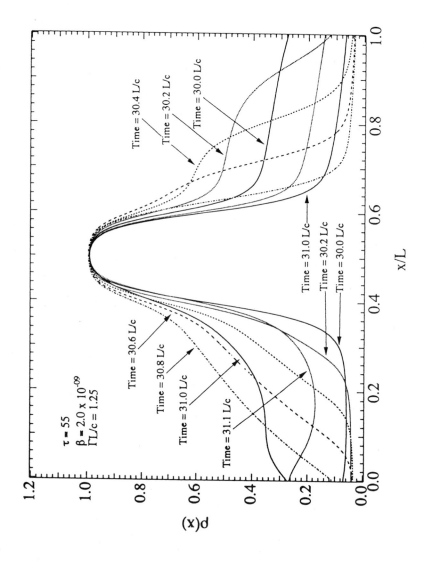

Figure 2: The quantity $\rho(x)$ which is the actual population difference between the masing states normalized by the population difference that would occur if there is no maser radiation, as a function of location x within the maser, for the calculation of Figure 1. These are the population differences when the various rays in Figure 1 are at the location x. For example, $\rho(x)$ for the curve labelled $30.8L/c$ is the value of $\rho(x)$ at a time $[30.8L/c - (L-x)/c]$.

A MONITORING PROGRAM OF INTERSTELLAR OH AND H_2O MASERS

John H. Simonetti
Virginia Polytechnic Institute and State University
Blacksburg VA 24061

P. J. Diamond
National Radio Astronomy Observatory
Socorro NM 87801

Jeffrey A. Uphoff, David Boboltz, and Brian Dennison
Virginia Polytechnic Institute and State University
Blacksburg VA 24061

ABSTRACT

We obtained spectra of a set of interstellar OH and H_2O masers at intervals of a few days to about a week throughout a period of about 5 months. The purpose was to determine whether interstellar scintillation and scattering processes contribute to maser variability. Some OH spectral features clearly varied, but their variation time scales ($\sim$10 days) are inconsistent with interstellar scintillation. In addition to some flares, nearly all spectral features of the H_2O masers showed $\sim$10–50% variations in flux density, with typical time scales of $\sim$1 month to $\gtrsim$ 5 months. These H_2O feature variations are consistent with refractive interstellar scintillations.

1. INTRODUCTION

Masers occasionally flare due to processes which are instrinsic to these sources (Reid & Moran 1988; Garay et al. 1989). However, other less dramatic intensity variations are not obviously intrinsic — some may be produced by propagation effects within the interstellar medium. The situation may be like that for compact extragalactic radio sources (cores of radio galaxies and quasars with intrinsic angular sizes $\lesssim$ 10 mas). Intrinsic outbursts of extragalactic sources occur at radio frequencies $\nu \gtrsim 1$ GHz (Kellermann & Owen 1988). However, a host of other variations of extragalactic sources are probably caused by refractive effects within the interstellar medium: low-frequency variability, flicker, and extreme scattering events (Rickett 1990). Interstellar propagation effects may account for some variability, other than flares, in masers (Rickett 1990).

2. OBSERVATIONS

We monitored interstellar masers with small angular spot sizes (and a few masers with large spot sizes as controls). The 1665-MHz line was observed for 9 OH masers (OH34.26, OH35.20, OH45.47, W3(OH), W33A, W49N, W49S, W75N, and W75S), and

311

the 22-GHz line for 4 H_2O masers (Sgr B2(N), W49N, W51M, and W75N). W3(OH) W49N, and W49S were also observed at 1667 MHz.

Observations were made in LCP with a single VLBA dish (25m) and the signal was digitized and recorded with the MkII VLBI system. On-source integration times were 20 minutes. Autocorrelations, obtained from the NRAO MkII correlator, were used to produce 288-channel spectra using NRAO's Astronomical Image Processing System (AIPS). After removing a polynomial baseline, the flux density scale in each spectrum for a given maser was adjusted to make the flux density integrated over the spectrum equal to the integrated flux density for a chosen nominal day. The flux density F at the peak of each feature in a spectrum was obtained from a quadratic fit through the upper 3 points in the feature.

3. RESULTS

Here we present representative results. A detailed discussion of the full results will appear elsewhere.

The results for the OH maser W33A are displayed in Figure 1. Each point in Figure 1b represents the statistics (rms/mean and mean) for the day-to-day variations in flux density of a single channel in the spectra for W33A. The solid line in Figure 1b is the expected dependency if the variations are due solely to system noise: the rms system noise for a channel at frequency ν is $\sigma(\nu) = \alpha(F_{rad} + F_{source}(\nu))/(\Delta\nu\,\Delta t)^{1/2}$. The flux

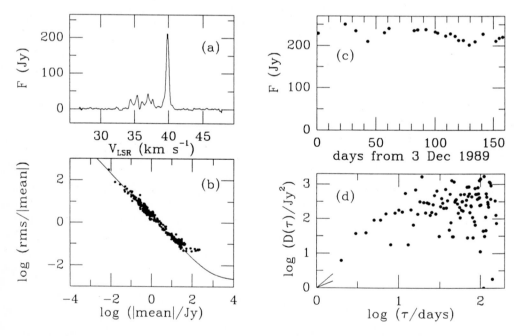

Fig. 1.— Results for the OH maser W33A. (a) Spectrum from 14 March 1990. (b) Statistics for each channel (see text). (c) Light curve for the feature at 39.8 km s^{-1} (mean = 225.5 Jy, rms/mean = 0.055). (d) Structure function for the light curve.

density F_{rad} (equivalent to the system noise temperature in the absence of a source) was determined from the residuals in the baseline fit for a typical W33A spectrum, and $\alpha \approx 1.5$ for 2-bit sampling. Clearly the day-to-day variations of most channels are consistent with system noise. However, the variations of the strongest feature (at $V_{LSR} = 39.8$ km s^{-1}) cause the points to deviate markedly from the curve at the largest flux densities in Figure 1b.

Figure 1c shows the light curve for the 39.8 km s^{-1} feature in W33A. Figure 1d displays the structure function $D(\tau) = \langle [F(t) - F(t + \tau)]^2 \rangle$ calculated from the light curve. For a long time series of a stationary process, the structure function and autocorrelation function $R(\tau) = \langle [F(t) - \langle F \rangle][F(t + \tau) - \langle F \rangle] \rangle$ are simply related by $D(\tau) = 2R(0)[1 - R(\tau)/R(0)]$ but the structure function is more accurately estimated (i.e., without bias) for a short time series. The correlation time scale (where $D(\tau)$ levels out) for the variations in this feature of W33A is ~20 days. Variations correlated over a number of days cannot be due to unrepeatable intensity-proportional errors such as pointing errors.

A spectral feature of the OH maser W75N also varied markedly, with rms/mean ~ 8%, and a correlation time scale of ~5 days.

Representative results for H_2O maser features are displayed in Figures 2 and 3. The structure function indicates the correlation time scale is probably ≳80 days for the −78.5 km s^{-1} feature of W49N. The correlation time scale is ≳40 days for the 43.2 km s^{-1} feature of W51M.

On time scales shorter than the characteristic time scales of the process, $D(\tau) \propto \tau^2$.

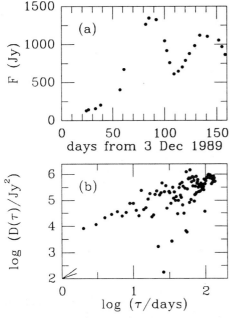

Fig. 2.— Results for the −78.5 km s^{-1} feature of the H_2O maser W49N (mean flux density = 786.4 Jy, rms/mean = 0.477).

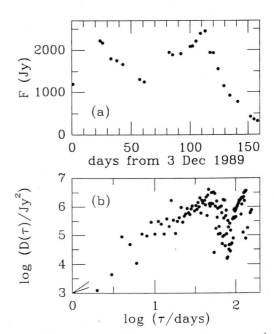

Fig. 3.— Results for the 43.2 km s^{-1} feature of the H_2O maser W51M (mean flux density = 1591 Jy, rms/mean = 0.401)

For example, a process with a gaussian power spectrum has one characteristic time scale, and its gaussian autocorrelation function would depend quadratically on τ at small lags, as would its structure function. For the W49N feature $D(\tau) \propto \tau^x$ where $x < 2$, indicating a range of time scales from $\lesssim 2$ to $\lesssim 80$ days contribute to the variability. The same may be true for the W51M feature for 5 days $\lesssim \tau \lesssim 40$ days.

4. DISCUSSION

For a scattering medium extended along the line of sight, the correlation time scale of refractive interstellar scintillations is $t_r \approx L\theta_o/V_\perp$ for a source of observed angular size θ_o and distance L, where $V_\perp$ is a relative transverse velocity (of the medium, source, or observer). $V_\perp$ should range from $\sim$30–100 km s^{-1} — we use $V_\perp$=60 km s^{-1} in the following calculations. The diffractive scintillation time scale is $t_d \approx \lambda/2\pi\theta_s V_\perp$, where θ_s is the scattered size of a point source (presumably $\theta_s \approx \theta_o$ for the masers).

Nominal distances and spot sizes for the OH masers W33A and W75N are 4 kpc and 20 mas, and 2.5 kpc and 4 mas, respectively. Thus refractive scintillation time scales for these masers would be about 6 and 0.8 years, respectively — too long compared to our observed time scales of $\sim$20 and $\sim$5 days. The diffractive time scales (about 100 and 500 seconds) are too short to explain our observations. Furthermore, for diffractive scintillation to occur, the intrinsic angular sizes of these masers must be very small ($\lesssim 10$ *nano*arcsec) requiring extremely large brightness temperatures $\gtrsim 10^{22}$ K.

The H_2O feature variations are apparently consistent with refractive interstellar scintillations. For W49N (L=14 kpc, $\theta_o \approx 0.2$ mas) and W51M (L=7 kpc, $\theta_o \approx 0.1$ mas), the refractive time scales are $\sim$80 and $\sim$20 days, respectively. Furthermore, the H_2O variation levels of $\sim$10–50% are what one would expect if the intrinsic angular sizes of the masers are comparable to or less than the scattering size. Finally, structure functions of flux density variations which rise less quickly than quadratic are observed for compact extragalactic source flicker and interpreted as produced by a scattering medium extended along the line of sight. Such behavior is common in the observed structure functions for the H_2O maser features.

In conclusion, water maser variations occurring on $\gtrsim$monthly time scales with variations of rms/mean $\sim$ 10–50% (and distinct from flares) are quite possibly refractive interstellar scintillations. It is important to note that simple scintillation theory predicts that refractive scintillations should be occurring at such levels and on such time scales, even if our observations can be explained by some intrinsic model.

REFERENCES

Garay, G., Moran, J.M., Haschick, A.D. 1989, ApJ, 338, 244
Kellermann, K.I. and Owen, F.N. 1988, in Galactic and Extragalactic Radio Astronomy, eds. G.L. Vershuur & K.I. Kellermann (New York: Springer), p. 563
Reid, M.J., and Moran, J.M. 1988, in Galactic and Extragalactic Radio Astronomy, eds. G.L. Vershuur & K.I. Kellermann (New York: Springer), p. 255
Rickett, B.J. 1990, ARA&A, 28, 561

Water Maser Monitoring of IRAS 16293-2422

Alwyn Wootten
National Radio Astronomy Observatory, 520 Edgemont Road, Charlottesville, VA
22903-2415 USA

ABSTRACT. Emission from the water maser associated with the primary component of the protobinary star IRAS16293-2422 has been monitored since 1986 Dec.

OBSERVATIONS.

The NRAO[1] Very Large Array (VLA) observed the IRAS16293-2422 region on 20 Jan 1989, with the array in its largest, 'A', configuration. The spectral line observations of the 6_{16}-5_{23} masing line of H_2O employed the "C" IF channel with 63 channels of 0.33 km s^{-1} resolution centered on the strongest hyperfine component at 22.235120 GHz which was placed at the nominal velocity of the ambient cloud, V_{LSR}=4.0 km s^{-1}. 20.3 km s^{-1} of velocity coverage was available. 3C286 observations set the flux scale (assumed flux $S_{22.24}$=2.55 Jy), and 1622-297 (measured flux $S_{22.46}$=1.77 Jy), served as phase calibrator. The final naturally weighted images reached sensitivities of 25000 μJy per channel at a resolution of 0."18 × 0."08. Figure 1 shows the water maser positions and velocities superposed upon the simultaneous 22.46 GHz continuum image of the region (Mundy, Wootten, Wilking, Blake and Sargent 1992).

Table I lists the positions, velocities and fluxes of the 12 distinct H_2O maser sources found.

Observations were made at times noted in Figure 2 at the 43m telescope of the National Radio Astronomy Observatory (NRAO) in Green Bank, West Virginia. The 1024 channel Mark IV autocorrelator provided a resolution of 0.15 km s^{-1} for most sources. The data were taken using a frequency-switched mode, and the autocorrelator was divided into two 512 channel sections sampling the output of the receiver. The two sections were averaged, and the data were shifted in frequency and summed to produce the final spectra. The beamwidth of the 43m telescope is 1.'3 at 22 GHz and the beam efficiency is 20%. Line intensities are expressed as observed antenna temperature. Owing to the short observing sessions, the variety of weather conditions encountered, and the variable high frequency performance of the 43m telescope, the calibration is insecure. Pointing was, however, checked at the beginning of each session.

The similarity of the velocity pattern of the masers with that of the CO outflow suggests that the masers and the outflow share a common origin. The locations of the

---

[1]   The National Radio Astronomy Observatory is operated by Associated Universities, Inc., under cooperative agreement with the National Science Foundation.

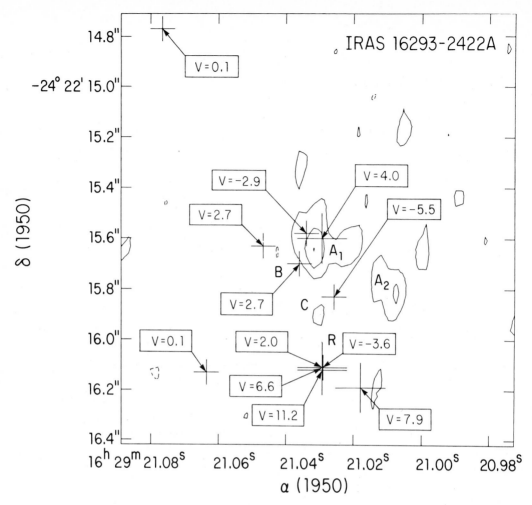

Fig. 1. A detail of the region around IRAS16293A, showing continuum emission at 22.46 GHz (Mundy et al. 1992). Contours are drawn at -2, 2, 4, 6, and 8 times 250 μJy. Positions of the water masers on 20 Jan 89 are plotted and tagged with their peak velocities. Large crosses mark masers at velocities above the ambient cloud velocity of $V_{LSR}=3.7$ km s^{-1}, and small crosses mark masers below that velocity. The geometric mean position of redshifted masers is marked by the letter R, and of the blueshifted masers by the letter B, with their centroid marked by a letter C. Each 0.″2 covers approximately 30 AU at the 160 pc distance of the cloud.

masers present further clues to the outflow origin. The median position of blueshifted maser features lies east of the A1 ionized gas structure, which may locate the 'A' star, or northeast of the mean A1/A2 position. The median position of redshifted features lies to the southwest. The axis of median water maser positions aligns with the axis of the A1/A2 ionized gas features, and in a general sense with the CO flow direction mapped on a much larger scale. No maser feature is found near the 'B' component of the region, which shows no evidence for an outflow. These conclusions, also reached in an 01 Jul 87 study of the region (Wootten 1990) which detected fewer masers, are strengthened by the evidence of the twelve masers located in Figure 1 .

TABLE I
Water Maser Positions near IRAS16293-2422 (1989.05)

Object :	$\alpha(1950)$	$\delta(1950)$	Flux (Jy)	V_{LSR} (km/s)
IRAS16293-2422 H_2O :	$16^h29^m21.0258^s$	-24°22'15.83"	0.394	-5.9
IRAS16293-2422 H_2O :	$16^h29^m21.0289^s$	-24°22'16.11"	0.316	-3.6
IRAS16293-2422 H_2O :	$16^h29^m21.0339^s$	-24°22'15.58"	0.301	-2.9
IRAS16293-2422 H_2O :	$16^h29^m21.0765^s$	-24°22'14.77"	0.531	0.1
IRAS16293-2422 H_2O :	$16^h29^m21.0636^s$	-24°22'16.13"	15.3	0.1
IRAS16293-2422 H_2O :	$16^h29^m21.0292^s$	-24°22'16.11"	0.614	2.0
IRAS16293-2422 H_2O :	$16^h29^m21.0468^s$	-24°22'15.63"	0.394	2.7
IRAS16293-2422 H_2O :	$16^h29^m21.0360^s$	-24°22'15.70"	0.771	2.7
IRAS16293-2422 H_2O :	$16^h29^m21.0293^s$	-24°22'15.60"	2.79	4.0
IRAS16293-2422 H_2O :	$16^h29^m21.0292^s$	-24°22'16.11"	3247.	6.6
IRAS16293-2422 H_2O :	$16^h29^m21.0179^s$	-24°22'16.19"	721.	7.9
IRAS16293-2422 H_2O :	$16^h29^m21.0292^s$	-24°22'16.12"	0.619	11.2

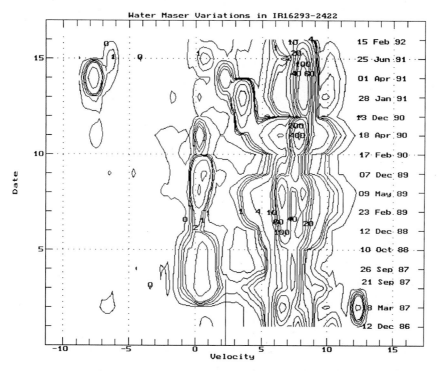

Fig. 2. Contours of water vapor spectra observed toward IR16293 from 12 Dec 86 (bottom) through 15 Feb 92 (top) with the NRAO 43m telescope. Velocities are plotted on the abcissa, and session numbers advancing with time are plotted on the ordinate. Sensitivity of the observations is typically 0.1K (several hundred mJy); the lowest contour unit shown is 0.2 K.

Long-term variations in velocity and intensity of maser features have been monitored using the 43m telescope. The higher velocity features tend to be weaker and more transitory than those within ± 4 km s^{-1} of the systemic velocity of 4.1 km s^{-1}.

Particularly notable higher velocity features occurred at 12.3 km s^{-1} in early 1987, and at -7.8 km s^{-1} during early 1991. Each of these features was characterized by an appearance and disappearance within six months.

Strong, low velocity features persist at all times. Emission has always been present between 6 and 8 km s^{-1}. During both epochs of interferometric measurements, features at this velocity lay just to the south and southwest of the continuum sources. A second set of prominent features occurs near 0 km s^{-1} and was quite prominent during 1987 through early 1990, reappearing during the latter half of 1991. The interferometry shows this set of features to be a blend of a number of components arising in spatially distinct regions. A third set of low velocity features, occurring very near the systemic velocity, became prominent during early 1989 and again during 1990-91. Despite the presence of this feature, Figure 2 shows that maser features generally avoid the range of 2-6 km s^{-1} within which gas rotating in the central structure occurs (Mundy, Wootten and Wilking 1990). For all of these lower velocity components, the true spatial velocities may be similar, and the distinction between them on the basis of radial velocity alone may be misleading.

The transverse velocities of the masers are still unknown. Assume half of the velocity spread of the low velocity masers occurs in a transverse component, or about 5 km s^{-1}. At the distance of Ophiuchus (160 pc), this results in a motion of about 0.''01 (1 AU) in one year. This study verifies that several masers are sufficiently long-lived for detections of motion over yearlong timescales. Such motions should be discernible with the VLBA, and images will be made in an attempt to measure them.

At both epochs of VLA measurements, a bipolar pattern has been apparent in the positions and velocities of the maser components. The mean position of the redshifted components lies 0.''4 southwest of the mean position of the blushifted components, along an axis parallel to that of the two continuum source components. The direction and sense of the shift mimics that seen on larger scales in the CO outflow from the source. The character of the flow is established, therefore, on scales of 40 AU, the scale of the orbits of the giant planets in the solar system.

Acknowledgements. This project arose out of discussions among a number of people, notably Priscilla Benson, Bruce Wilking, Mark Claussen, Phil Myers, Sue Tereby and Stu Vogel. This paper serves as a progress report for the continuing monitoring program underway at the 43m and VLA telescopes at NRAO.

References

Mundy, L. G., Wootten, H. A. and Wilking, B. A., Blake, G. A. and Sargent, A. I. 1992, Ap. J., 385,, 306.

Mundy, L. G., Wootten, H. A. and Wilking, B. A. 1990, Ap. J., 352,, 159.

Wootten, A. 1990, Ap. J., 337,, 858.

12. CIRCUMSTELLAR MASERS (GENERAL)

RECENT STUDIES OF CIRCUMSTELLAR MASERS

P. F. Bowers[1]
SFA, Inc., 1401 McCormick Dr., Landover, MD 20785

1. INTRODUCTION

Masers often are detected from star formation regions and evolved stars. Conventionally, the term circumstellar masers refers to those found from post-main sequence stars: red giants or supergiants with rates of mass loss $\approx 10^{-7}$ to 5×10^{-4} $M_\odot$ y^{-1}, protoplanetary nebulae (former red giants surrounded by shells produced in the mass loss phase), and young, ionization-bounded planetary nebulae. Masers are valuable probes of the envelope structures, and conversely, the envelopes serve as a good laboratory to study maser phenomena since the stars usually are not embedded in complex ambient environments. The field is very active. Numerous results have been published in the past few years, leading to detection of many new sources, new species of circumstellar masers, and much more information about characteristics of the masers. Developments since about 1987 are briefly discussed, including the data base, properties of the more common masers, and efforts to model the profiles and angular distributions. Other recent reviews are given by Cohen (1989; good primer) and Elitzur (1992).

2. THE DATA BASE

Source catalogs have been published by Engels & Heske (1989), te Lintel Hekkert et al. (1989), Comoretto et al. (1990), and Benson et al. (1990) for OH, H_2O, and SiO masers discovered prior to about 1988. Because maser stars are located in relatively well defined regions of the IRAS color-color plot, many sources have since been discovered with a high detection efficiency by sensitive 1612 MHz OH surveys of stars in the Point Source Catalog [e.g., Eder et al. 1988 (169); Gaylard et al. 1989 (9); Lewis et al. 1990 (79); Sivagnanam et al. 1990a (36); te Lintel Hekkert et al. 1991 (597)]. Several programs have been focused on cold (T_c < 200 K) IRAS sources in searches for masers from protoplanetary or planetary nebulae [Likkel 1989 (10); Zijlstra et al. 1989 (10); te Lintel Hekkert 1991 (3)], while other programs have been guided by the presence of the 9.7 μm silicate feature indicated in the IRAS Low Resolution Spectra Catalog [Gaylard & Whitelock 1988 (18); Galt et al. 1989 (25); Le Squeren et al. 1992 (74)]. New OH maser stars also have been found from surveys of nearby Miras [Sivagnanam et al. 1988 (14)], singly peaked 1612 MHz sources [Bowers & Knapp 1989 (11)], the galactic center region [Lindqvst et al. 1992 (120)], and serendipitously [Becker et al. 1992 (14)]. There are about 1700 - 1800 published detections of OH maser stars. As indicated in many of these papers, observations at 1612 MHz are becoming increasingly difficult because of frequent interference by the GLONASS satellites.

[1]Work performed at the Center for Advanced Space Sensing, Naval Research Laboratory, Washington, DC, under contract # N00014-89-C-2398.

Using the IRAS survey or lists of known OH maser stars, many observers
have conducted searches for other masers, including mainline (1665/1667
MHz) OH [Dickinson & Turner 1991 (41)], H_2O at 22 GHz [Deguchi et al. 1989
(27); Lindqvst et al. 1990 (4); Gómez et al. 1990 (4)], vibrationally excited (v = 1 - 3) ^{28}SiO [Allen et al. 1989 (12); Alcolea et al. 1989, 1990
(10); Gómez et al. 1990 (8); Le Bertre & Nyman 1990 (13); Hall et al. 1990
(31); Jewell et al. 1991 (55)], and (v = 4) SiO (Cernicharo et al. - these
proceedings). There are about 400 published detections of 22 GHz H_2O stars
and 300 published SiO sources, but many more have been discovered (Engels &
Lewis; Haikala - these proceedings). Other searches have led to the discoveries of OH or H_2O masers from a few carbon stars (and a carbon-rich
planetary nebula) surrounded by oxygen-rich material (Barnbaum et al. 1991;
Zijlstra et al. 1991), additional ground- and excited-state lines of H_2O
(Cernicharo et al. 1990; Menten & Melnick 1991), ground-state masers from
^{29}SiO and ^{30}SiO (Alcolea & Bujarrabal 1992) and ^{28}SiO (e.g., Jewell et al.
1991), vibrationally excited ^{29}SiO (Cernicharo et al. 1991), excited- and
possibly ground-state HCN and ground-state $H^{13}CN$ masers in oxygen-rich or
carbon-rich envelopes (Izumiura et al. 1987; Lucas et al. 1988; Nguyen-Q-
Rieu et al. 1988), ground-state SiS masers in carbon-rich envelopes
(Nguyen-Q-Rieu et al. 1984; Sahai et al. 1984), and recombination line
masers H29-31α from the binary system MWC349 (Martin-Pintado et al. 1989).

A claim for a ground-state CO maser was made and then retracted by
Zuckerman & Dyck (1986, 1989). The basis for the claim was a narrow line
and apparent time variability. For some of the above species, claims for
masers rest only on a narrow line. Maser emission from the vibrational
ground-state of diatomic molecules may commonly be produced in low-density
molecular envelopes (Schönberg 1988), but attempts to measure the brightness temperature or polarization of at least some of the putative ground-
state masers are needed, since narrow, presumably thermal lines have been
found in various stars (e.g., CO data of Margulis et al. 1990) and may be
formed in complex envelope structures or velocity fields. Efforts to
determine if narrow CO lines are indeed thermal also might be interesting
in view of proposed pumping schemes for CO (e.g., Piehler et al. 1991).

3. SOME GENERAL PROPERTIES OF SIO, H_2O, AND OH

3.1 <u>SiO</u>.   The vibrationally excited (v = 1: J = 1 - 0; J = 2 - 1)
masers are the most studied and may be present in the oxygen-rich envelopes
of all Miras ($\Delta m \geq 2.5$ mag) with $\dot{M} \geq 10^{-7} M_\odot y^{-1}$ and spectral classes
later than M4 (Heske 1989; Hall et al. 1990; Jewell et al. 1991); semi-
regular giants or supergiants with $\Delta m \geq 2.5$ mag also tend to have stronger
SiO emission (Alcolea et al. 1990). The masers are located close to the
star (3 - 5 R_*) in clumpy distributions (e.g., McIntosh et al. 1989) with
spot sizes $\approx 10^{12} - 10^{13}$ cm (Colomer et al. 1992), and maps or profiles at
different light cycles show little similarity. Emission can exhibit substantial linear or circular polarization. Hall et al. (1990) find similar
position angles of polarized features in sequential light cycles of the
symbiotic variable R Aqr, suggesting that some cycle-to-cycle properties of

the masers may not be completely random and that the linear polarization may be produced by a magnetic field rather than anisotropies in the out-flow. The binary nature of the star may influence their results, but indications of strong (10 - 100 G) magnetic fields are derived from circular polarization data for several red giants and supergiants (Barvainis et al. 1987). An interpretation of strong fields is supported by Elitzur (1991); other possibilities are discussed by Nedoluha & Watson (these proceedings).

The SiO luminosity is not strongly dependent on $\dot{M}$ (Bowers 1985; Le Bertre & Nyman 1990; Jewell et al. 1991), but a weak dependence is possible (Alcolea et al. 1990; Hall et al. 1990). Bujarrabal et al. (1987) find a tight correlation between the SiO and 8 μm luminosities at maximum light 1987), suggesting radiative pumping of saturated masers, but collisional pumping also is viable (Lockett & Elitzur, these proceedings). Monitoring observations do not resolve the issue (Nyman & Olofsson 1986; Martinez et al. 1988). Epochs of SiO and optical maxima are sometimes well correlated (phase lag $\approx$ 0.0 - 0.2 P), but there can be erratic variations and little correlation between the SiO and optical amplitudes. Bowers (1992) proposes that a tendency for the velocity ranges of SiO emission from Mira variables to be slightly red-shifted relative to V_* can be explained if some of the (v = 1) masers are radiatively pumped and located in the outflow.

3.2 $\underline{H_2O}$. Maser emission from the ground-state rotational ($6_{16} - 5_{23}$) transition at 22 GHz is also commonly associated with Mira variables ($\approx$ 75%; Bowers & Hagen 1984) and probably with most OH/IR stars (Engels et al. 1986). These authors find that the maser luminosity is proportional to $\dot{M}$. Nyman et al. (1986) question the correlation, but the distances of the stars in their sample are not well determined. Various authors find that the detection rate of H_2O masers decreases for the reddest IRAS stars (Likkel 1989; Gómez et al. 1990; Lewis & Engels 1991). Suggested causes include an insufficient gas density in the inner region of the envelope (e.g., a protoplanetary nebula) or an insufficient excitation temperature beyond the quenching radius.

The masers are located between about 6 and 30 R_* for both Mira variables and supergiants, with the shell radius depending on $\dot{M}$ (cf. Lane et al. 1987; Bowers 1991). Spot sizes range from about 0.5 x 10^{13} - 4 x 10^{13} cm (Diamond et al. 1987). Properties of the H_2O masers tend to be less random than those of SiO but more so than OH. Individual spectral features appear to be stable over a few months for semiregular variables, a few yr for Mira variables, and $\geq$ 15 yr for supergiants (Engels et al. 1988), but different features can have different lifetimes (Little-Marenin et al. 1991). Similar results may apply for the angular distributions. Over an interval of about one yr, Claussen et al. (1992) find agreement for the H_2O distributions of IK Tau, but there is no correspondence of the distributions for Miras observed in intervals larger than a few yr (R Aql - Lane et al. 1987; W Hya - Claussen et al. 1992). For some cases, the angular distributions indicate a probable component of outflow in the H_2O regions (S Per - Diamond et al. 1987; W Hya - Reid & Menten 1990; IK Tau - Claussen et al. 1992), but shell structures are complex.

Profile shapes are not easily classified. They can range from one or two narrow features within a few km s^{-1} of V_* (more typical for classical Miras with low $\dot{M}$) to complex, multiply peaked structures (IK Tau - Lane et al. 1987; S Per - Little-Marenin et al. 1991). There is a larger tendency for OH/IR stars (high $\dot{M}$) to show roughly doubly peaked profiles (Engels et al. 1986), but profile shapes can change strongly with time, possibly because of temperature changes in the maser shell (Lewis & Engels 1991). Quasi-doubly peaked structures with velocity ranges ≥ 200 km s^{-1} are seen for two bipolar outflows (Likkel & Morris 1988). For Miras, observations with high spectral sensitivity indicate that low-level, plateau emission can be present and that the maximum velocity range of the H_2O is comparable to that of the SiO and OH masers (Bowers 1992). The average velocity of the emission tends to be blueshifted relative to V_*, possibly indicating that some components in the H_2O shell are amplifying the stellar continuum. Searches for linear or circular polarization yield limits < 2 - 3% unless the line is very intense (Barvainis & Deguchi 1989; Reid & Menten 1990), indicating that the emission typically is not strongly saturated.

Closely sampled (≈ 30 d) monitoring results are now available (Little-Marenin & Benson 1991; Little-Marenin et al. 1991). For Miras, individual features have different phase lags relative to the optical maximum, varying independently of each other, but the dominant feature mimics the optical light curve with a phase lag $\approx 0.3 - 0.4$ P, providing good evidence for a collisional pumping mechanism. For the supergiant S Per, variation of the redshifted features (relative to V_*) is uncorrelated with the optical variation. Blueshifted features reach maximum in phase with the optical, implying radiative coupling, but there is a secondary maximum about ten weeks later for some features. Comparison with previous data indicates that the correlation of maser activity with the visual light curve is different at different epochs. Little-Marenin et al. conclude that the H_2O variations for S Per can be interpreted in terms of an asymmetric distribution of the masers and/or different pumping schemes. One pumping scheme (collisional) may be adequate, however, if radiative coupling of the features in phase with the optical cycle is caused by amplification of the stellar continuum. Recent calculations for excitation of the 22 GHz and other transitions are provided by Deguchi & Nguyen-Q-Rieu (1990) and Neufeld & Melnick (1991).

3.3 OH. It is estimated that OH maser stars comprise about 50% of Miras with spectral types of M5.5 or later (Sivagnanam et al. 1988). For $\dot{M} \approx 10^{-7}$ to 10^{-6} $M_\odot$ y^{-1}, the 1665/67 MHz lines are dominant, but if $\dot{M} \geq 10^{-5}$ $M_\odot$ y^{-1}, the 1612 MHz line usually is stronger, with a luminosity proportional to $\dot{M}^2$, enabling detection of OH/IR stars throughout the Galaxy (reviews of Bowers 1985; Herman & Habing 1985). Comparisons of OH and IRAS properties are numerous (Sivagnanam et al. 1988; Lewis 1990; citations in § 2). A scenario for how SiO, H_2O, and OH maser properties depend on the evolutionary phase and infrared colors is presented by Lewis (1989). A limitation to this scheme is provided by the well known bipolar outflow OH231.8+4.2 ($\dot{M} \geq 10^{-5}$ $M_\odot$ y^{-1}). Strongest OH emission occurs at 1667 MHz in the cold shell ($T_c \approx 100$ K, typical of a protoplanetary nebula), but in contrast to the Lewis scenario, there are SiO and H_2O masers (Morris et al.

1987), indicating that mass loss is ongoing and consistent with classification of the star as an M9 Mira. A strong concentration of matter near the
equatorial plane might account for the anomalously cold shell. Possibly
similar objects are discussed by te Lintel Hekkert et al. (1988).

Comparisons of OH and IRAS flux densities indicate that the 1612 MHz
masers from OH/IR stars are probably pumped by means of 53 μm as well as
35 μm photons (Röttgering 1989; Gaylard et al. 1989; Dickinson 1991), and
this also may apply for the main lines (Sivagnanam et al. 1989). In cold
shells pumping via 80 and 120 μm photons may be important (Gaylard & Whitelock 1988). Dickinson & Turner (1991) find that the 1667/1665 MHz flux
ratio for OH/IR stars tends to be larger than for Miras, perhaps because of
different dust temperatures. The 1667 MHz line can be stronger than the
1612 MHz line if $T_c \le 100$ K (Morris & Bowers 1980). For various cases,
the latter authors and others (Sivagnanam et al. 1989; Nedoluha & Bowers
1992) suggest that nonlocal, line-overlap effects may be operative. A
recent model for these effects is given by Cesaroni & Walmsley (1991).

The 1612 MHz masers from OH/IR stars are thought to be saturated, but
a 60-fold increase for the strength of some features from IRC+10420 demonstrates that this is not always the case (Nedoluha & Bowers 1992). The
magnitude of the increase is comparable with that for the well known 1612
MHz flare from U Orionis. Large amplitude variations (≥ 10) and changes of
profile shapes are more common for the main lines; this emission probably
is not highly saturated if $\dot{M} \le 10^{-6}$ $M_\odot$ y^{-1}. In addition to U Ori (cf.
Nedoluha & Bowers), examples include R Leo (Le Squeren & Sivagnanam 1985),
R Cnc and V Cam (Sivagnanam et al. 1988), and S Per (Bowers et al. 1989).

Significant polarization can be detected if data are obtained with a
high velocity resolution (0.1 km s^{-1}). Magnetic field strengths ≈ 1 mG at a
few x 10^{16} cm have been derived for OH supergiants (Cohen et al. 1987;
Nedoluha & Bowers 1992). Zell & Fix (1991) find weaker fields (0.001 - 0.1
mG) for OH/IR stars. Substantial polarization also is found from OH Miras
(Ukita & LeSqueren 1984; Chapman et al. 1991) and protoplanetary or planetary nebulae (Zijlstra et al. 1989). The field may influence the geometry
of the outflow in some cases, but interpretation is difficult (§ 3.1).

Many interferometric results have been published. At 1612 MHz, shell
radii can range from about 50 AU (classical Miras) to 10^4 AU (OH/IR stars),
depending on $\dot{M}$. Relative to the 1612 MHz radius, mainline emission can
occur at a smaller radius (OH127.8-0.0 - Diamond et al. 1985; VX Sgr -
Chapman & Cohen 1986), at the same radius (U Ori - Bowers & Johnston 1988),
or at a larger radius (IRC+10420 - Bowers 1984). First measurements of
proper motions of circumstellar masers (and the transverse expansion
velocity) have been presented by Chapman et al. (1991) for U Ori. Several
Miras have been mapped at 1667 or 1612 MHz by Bowers et al. (1989), who
find shell radii from about 40 to 400 AU and indications of axially symmetric or asymmetric outflows. Sivagnanam et al. (1990b) have mapped 1665
and 1667 MHz emission from U Her and propose that the extreme blueshifted
component amplifies the stellar continuum at 18 cm. High-quality 1612 MHz

maps have been published by Bowers & Johnston (1990) for OH127.8-0.0 (best known example which conforms to the standard expanding shell model) and OH26.5+0.6 (one of the less symmetric expanding shells for an OH/IR star). Maps and a detailed analysis of the 1612 and 1665 MHz shell structures for IRC+10420 are presented by Nedoluha & Bowers (1992). The bipolar outflows of OH19.2-1.0 and HD 101584 have been respectively mapped at 1612 and 1667 MHz by Chapman (1988) and te Lintel Hekkert et al. (1992). Maps at 1612 or 1667 MHz also have been obtained for several planetary nebulae with OH (NGC 6302 - Payne et al. 1988; Vy 2-2 and OH0.9+1.3 - Shepherd et al. 1990; M1-92 - Seaquist et al. 1991). From measurements of 1612 MHz spot sizes for stars near the Galactic center, van Langevelde & Diamond (1991) conclude that the spots are broadened by interstellar scattering and that phase-lag distances cannot be obtained for such stars. On a related note, van Langevelde et al. (1990) have reassessed phase-lag data for OH/IR stars and revise some previously published values by as much as a factor of 2.

4. MODELING MASER PROFILES AND DISTRIBUTIONS

There have been several recent attempts to develop more realistic models of maser shells in order to explain complexities in profile shapes and angular distributions. A stochastic model is considered by Zell & Fix (1990), who find that randomly distributed, discrete emitting elements (blobs) can produce fine structure seen in the 1612 MHz profiles of OH/IR stars if the number of blobs is large (10^3 - 10^4), or large differences between the flux densities of the blue- and redshifted peak features if the number of blobs is small (25). They cannot model profiles with both fine structure and unequal flux densities of the peaks. A similar approach is used by Szymczak (1989, 1990), who finds that unequal peak flux densities can result from small-amplitude density fluctuations with a size comparable to the shell radius, from small fluctuations of the outflow velocity, or from spatial fluctuations of the pump rate. Systematic effects (blue peak > red peak) can be produced by amplification of stellar emission (Szymczak 1988) or by absorption of redshifted emission through ionized gas (see references for OH-planetary nebulae in previous paragraph).

Others have considered properties of aspherical outflows. From a dynamical model, Collison & Fix (1992) demonstrate that an axially symmetric outflow with a smooth wind can produce many aspects of observed profiles and interferometer maps which neither the spherical shell model nor the blob model can explain. Bowers (1991) computes profiles and maps from kinematical models of ellipsoidal (three-dimensional) shells. This approach can be used for any maser species and provides a way to parameterize aspherical outflows by fitting simultaneously to the observed profiles and the angular distributions. Sources modeled with this technique include OH231.8+4.2 and Orion-IRc2 (Bowers 1991) and IRC+10420 (Nedoluha & Bowers 1992).

Although it is common practice to attempt to deduce the geometry or velocity field of the maser shell solely from analysis of profile shapes,

326

results from kinematical models indicate that: 1) similar profile shapes can be produced by completely different velocity fields, and 2) for a given geometry/velocity field, different profile shapes can be formed, depending on the Doppler width and outflow velocity. The latter point is particularly important when comparing maser profiles from different regions of the envelope, and it applies even in the case of an expanding, spherical shell.

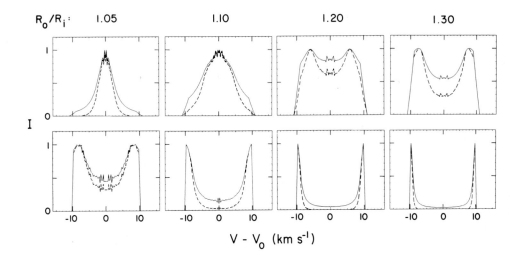

Figure 1: Intensity profiles for a spherical shell with a constant outflow velocity of 10 km s^{-1}. The FWHM Doppler width is 2.0 km s^{-1} for the upper row of figures and 0.35 km s^{-1} for the lower row. The ratio of the outer radius R_o to the inner radius R_i of the shell is indicated. Gas is distributed uniformly between these radii and the emission is saturated (no line narrowing). Dashed curves illustrate the effects of a minimum column density for saturated emission, where the minimum velocity-coherent pathlength is 0.25 times the maximum at any velocity.

The influence of the Doppler width is demonstrated in Figure 1, which shows computed profiles for a spherical shell as a function of the shell thickness. Profiles in the upper row of panels are computed for a large Doppler width, and profiles in the lower row are computed for a small Doppler width. The outflow velocity is identical for all cases (10 km s^{-1}). It is evident that if the shell is sufficiently thin, profile shapes can differ significantly as the Doppler width changes. It also is evident that the profile is no longer doubly peaked if the shell is very thin and if the gas is very warm, because the intensity in the tangential direction integrated over a spherical distribution can exceed the intensity in the radial direction. Profiles with a dominant peak near the stellar velocity are not uncommon in circumstellar shells, but interferometric data are needed to differentiate between the thin-shell model and more complex cases (e.g., accelerative or anisotropic outflows) which can produce similar profile shapes (Bowers 1991, 1992).

REFERENCES

Alcolea, J., & Bujarrabal, V. 1992, A&A, 253, 475
Alcolea, J., Bujarrabal, V., & Gallego, J.D. 1989, A&A, 211, 187
Alcolea, J., Bujarrabal, V., & Gómez-González, J. 1990, A&A, 231, 431
Allen, D.A., et al. 1989, MNRAS, 236, 363
Barnbaum, C., Morris, M., Likkel, L, & Kastner, J.H. 1991, A&A, 251, 79
Barvainis, R., & Deguchi, S. 1989, AJ, 97, 1089
Barvainis, R., McIntosh, G., & Predmore, C.R. 1987, Nat, 329, 613
Becker, R.H., White, R.L., & Proctor, D.D. 1992, AJ, 103, 544
Benson, P.J., et al. 1990, ApJS, 74, 911
Bowers, P.F. 1984, ApJ, 279, 350
Bowers, P.F. 1985, in Mass Loss from Red Giants (Dordrecht:Reidel), p.189
Bowers, P.F. 1991, ApJS, 76, 1099
Bowers, P.F. 1992, ApJ, 390, L27
Bowers, P.F., & Hagen, W. 1984, ApJ, 285, 637
Bowers, P.F., & Johnston, K.J. 1988, ApJ, 330, 339
Bowers, P.F., & Johnston, K.J. 1990, ApJ, 354, 676
Bowers, P.F., Johnston, K.J., & de Vegt, C. 1989, ApJ, 340, 479
Bowers, P.F., & Knapp, G.R. 1989, ApJ, 347, 325
Bujarrabal, V., Planesas, P., & del Romero, A. 1987, A&A, 175, 164
Cernicharo, J., Bujarrabal, V., & Lucas, R. 1991, A&A, 249, L27
Cernicharo, J., et al. 1990, A&A, 231, L5
Cesaroni, R., & Walmsley, C.M. 1991, A&A, 241, 537
Chapman, J.M. 1988, MNRAS, 230, 415
Chapman, J.M., & Cohen, R.J. 1986, MNRAS, 220, 513
Chapman, J.M., Cohen, R.J., & Saikia, D.J. 1991, MNRAS, 249, 227
Claussen, M.J., Bowers, P.F., & Johnston, K.J. 1992, AJ, submitted
Cohen, R.J. 1989, Rep.Prog.Phys., 52, 881
Cohen, R.J., et al. 1987, MNRAS, 225, 491
Collison, A.J., & Fix, J.D. 1992, ApJ, 390, 191
Colomer, F., et al. 1992, A&A, 254, L17
Comoretto, G., et al. 1990. A&AS, 84, 179
Deguchi, S., Nakada, Y., & Forster, J.R. 1989, MNRAS, 239, 825
Deguchi, S., & Nguyen-Q-Rieu. 1990, ApJ, 360, L27
Diamond, P.J., et al. 1987, A&A, 174, 95
Diamond, P.J., Norris, R.P., & Booth, R.S. 1985, MNRAS, 216, 1p
Dickinson, D.F. 1991, ApJ, 379, L29
Dickinson, D.F., & Turner, B.E. 1991, ApJS, 75, 1323
Eder, J., Lewis, B.M., & Terzian, Y. 1988, ApJS, 66, 183
Elitzur, M. 1991, ApJ, 370, 407
Elitzur, M. 1992, ARAA, 30, in press
Engels, D., & Heske, A. 1989, A&AS, 81, 323
Engels, D., Schmid-Burgk, J., & Walmsley, C.M. 1986, A&A, 167, 129
Engels, D., Schmid-Burgk, J., & Walmsley, C.M. 1988, A&A, 191, 283
Galt, J.A., Kwok, S., & Frankow, J. 1989, AJ, 98, 2182
Gaylard, M.J., et al. 1989, MNRAS, 236, 247
Gaylard, M.J., & Whitelock, P.A. 1988, MNRAS, 235, 123
Gómez, Y., Moran, J.M., & Rodríguez, L.F. 1990, Rev.Mex.A&A, 20, 55
Hall, P.J., et al. 1990, MNRAS, 243, 480
Herman, J., & Habing, H.J. 1985, Phys.Rep., 124, 255
Heske, A. 1989, A&A, 208, 77
Izumiura, H., et al. 1987, ApJ, 323, L81
Jewell, P.R., et al. 1991, A&A, 242, 211
Lane, A.P., et al. 1987, ApJ, 323, 756
Le Bertre, T., & Nyman, L.-Å. 1990, A&A, 233, 477

Le Squeren, A.M., et al. 1992, A&A, 254, 133
Le Squeren, A.M., & Sivagnanam, P. 1985, A&A, 152, 85
Lewis, B.M. 1989, ApJ, 338, 234
Lewis, B.M. 1990, AJ, 99, 710
Lewis, B.M., Eder, J., & Terzian, Y. 1990, ApJ, 362, 634
Lewis, B.M., & Engels, D. 1991, MNRAS, 251, 391
Likkel, L. 1989, ApJ, 344, 350
Likkel, L., & Morris, M. 1988, ApJ, 329, 914
Lindqvst, M., et al. 1992, A&AS, 92, 43
Lindqvst, M., Winnberg, A., & Forster, J.R. 1990, A&A, 229, 165
te Lintel Hekkert, P. 1989, A&AS, 78, 399
te Lintel Hekkert, P. 1991, A&A, 248, 209
te Lintel Hekkert, P., Chapman, J.M., & Zijlstra, A.A., 1992, ApJ, 390, L23
te Lintel Hekkert, P., et al. 1988, A&A, 202, L19
te Lintel Hekkert, P., et al. 1991, A&AS, 90, 327
Little-Marenin, I.R., & Benson, P.J. 1991, ASP Conf. Ser. 16, 73
Little-Marenin, I.R., et al. 1991, A&A, 249, 465
Lucas, R., Guilloteau, S., & Omont, A. 1988, A&A, 194, 230
Margulis, M., et al. 1990, ApJ, 361, 673
Martinez, A., Bujarrabal, V., & Alcolea, J. 1988, A&AS, 74, 273
Martin-Pintado, J., Thum, C., & Bachiller, R. 1989, A&A, 222, L9
McIntosh, G.C., et al. 1989, ApJ, 337, 934
Menten, K.M., & Melnick, G.J. 1991, ApJ, 377, 647
Morris, M., & Bowers, P.F. 1980, AJ, 85, 724
Morris, M., Guilloteau, S., Lucas, R., & Omont, A. 1987, ApJ, 321, 888
Nedoluha, G.E., & Bowers, P.F. 1992, ApJ, 392, 249
Neufeld, D.A., & Melnick, G.J. 1991, ApJ, 368, 215
Nguyen-Q-Rieu, et al. 1984, ApJ, 286, 276
Nguyen-Q-Rieu, et al. 1988, ApJ, 330, 374
Nyman, L.-Å., Johansson, L.E.B., & Booth, R.S. 1986, A&A, 160, 352
Nyman, L.-Å., & Olofsson, H. 1986, A&A, 158, 67
Payne, H.E., Phillips, J.A., & Terzian, Y. 1988, ApJ, 326, 368
Piehler, G., Kegel, W.H., & Tsuji, T. 1991, A&A, 245, 580
Reid, M.J., & Menten, K.M. 1990, ApJ, 360, L51
Röttgering, H.J.A. 1989, A&A, 222, 125
Sahai, R., Wootten, A., & Clegg, R.E.S. 1984, ApJ, 284, 144
Schönberg, K. 1988, A&A, 195, 198
Seaquist, E.R., Plume, R., & Davis, L.E. 1991, ApJ, 367, 200
Shepherd, M.C., Cohen, R.J, Gaylard, M.J., & West, M.E. 1990, Nat, 344, 522
Sivagnanam, P., et al. 1989, A&A, 211, 341
Sivagnanam, P., et al. 1990a, A&A, 233, 112
Sivagnanam, P., et al. 1990b, A&A, 229, 171
Sivagnanam, P., Le Squeren, A.M., & Foy, F. 1988, A&A, 206, 285
Szymczak, M. 1988, Ap & Space Sci., 141, 353
Szymczak, M. 1989, MNRAS, 237, 561
Szymczak, M. 1990, MNRAS, 243, 375
Ukita, N., & LeSqueren, A.M. 1984, A&A, 138, 343
van Langevelde, H.J., & Diamond, P.J. 1991, MNRAS, 249, 7p
van Langevelde, H.J., et al. 1990, A&A, 239, 193
Zell, P.J., & Fix, J.D. 1990, AJ, 99, 314
Zell, P.J., & Fix, J.D. 1991, ApJ, 369, 506
Zijlstra, A.A., et al. 1989, A&A, 217, 157
Zijlstra, A.A., et al. 1991, A&A, 243, L9
Zuckerman, B., & Dyck, H.M. 1986, ApJ, 311, 345
Zuckerman, B., & Dyck, H.M. 1989, A&A, 209, 119

OBSERVATIONS OF MASER RRLs FROM THE STAR MWC 349A

M. A. Gordon

National Radio Astronomy Observatory*

Tucson, Arizona USA 85721-0655

ABSTRACT

Observations of H29α, H30α, and H31α radio recombination lines (RRLs) with rest frequencies 256 GHz, 232 GHz, and 210 GHz, respectively, from the star MWC 349A show variations in intensity and radial velocity over a period of 2.5 yrs. The data are consistent with a model of a star with a circumstellar disk embedded within an expanding, ionized atmosphere. Presumably, outbursts on the surface of the the star itself would cause changes in the circumstellar disk, which would affect the intensities of the RRLs. In this model the double-peaked profiles seen at mmwave frequencies would originate from 2 regions on opposite sides of a circumstellar disk in Keplerian motion, each region lying at a radius of $\approx$ 30 AU from the star. The separation of the emitting regions would then be $\approx$ 60 AU. The average radial velocity of the 2 peaks marks the radial velocity of the central star MWC 349A. At lower frequencies, the masing would subside. The increased opacity of the ionized wind would absorb the weakened RRLs from the disk and, correspondingly, generate increasingly visible RRLs from the stellar wind.

1. INTRODUCTION

Optically, MWC 349 is an emission star long known to have peculiar properties. First discovered by Merrill and Burwell (1933), it consists of a double system separated by 2.″4 and containing a B0 III star (MWC 349B) and a peculiar star (MWC 349A) with so many emission lines that it defies a simple classification (Brugel and Wallerstein 1979). Furthermore, MWC 349A lies at the center of a 20″ double-lobed emission that can be seen optically (Ney et al. 1975). It lies at a distance $\approx$ 1.2 kpc from the Sun.

MWC 349A is also a peculiar object with respect to its radio and IR spectra. It is one of the strongest radio stars known, with the emission coming from the same nebulosity seen optically (Cohen et al. 1985, White and Becker 1985). The continuum flux density varies as $\lambda^{-0.6}$, and its spectrum has been successfully modeled by a star with an ionized, stellar wind (Olnon 1975).

Perhaps most surprising was the discovery of unexpectedly intense, double-peaked radio recombination lines (RRLs) at millimeter wavelengths from MWC 349A (Martín-Pintado et al. 1989), whereas only weak, normal RRLs have been observed at centimeter wavelengths (Altenhoff et al. 1981). "Normal" RRLs at these frequencies have line-to-continuum ratios of $\approx$ 1 (see Fig. 1 of Gordon 1990), whereas these RRLs have line-to-continuum ratios of $\approx$ 50. The prima facie conclusion is that these lines are masing.

* Operated by Associated Universities, Inc., under cooperative agreement with the National Science Foundation

2. DATA

The line profile of the mmwave RRLs appear to be a composite of 2 narrow Gaussian components superimposed upon a broad Gaussian component. Fig. 1 shows a sample H31α RRL from MWC 349A observed with the NRAO 12-m telescope. The upper panel shows how the "pedestal" RRL was fitted; the lower panel, the line profile after removal of the pedestal RRL. Note that the fitting is imperfect. While the profile of the red (R) component fits a Gaussian well, the violet (V) component is slightly skewed. Furthermore, the baseline between the peaks after removal of the pedestal feature still shows a dip, indicating that the broad emission is not a true Gaussian. In the discussion that follows, the reader should note that the word "data" refers to the Gaussian fits rather than to the actual lines.

The top panel of Fig. 2 shows the variations of the violet/red intensity ratio (V/R) as a function of time. Panels 2 and 3 show the variations of the radial velocities of the 2 components to vary in opposite senses. Panel 4 shows the average radial velocity of the compoent to remain constant – thereby giving the radial velocity of the entire system. Thum et al. (1992) have made similar observations but do not report the variations in radial velocities. The average radial velocity of the R and V components differs from that observed from the cmwave lines observed by Altenhoff et al. (1981), implying that the low frequency lines have a different origin than the mmwave lines.

Both the velocity separation of the peaks and the intensity ratio V/R vary in a way similar to what has been observed in optical and IR emission lines from Be stars (see Fig. 4 of Cowley and Gugula 1973). However, the variations in RRLs reported here are much less regular and may be due to a different mechanism. In particular, the late 1991 and early 1992 data obtained after submission of Gordon (1992) shown in Fig. 2 do not confirm a regular, periodic variation in the line properties that would be seen in emission lines from a Be star.

3. MODEL

Martín-Pintado et al. (1989) proposed a model in which the λ1mm lines were masing within the 8500 K expanding stellar envelope. The model required a radial temperature gradient within the envelope to account for the line intensity. Although their model could account for the observed line intensities, it did not produce the double-peaked line profile.

To account for the double-peaked profile, Gordon (1992) proposed that the masing lines originate in a circumstellar disk in Keplerian rotation embedded within the large, ionized stellar envelope of radius $\approx$ 200 AU. The V component would come from the disk on one side of the star; the R, component from the other side. The time variations would then arise from changing excitation and kinematic conditions within the disk. This model is a slight modification of one proposed for optical and IR lines from MWC 349A by Hamann and Simon (1986, 1988). At short millimeter wavelengths, the observer sees primarily the masing RRLs; the normal RRLs emitted from the stellar envelope are too weak to contribute. As the wavelength increases from 1 mm, the masing RRLs from the inner disk weaken, and the free-free opacity of the stellar envelope increases. The composite RRL line profiles seen by an observer become increasingly dominated by the weak RRLs emitted by the stellar envelope as the free-free absorption blocks the RRLs emitted by the disk.

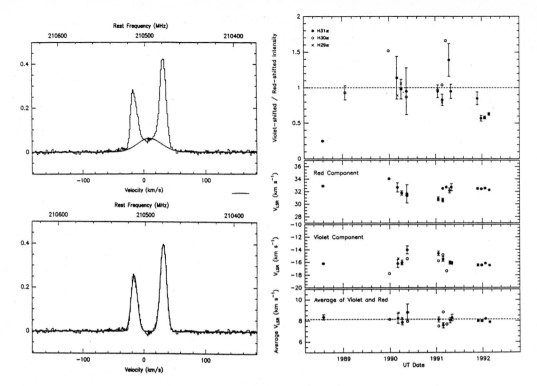

(*Left:*) **Fig. 1**—A spectrum of the H31α line at 210 GHz showing the violet (left) and the red (right) components superimposed upon a "pedestal" line that has been fitted with the Gaussian line shown. The bottom panel shows Gaussian fits to the V and R components after subtraction of the fit to the pedestal. (*Right:*) **Fig. 2**— Time variations of the V/R intensities, radial velocities of the V and R components, and the average of the V and R radial velocities. The 1988 point is from Martín-Pintado (1989).

Based upon the radial velocities observed for the optical and IR lines by Hamman and Simon (1986,1988), Gordon (1992) predicted that source of the masing lines should lie within the circumstellar disk at a radius of about 30 AU as shown in Fig. 3, or a separation of about 60 AU. Recently Planesas et al. (1992) reported interferometric observations of the H30α line at 232 GHz that showed the maser regions to be separated by 80 AU along a line perpendicular to the optical nebulosity, thereby confirming the suggestion that the double-peaked profile arises in the circumstellar disk.

Although the disk lines could well be maser lines with true inversions in level populations, additional observations of—for example, the polarization and the variation of the line intensities with frequency—could be helpful to clarifying the physical mechanism causing the line enhancement.

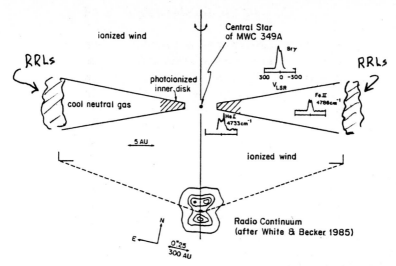

Fig. 3—Cartoon adapted from Hamann and Simon (1986) shows the proposed location of the emitting regions of the masing RRLs. The circumstellar disk at top is embedded in the stellar wind shownat the bottom

REFERENCES

Altenhoff, W. J., Strittmatter, P. A., & Wendker, H. J. 1981 A&A, 93, 48

Brugel, E. W., & Wallerstein, G. 1979, ApJ, 229, L23

Cohen, M., Bieging, J. H., Dreher, J. W., & Welch, W. J. 1985, ApJ, 292, 249

Cowley, A. P., & Gugula, R. 1973, A&A, 22, 203

Gordon, M. A. 1990, in *Radio Recombination Lines: 25 Years of Investigation*, edited by M. A. Gordon and R. L. Sorochenko, (Kluwer: Dordrecht), pp 93-104

———— 1992, ApJ, 387, 701

Hamann, F., & Simon, M. 1986, ApJ, 311, 909

———— 1988, ApJ, 327, 876

Martín-Pintado, J., Bachiller, R., Thum, C., & Bachiller, R. 1989, A&A, 215, L13

Merrill, P. W., & Burwell, C. G. 1933, ApJ, 78, 87

Ney, E. P., Merrill, K. M., Becklin, E. E., Neugebauer, G., & Wynn-Williams, C. G. 1975, ApJ, 198, L129

Olnon, F. M. 1975, A&A, 39, 217

Planesas, P., Martín-Pintado, J., Serabyn, E. 1992, ApJ, 386, L23

Thum, C., Martín-Pintado, J., & Bachiller, R.,1992, A&A, in press

White, R. L., & Becker, R. H. 1985, ApJ, 262, 657

THE COLOR MIMICS OF OH/IR STARS

B. M. Lewis
Arecibo Observatory
PO Box 995, Arecibo PR00612

ABSTRACT

OH / IR star candidates are readily identified by color selection from the IRAS Point Source Catalog; they are confirmed by detecting a 1612 MHz maser. However, $\geq 40\%$ of color selected sources do not exhibit masers, and most of these do not exhibit water or mainline OH masers either, even though their low resolution IRAS spectra may show the 9.7 μm silicate feature that flags an oxygen-rich status. These IR sources are therefore generally not associated with carbon stars, though most are generated by circumstellar shells. The most likely explanation for these *"OH / IR star color mimics"* is that mimics have a degenerate companion collecting an accretion disk from their winds, which provides them with an extra local source of UV to dissociate molecules from within their dust shells. In some cases the extra UV is sufficient to excise all molecules from a shell, as happens with symbiotic novae; in some cases it merely reduces their number and the ability of a shell to support a maser.

1. INTRODUCTION

One of our most general understandings about the Galaxy is that its interstellar UV intensity destroys simple molecules quickly. Molecules only continue to exist in this hostile environment when protected by dust screening from UV and X-rays. And without molecules we have no celestial masers. The detection of circumstellar masers about late-type stars implies the presence of both molecules and dust screening there, so it is not surprising that observers can use the IR signatures of dusty circumstellar shells to guide them to objects with masers. What has been more surprising, however, is that this approach has brought the existence of a numerous set of dusty sources without 1612 MHz masers to our notice. These sources have the requisite dust to protect molecules from rapid degradation by interstellar UV, but lack the usual complement of masers. For convenience we call them "OH / IR star color mimics", or mimics henceforth. This paper summarizes the evidence for mimics and discusses their probable nature.

Many color selected IRAS sources exhibit 1612 MHz masers, though their detection rate is color dependent. But almost half of those selected have no 1612 MHz masers (Lewis 1992). This lack is more significant than similar detection statistics from mainline, water, or SiO masers, as these other masing species show larger intensity changes over the pulsation cycle, and the high frequency masers are observed with less sensitive receivers, are affected by the weather, and are often polarized. We also know that the detectability of water masers is affected by their particular excitation requirements (Lewis and Engels 1991). On the otherhand, the robust 1612 MHz maser is only known to require a suitably long column density of OH molecules and a flux of 35 and 53 µm pump photons: these needs are usually satisfied in sources with the selected colors. The absence of masers from so many IR sources has therefore been puzzling.

The multitude of sources without 1612 MHz masers was first noticed by Lewis *et al* (1987) in discussing the initial segment of the Arecibo survey. It is confirmed by all other observers. It is easy to see that mimics are not artifacts of deployed sensitivity, as the 1612 MHz detection rate is constant at ≈55% for sources with $-0.55 \leq (25\text{-}12)$ µm $\leq +0.05$, when the 25 µm flux $S(25) > 25$ Jy; it drops smoothly thereafter with $S(25)$. In practise the only benefit of the factor of 6 better Arecibo sensitivity in detection surveys is the maintenance of the constant detection rate down to $S(25) \approx 6$ Jy before showing any decline (Fig. 1). So the sources with the largest fluxes and the telescopes with the best sensitivity are both limited to the same detection rate.

OH / IR stars are variables with more than a bolometric magnitude of range, and a diversity of periods from 250 to ≈2000 days. We therefore have to consider the possibility that mimics are artifacts of variability. But a $\sqrt{2}$ more sensitive second look at 293 first epoch nondetections results in <2% (4) new detections (Lewis 1992); most mimics are not artifacts of variability. They are also more frequent in the galactic plane (| b | < 2°) at blue colors, and among sources which IRAS found to have a small probability of being variable. On the other hand, all identification surveys of color selected sources have found few examples of any other type of astronomical object in the color range occupied by circumstellar shells (e.g. Walker *et al* 1989). The most numerous of these "confusing objects" are T Tauri stars, which share a portion of the red range of the most massive OH / IR stars; but the majority of mimics lie at bluer colors with optically thinner shells.

Many of the mimic features are expected from carbon stars. These have carbon-rich shells in which most of the oxygen is bound into CO; lacking water they lack both water and OH masers too. But identifiably carbon-rich shells are characterized by a SiC feature in their IR spectra at 11.3 µm, while the majority of mimics with low resolution spectral types exhibit a 9.7 µm silicate feature. About 22% of the mimics that were examined for

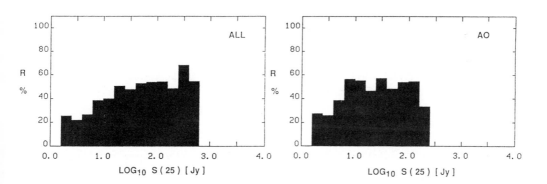

Fig 1: The dependence of the 1612 MHz detection rate on S(25) from ALL observed color selected PSC sources, and from the Arecibo sample alone, over the color range -0.55 < (25-12) µm < +0.05.

OH mainlines, exhibit them, even though they do not exhibit the 1612 MHz line; ≈24% of the searched mimics have water masers. Clearly none of these objects is carbon-rich, nor is it easy to see how a carbon-star explanation for the generality of mimics can generate their variability and latitude dependent signatures. I conclude that mimics are real.

2. A WHITE DWARF COMPANION

Binary stars with circumstellar shells rarely exhibit masers, and many are known to be D(usty)-type symbiotic stars. The most extreme subset of these objects are the symbiotic novae, which have the IR colors of a circumstellar dust shell, *but no masers.* They are also strong UV and X-ray sources (Bryan and Kwok 1991). These properties are causally related, as the intense high energy radiation from the nova outburst, which we measure directly, dissociates most if not all of their shell molecules. These outbursts photoionize the innermost shells, and extend their inner dustless regions to 2-6 times the usual radius. It is easy to appreciate that circumstellar dust is unable to protect embedded molecules against degradation in circumstances where the high energy flux reaches us in abundance. So symbiotic novae, such as RR Tel, V1016 Cyg, and HM Sge provide us with a well-founded model for a system with the colors of a dusty circumstellar shell, and an expectation of few molecules. The most likely explanation for our mimics, therefore, is that they are disguised symbiotic stars.

Our working model supposes that mimics have a degenerate stellar companion, which is usually a white dwarf. This inevitably collects an accretion disk from the outflowing red giant wind, to become a source of molecule destroying UV tethered within the shell. The intensity of the extra UV flux depends on the separation of the companion, the mass-loss rate from the red giant, and the extent to which the white dwarf has cooled. The efficiency of molecule destruction then varies from system to system. Some systems have low mass-loss rates and widely separated companions, implying a moderate UV flux and the preservation of most molecules; Mira is a well-studied example. On the other hand, there will also be systems with larger mass-loss rates, and / or closer companions, with more intense UV fluxes, which destroy most of the water and OH molecules that sustain observable masers. These are the mimics.

A partial rather than total loss of the molecular component occurs in systems like Mira, which has a known white dwarf companion, with an effective temperature below 14,000 K (Reimers and Cassatella 1985). Mira has detectable water masers, but has only once before this Meeting been seen with a mainline maser; its shell clearly has some molecules. But CO maps of the system are asymmetric (Planesas *et al* 1990), which leads us to suppose that the portions of the CO shell near the companion are extensively photodissociated. The presence of mainline and water masers in just some of the mimics is thus an expected result of the range in the intensity of their self-generated UV.

Symbiotic stars incorporate features from two identifiably different temperature regimes in their spectra. When one component has a much hotter temperature than a red giant, is at a radius beyond that for Roche lobe interchanges, and accretes mass from the red giant's wind, it is a D-type symbiotic (e.g. Allen 1988). These may be obviously active, with intermittent dwarf nova-like outbursts when the accreted mass rate is high, or rather passive sources of UV, when the accretion rate is low. The symbiotic status of mimics has to be inferred, as their optically thick shells preclude direct inspection of their spectra. But the presence of a cool red giant within a dust shell is safely inferred from the characteristic IR colors of the dusty shell, which every mira has. We can now use the persistent absence of the regular complement of circumstellar masers from O-rich shells to infer the presence of molecule destroying UV, and hence the presence of a degenerate companion. This explanation for the mimics suggests that they are all D-type symbiotic stars. A wider implication of this result is that it also suggests that widely separated white dwarf companions accompany more than 40% of all demonstrably old stars.

The symbiotic model for mimics explains the surprisingly wide IR color range occupied by some of the IRAS Low Resolution Spectral types. There is a progression in IR color with mass-loss rate, such that the IR sources with the largest rates have the reddest colors. They also exhibit the 9.7 μm silicate feature in absorption, and have LRS types of 38 or 39. Most of these fall where they should at the red edge of the OH / IR star color distribution. But 24% of the type 38-39 sources have distinctly bluer colors, which need explanation. I can reconcile these color displaced objects with their spectral type, if their IR colors are modified by the dissipation of the extra energy generated in their accretion disks; dust in the environs of a disk will radiate at a higher temperature and at a radius less subject to the usual dust absorption. The IR colors of dusty shells with unocculted accretion disks are always bluer than they would otherwise be.

The symbiotic model for mimics also draws our attention to the changes in the complement of masers, that follows on a change in available UV. Increases in the ambient UV radiation field lead to increases in the rate of photodegradation of molecules at the edge of a shell, and to a contraction in their radial distribution. From this point of view an extra source of UV is equivalent to a larger ambient UV intensity. The most fragile maser in these circumstances is observationally seen to be the 1612 MHz maser. If the column density of OH molecules is too low, the 1612 MHz maser is suppressed, and the OH molecule radiates through its 1665 / 1667 MHz lines instead. Some mimics have enough molecules left to support a water maser, and therefore an OH mainline maser too; some only have enough molecules to support a mainline maser, and some do not have enough water or OH molecules left to support any of the usual masers.

This work is supported by NAIC, which is operated by Cornell University under a cooperative management agreement with the National Science Foundation.

Allen, D. 1988, "The Symbiotic Phenomenon", ed J. Mikolajewska, M. Friedjung, S. J. Kenyon, and R. Viotti, ASSL 145, p3 [Kluwer: Dordrecht]
Bryan, G. L., and Kwok,S., 1991, *Ap.J.*, **368**, 252
Joint *IRAS* Working Group, 1985, *IRAS Point Source Catalog* (U.S. GPO, Washington, DC) [PSC]
Lewis, B. M. 1992, *Ap.J. (in press, Sep 1)*
Lewis, B. M., and Engels, D., 1991, *M.N.*, **251**, 391
Lewis, B. M. Eder, J., and Terzian, Y. 1987, *Astr.J.*, **94**, 1025
Planesas, P., Kenney, J. D. P., and Bachiller, R. 1990, *Ap.J.*, **364**, L9
Reimers, D., and Cassatella, A., 1985, *Ap.J.*, **297**, 275
Walker, H. *et al.*, 1989, *Astr.J.*, **98**, 2163

THE MASERS OF GY AQL

B. M. Lewis
Arecibo Observatory
PO Box 995, Arecibo PR00612

ABSTRACT

The chronological sequence of circumstellar masers describes the addition of masers as the stellar mass-loss rate changes; SiO, water, OH mainline, and 1612 MHz masers then appear in order as the (25-12) μm color reddens. The most notable exception has been GY Aql, which has SiO and mainline masers, but has not previously been seen with a water maser despite ≥10 searches. A few other stars have these characteristics too. This pattern of masers is explicable, however, if these stars have a white dwarf companion collecting an accretion disk from their winds, which becomes a source of molecule destroying UV within their dust shells. So these stars are not counterexamples to the chronological paradigm. In any case GY Aql itself is no longer an obvious anomaly, as its water maser has now been detected.

1. INTRODUCTION

The chronological sequence of circumstellar masers describes the changes to the set of masers exhibited by circumstellar shells as their rates of mass-loss evolve. Initially SiO, water, the OH mainlines, and 1612 MHz masers are added in order, as a shell grows in thickness and increases its ability to shield molecules against interstellar UV (Lewis 1989; 1990). This scenario explains the systematic addition of masers, when the only source of energetic photons for molecular dissociation is interstellar. Though successful in describing most stars with detected masers, the number with well observed SiO, water and OH is still small, so further consolidation depends on increasing their number and on periodic reexamination of exceptions.

The most notable exception to the chronological paradigm has been GY Aql. This SR star has SiO and mainline masers (Fillet *et al* 1972; Dickinson *et al* 1986), but has never before been seen with a water maser. There is always some residual uncertainty about a reported masing nondetection being the final word on a system, as the variability of masers over the pulsation cycle may contribute to the lack of a detection: Little-Marenin and Benson (1989) find that water emission from miras is most frequent in the 30% of the cycle following peak luminosity. But many searches (≥10) for water masers in GY Aql have been fruitless (Dickinson and Dinger 1982; Benson *et al* 1990; Lewis and Engels 1991a), despite its strong 271 Jy IR flux at 25 μm. This lack is anomalous in the sense that the existence of a mainline maser shows that the shell is O-rich, and has OH molecules which are a daughter product of the photodegradation of water by interstellar UV. They therefore lie in an annulus around the outside of a circumstellar shell (CS) with water molecules inside. Shells like that of GY Aql, with (25-12) μm = –0.64, normally support a water maser during at least part of its pulsation cycle.

GY Aql is particularly anomalous in its lack of a water maser when compared with other low mass (<1.5 $M_\odot$) stars, which it seemed to be with its b = -16°, period ~204 days (GCVS), and small reported expansion velocity ~5.5 km s^{-1} (Fillet *et al* 1972) from both the SiO and OH masers. Almost all such stars within the Arecibo sky with 1612 MHz masers are found to exhibit mainline and water masers after one inspection (Lewis and Engels 1991). Nor can this difference be attributed to GY Aql being smaller than other stars in the group, as it was equally anomalous in the more limited category of short period SR stars, where Dickinson and Dinger (1982) have 5 examples with water masers.

I reconciled GY Aql to the chronological scenario in the past by assuming that as the only known case it was unique. It might then be explained by invoking a rare occurrence (Lewis 1990). In particular, if it is a star recovering from a thermal pulse, its outer envelope might support an OH maser from past mass-loss, while it is temporarily bereft of an inner shell to support a water maser, as a result of the interruption to mass-loss following on a thermal pulse. This putative conjunction of affairs would be necessarily brief, as its low resolution spectral type of 28 and IR colors (–0.64,–1.18) suggest that GY Aql has an inner shell and ongoing mass-loss. But this approach does not lead us to expect many of these objects, and we are starting to accumulate more.

2. A COMPANION-STAR RESOLUTION

A general explanation for the system of masers exemplified by GY Aql can be reached by considering the changes wrought on a CS by an extra source of UV. Water masers are collisionally excited, with an innermost masing radius, r_q, set by quenching, that only depends on the mass-loss rate (Cooke and Elitzur 1985). However, their outermost radius, r_d, is usually set by the photodegradation of water molecules by interstellar UV. Consequently if the external radiation field is unusually large, or there is an extra internal source of UV to destroy a proportion of the molecules, r_d contracts. When the extra source of UV causes $r_d \leq r_q$, all possibility of a water maser vanishes. Yet the dissociation of water produces OH as a daughter, and if the UV flux permits OH molecules some longevity, OH masers may still flourish in the absence of water.

GY Aql is probably immersed in a weaker than ambient UV flux, as it has a b of –16°. A sufficient model for explaining its masers, therefore, is based on the assumption that it has a white dwarf companion, which would inevitably garner an accretion disk from its outflowing wind to become a significant source of UV. The intensity of this extra UV depends on both the separation of the companion and GY Aql's mass-loss rate. An accretion disk can readily supply the extra UV needed for destroying a significant number of water molecules before they are exposed to interstellar UV. This postulate is in any case already needed as an explanation for the large number of color selected O-rich shells without 1612 MHz masers (Lewis 1992), and may be applicable to ≥40% of all CSs. For GY Aql the extra UV only needs to ensure $r_d \approx r_q$. There is then almost no zone in which a water maser can thrive, while there are still plenty of OH molecules at $r > r_d$ to nourish a radiatively pumped OH maser.

This general approach to the masers exhibited by GY Aql leads us to expect more of these objects than a thermal-pulse explanation allows. Indeed ≈90 color selected objects from the Arecibo survey, that do not have 1612 MHz masers, are known to have OH mainline masers (Lewis 1992). Four of these have been fruitlessly searched 3 times for water (BI Cyg, BM Gem, IRC+30074, RV Boo); these are potentially further examples of the GY Aql syndrome and may perhaps be detectable UV sources. We may also expect some of these objects without 1612 MHz masers, but with water masers, to be at a "pre GY Aql stage", where they have weak tangential masers like W Hya (Reid and Menton 1990), because their capacity to support water masers has been marginalized.

3. NEW OBSERVATIONS

The need for extra flexibility in the chronological paradigm stems partly from the persistent lack of water masers from GY Aql, and partly from a developing realization that there are more systems like it. However after this paper was prepared, and en route to this Meeting, I had the opportunity to search for water yet again in GY Aql. I have now to report that Engels and I detected a weak (0.3 K) water maser from it on the 4th and 7th of March 1992 at Effelsberg. These are ~6σ detections, at a velocity of 33.6 km s^{-1}, in each of three independent spectra, which somewhat spoil GY Aql as a prototype.

Another new finding has a bearing on our discussion of GY Aql. Dr. Bujarrabal reports a longer (~450 rather than 204 day) period at this Meeting, from SiO monitoring over more than 5 well sampled cycles. GY Aql is therefore not a low mass star. The doubling of the period also suggests that the previously adduced expansion velocity is low, and a search through all of its OH detections shows that at 1665 MHz its low velocity peak is ~26-28 km s^{-1}, while its high velocity peak is ~40-41 km s^{-1} (Olnon *et al* 1980), giving an expansion velocity of ~7.5 km s^{-1}. Our water maser is then indeed near the stellar velocity, where the optimal gain is along a tangential ray, and the maser is only marginally excited. The SiO light-curve incidentally locates the detection at an epoch when the SiO signal is near its minimum intensity.

4. CONCLUSIONS

GY Aql was difficult to reconcile with the chronological paradigm, because it had SiO and OH masers, but no water masers. We have found that this pattern of masers does not constitute a counter-example to the paradigm, if the star has a white dwarf companion. GY Aql itself is in any case no longer directly at variance with the paradigm, as a weak water maser has just been detected after numerous previous failures. This is all the more surprising in coming as it does now near the minimum in its light curve. It is still likely that GY Aql has a white dwarf companion, that is responsible for the history of weakness in its water masers. Their past absence may have coincided with the passage of the companion through periastron, and an ability to garner a more massive accretion disk there, while it is now moving away on an elliptical orbit, with a reduced flow of material into its accretion disk, which provides a lower UV flux, and a lowered destruction rate for circumstellar molecules. This possibility can be clarified by an IUE search for accompanying UV. Alternately, it is perhaps possible that we are seeing GY Aql emerge from a phase where its water masers have for other reasons been suppressed for ≥ 20 years, so they may strengthen markedly over the current cycle, as the star moves to a regular maximum. This supposition is based on the somewhat reminiscent longterm behaviour of the mainline masers of R Leo (Le Squeren *et al* 1985), which were also much weaker than usual for several pulsation cycles. If the water masers do strengthen quickly, we should expect them to increase their gain most notably along the more nearly radial rays, with a resulting increase in the width of the water profile.

A lasting result of this exercise is the realization that when a circumstellar shell is significantly modified by self-generated UV, this first eliminates the possibility of a 1612 MHz maser from an IR source that would otherwise exhibit one, and then suppresses the occurrence of a water maser, while it may still permit a mainline OH maser. UV-rich sources may depart from the expectations of the chronological scenario without providing counter-examples to it. This paradigm is now clearly seen to be most apposite to solitary sources, and so needs a codicil added to it. In circumstances where there is an extra source of UV, or a strong ambient external UV flux, we should expect the 1612 MHz masers to be fragile. They are followed in decline by the water masers. And viceversa, when the external UV field is weak, as it may be in the SMC, the LMC, and the outer Galaxy, OH masers may be weak or absent, while the outer radius, r_d, of the water masing zone is enlarged. In those with the largest mass-loss rates, and those only exposed to extremely weak UV fluxes, r_d may even be set by the radial decrease in shell temperature. In these circumstances water masers should be strengthened by their increased path length.

I am grateful to Dr. V. Bujarrabal for faxing a copy of his SiO light curve for GY Aql after the Meeting. This work is supported by NAIC, which is operated by Cornell University under a management agreement with the National Science Foundation.

Benson, P. J., *et al,* 1990, *Ap.J.Suppl..,* **74**, 911

Cooke, B. and Elitzur, M. 1985, *Ap.J.,* **295**, 175

Dickinson, D.F., Turner, B. E., Jewell, P. R., and Benson, P. J. 1986, *A.J.,* **92**, 627

Dickinson, D.F., and Dinger, A.St. C. 1982, *Ap.J.,* **254**, 136

Fillit, R., Gheudin, M., Rieu, Ng-Q., Paschenko, M., and Slysh, V. 1972, *A&A.,* **21**, 317

Kolopov, P. N., *et al* 1985, *General Catalog of Variable Stars,*(Moscow; Nauka) **[GCVS]**

Le Squeren, A. M., and Sivagnanam, P. 1985, *A.&A.,***152**, 85

Lewis, B. M. and Engels, D. 1991, *MN,* **251**, 391

Lewis, B. M. and Engels, D. 1991a *(unpublished observations)*

Lewis, B. M. 1992, *Ap.J.,(in press, Sep. 1)*

Lewis, B. M. 1990, *A.J.,* **99**, 710

Lewis, B. M. 1989, *Ap.J.,* **338**, 234

Little-Marenin, I. R. and Benson, P. 1988, *BAAS,* **20**, 698

Olnon, F. M., *et al* 1980, *A.&A.Suppl..,* **42**, 119

Reid, M., and Menton, K. M. 1990, *Ap.J.,* **360**, L51

13. CIRCUMSTELLAR OH MASERS

OH/IR STARS AND THE GALACTIC CENTRE DISTANCE

Jessica M. Chapman[1], N.E.B. Killeen[1], P. te Lintel Hekkert[2], J.L. Caswell[1], & J. Harnett[3]

[1] Australia Telescope National Facility, CSIRO, P.O. Box 76, Epping, NSW 2121, Australia
[2] Mount Stromlo & Siding Spring Observatories, Australian National University, Private Bag, Weston Creek P.O., ACT 2611, Australia
[3] School of Physics, University of Sydney, NSW 2006, Australia

ABSTRACT

Since December 1988, we have been using the Parkes 64-m radio telescope to monitor the OH 1612-MHz maser emission from 95 OH/IR stars in the disc, halo and bulge of our Galaxy. The observations are used to determine stellar periods and linear diameters of the circumstellar envelopes. Here we give a progress report on the monitoring programme and show example phase-lag results. At a later date the linear diameters will be combined with angular diameters to determine stellar distances and the distance to the Galactic centre.

1. INTRODUCTION

OH/IR stars provide an excellent means of determining accurate, geometric stellar distances. The distance to an OH/IR star may be obtained by combining the angular and linear diameters of the circumstellar envelope. The angular diameter is obtained from aperture synthesis imaging of the OH maser emission; the linear diameter is determined by measuring the time-delay between the OH light curves from the front and back of the circumstellar envelope. Typically, the angular diameters of OH/IR stars are between 0.3 and 2.0 arcsec, the stellar periods are between 400 and 1600 days and the front-back phase-lags are between 5 and 50 days. Recent phase-lag results are given by van Langevelde, van der Heiden & Schooneveld (1990) and by Gaylard et al. (1992).

Since December 1988, we have been using the Parkes 64-m radio telescope to monitor the OH 1612-MHz maser emission from 95 southern OH/IR stars. The principal aim of this project is to determine the Galactic centre distance. The Parkes sample includes 39 stars which are likely to be either on the solar circle or at tangential point distances, and 28 stars in the Galactic bulge. In addition, we are monitoring sources at high Galactic latitude and sources of special interest such as the bipolar source OH19.2-1.0 (Chapman 1988). To avoid the problem of interstellar scattering, we are not monitoring stars close to the Galactic centre. The single-dish observations are used to determine stellar periods and linear diameters of the circumstellar envelopes. Angular diameters will be obtained at a later date and the angular and linear diameters will then be combined to give stellar distances.

2. RESULTS FROM THE MONITOR PROGRAMME

The Parkes observations began in December 1988 and the 95 OH/IR stars have been observed at regular intervals of 4 to 6 weeks since September 1989. From the OH data obtained so far we find that 89 of the 95 sources have sinusoidal-like OH flux density variations with maximum-flux/minimum-flux ratios in the range 1.1 to 3.8 and stellar periods > 350 days. The other six sources show no significant variations. These sources either have intrinsically small amplitudes or very long stellar periods. Fig. 1 shows the distribution of stellar periods determined from the Parkes observations for 79 stars with periods below 1300 days. The ten longest period sources have not been included in this figure as their periods are not well determined from the data obtained so far.

PERIOD DISTRIBUTION – 79 OH/IR STARS

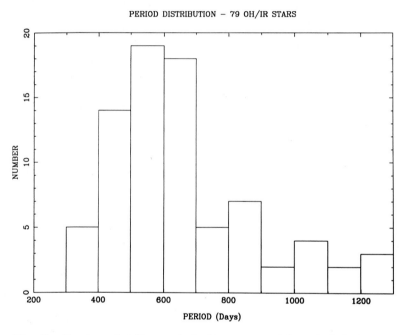

Fig. 1 The distribution of stellar periods for sources with periods below 1300 days.

In Fig. 2 we plot the stellar periods against the velocity range of the OH maser emission, for the 79 sources with periods < 1300 days. As first noted by Dickinson & Chaisson (1973) there is an approximately linear correlation between the stellar period and the OH velocity range for sources with stellar periods below ~ 800 days.

By fitting light curves separately to the extreme blue- and red-shifted emission peaks we obtain the front-back phase-lags and hence the linear diameters of the circumstellar envelopes. So far, we have obtained preliminary phase-lags for 49 of the shorter period sources (P < 700 days). The best-fit phase-lags are between 3 and 50 days corresponding to linear diameters in the range 0.8 x 10^{16} cm to 12.4 x 10^{16} cm. In the best cases we obtain linear diameters to a precision of $\sim$ 15 %. More typical errors are $\sim$ 30%.

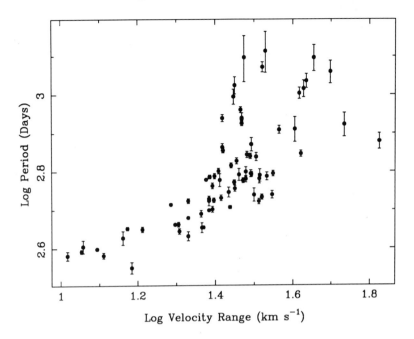

Fig. 2 The best-fit stellar periods plotted against the velocity separation of the extreme emission peaks for 79 OH/IR stars with periods below 1300 days.

To illustrate the phase-lag technique, Fig. 3 shows OH flux density variations for the two sources IRAS 16105−4205 (OH 338.1+6.4) and IRAS 18467−4802 (OH 348.2−19.6). In each case we plot the data for the two channels containing the extreme blue- and red-shifted emission peaks. The solid lines show best-fit light curves determined using a two Fourier-component model which allows for asymmetries in the shapes of the OH light curves. In general, the OH maser intensity increases more steeply towards light maximum and decreases more gradually after light maximum. For IRAS 16105−4205, the best-fit stellar period is 628.6 ± 11.1 days and the front-back phase-lag between the outer peaks is 22.9 ± 4.7 days, corresponding to a linear diameter of (5.3 ± 1.0) × 10^{16} cm. For IRAS 18467−4802, the stellar period is 596.3 ± 8.8 days and the front-back phase-lag of 10.7 ± 2.8 days corresponds to a linear diameter of (2.8 ± 0.7) × 10^{16} cm.

We intend to continue monitoring until approximately December 1993. This will allow us to obtain $\sim$ 40 points on the OH curves for each source. From these data we expect to obtain linear diameters, for most sources, to a precision of 10 to 15 %. The angular diameters of the circumstellar envelopes will be determined during 1992 and 1993 from long baseline interferometry observations which will be taken with the MERLIN and VLBA arrays. The distance estimates for individual sources will then be combined to determine the Galactic centre distance to an expected accuracy of between 5 and 10%.

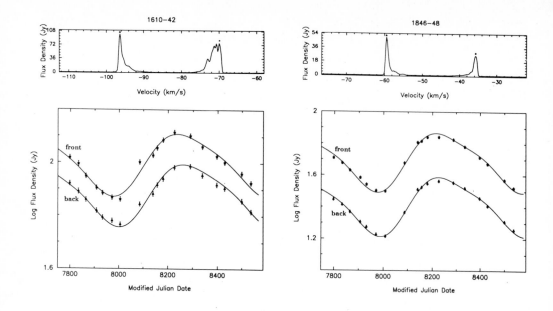

Fig 3 Averaged OH spectra and OH flux densities obtained for the extreme blue-and red-shifted emission peaks for the two OH/IR stars IRAS 16105−4205 and IRAS 18467−4802, together with model light curves through the data. The front-back phase-lags are 22.9 days and 10.7 days respectively. Velocities are with respect to the local standard of rest.

REFERENCES

Chapman, J.M., 1988, MNRAS, 230, 415
Dickinson, D.F. & Chaisson E.J., 1973, ApJ, 181, L135
Gaylard, M.J., West, M.E., Shepherd, M.C. & Cohen, R.J., 1992, these proceedings
van Langevelde H.J., van der Heiden, R. & van Schooneveld, C., 1990, A&A, 239, 193

OH MASER EMISSION FROM F- AND G-TYPE SUPERGIANT STARS

Jessica M. Chapman[1], P. te Lintel Hekkert[2] & N.E.B. Killeen[1]

[1]Australia Telescope National Facility, CSIRO, P.O. Box 76, Epping NSW 2121, Australia
[2]Mount Stromlo & Siding Spring Observatories, Australian National University,
Private Bag, Weston Creek P.O., ACT 2611, Australia

ABSTRACT

We have detected OH maser emission from four stars classified at optical wavelengths as F- or G-type supergiants. The spectral characteristics of the OH masers are described. For the source HD 101584, images of the OH 1667 MHz maser emission obtained with the Australia Telescope show that the masers are located in a bipolar outflow. The stars are likely to be proto-planetary nebulae.

1. PARKES OBSERVATIONS

From the literature, we have compiled a list of approximately 150 stars which are classified at optical wavelengths as F- or G-type supergiants but which also show a strong excess of far-infrared emission. This list includes eight stars for which detections of OH maser emission have been published, including the well-known stars IRC 10420, Roberts 22 and HD 161796. We have searched southern sources in the list for OH maser emission at 1612, 1665 and 1667 MHz, using the Parkes 64-m radio telescope in 1990 September, 1991 February and 1992 February. So far we have detected OH maser emission from a further four of the supergiant stars. The sources are listed in Table 1 which gives for each source the the OH lines detected, the mean OH velocity, the total velocity range of the OH emission and the peak OH flux density.

Table 1: Parkes Detections of OH Maser Emission from Early Type Supergiant Stars

IRAS	Other Name	Class	l (°)	b (°)	OH line (MHz)	$<V_{OH}>$ (km s^{-1})	ΔV_{OH}	Peak_OH (Jy)
07471-2443	HD 63700	G6Ia	97	19	1612	11.5	23	0.15
11385-5517	HD 101584	F2Iape	293	6	1667	42	84	0.54
12022-6143	GC 16500	F0Ib/II	297	0	1667	-36.5	40	0.14
12022-6143	GC 16500	F0Ib/II	297	0	1665	-36.5	15	0.09
16122-5128	CD 519943	F5Ia/Iab	332	-1	1667	-80	75	0.12

Fig. 1 shows spectra obtained at Parkes for the sources IRAS 07471-2443 (HD 63700), IRAS 12022-6143 (GC 16500) and IRAS 16122-5128 (CD 519943).

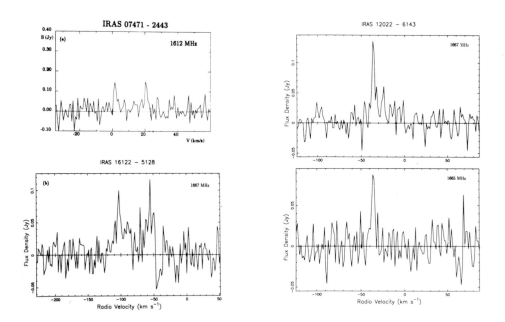

Fig. 1 OH spectra obtained with the Parkes 64-m radio telescope for the three sources IRAS 07471-2443, IRAS 12022-6143 and IRAS 16122-5128. The OH 1612 MHz spectrum of IRAS 07471-2443 was taken in 1991 February with a velocity resolution of 0.8 km s^{-1}. The spectra of IRAS 16122-5128 and IRAS 12022-6143 were obtained in 1992 February with a velocity resolution in each case of 1.7 km s^{-1}.

2. THE OH MASER EMISSION FROM HD 101584

The OH 1667 MHz maser emission from the F-type supergiant star HD 101584 (IRAS 11385-5517) was first discovered at Parkes in 1990 September. Higher sensitivity observations of the OH 1667 MHz emission were taken in 1991 August using the Australia Telescope Compact Array (ATCA). No emission has been detected from the other OH lines at 1612, 1665 or 1720 MHz. Full details of the Parkes and ATCA observations for this source are given by te Lintel Hekkert, Chapman & Zijlstra (1992). Fig. 2a shows the total intensity spectrum of the OH 1667 MHz maser emission of HD 101584 obtained from the ATCA observations by vector averaging the source visibilities over the total integration time of 4.5 hours.

With the 9.0 x 6.2 arcsec^2 beam of the ATCA observations, the OH maser emission from HD 101584 is partially resolved. Each channel map shows a single point-like feature. The relative positions of the emission features are plotted in Fig. 2b. The origin of the figure is the unweighted emission centroid of the maser positions. This position coincides within 0.5 arcsec with the optical position of the star.

The OH masers of HD 101584 are located within a bipolar outflow with the red- and blue-shifted emission features appearing on opposite sides of the stellar position. Each of the two emission regions is elongated along a bipolar axis at a position angle of $\sim -60^\circ$. In Fig. 2c, the velocity of each maser spot is plotted against the radial offset from the origin. Along the bipolar axis there is a strong velocity gradient with velocities increasing outwards. At the inner edge of each lobe the expansion velocity is 9 km s^{-1} and the radial offset is 1.0 arcsec, whilst at the outer edges, the expansion velocity is 40 km s^{-1} and the radial offset is 2.2 arcsec. For a distance of 1.0 kpc, the outer radius of the OH maser emission corresponds to a linear size of 3.3 x 10^{16} cm.

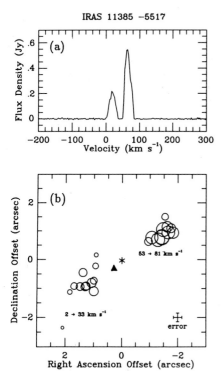
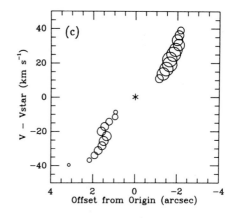

Fig. 2

(a) - The OH 1667 MHz total intensity maser spectrum of HD 101584 obtained with the Australia Telescope in 1991 August. The velocity resolution is 3.4 km s^{-1} and the rms noise level in the spectrum is 4 mJy. The optical stellar position is indicated by the triangle. (b) - The relative positions of the OH 1667 MHz masers of HD 101584 in each of 28 spectral channels. The areas of the circles are proportional to the integrated emission at each velocity. The error bar shows the typical precision of the relative positions.
(c) - The velocity of each maser feature, relative to the stellar velocity of 42 km s^{-1}, is plotted against the radial offset from the origin.

3. DISCUSSION

The OH maser spectra of the F- and G-type supergiants are strikingly different to those of OH/IR stars. In particular, the OH maser emission for these supergiants is generally stronger at 1667 MHz than at 1612 MHz and in some cases, the total velocity range of the OH emission is significantly larger than that observed in OH/IR stars. For two of the sources shown here, HD 101584 and IRAS 16122-5128 (CD 519943), the total velocity range of the OH emission is exceptionally large with values of 84 and 75 km s^{-1} respectively. For each of these sources the maser spectrum reveals two broad emission features each of FWHM $\sim$ 20 km s^{-1}. For HD 101584, the maximum observed outflow velocity of the OH molecules of 42 km s^{-1} is considerably smaller than the outflow velocity of CO molecules, determined by Trams et al. (1990) and by Loup et al. (1990) to be at least 150 km s^{-1}. The presence of high velocity molecular gas appears to be a common characteristic of these supergiants. For six of the twelve sources now detected in OH, the velocity range is $\geq$ 60 km s^{-1}. In contrast, less than 1% of OH/IR stars have velocity ranges above this value with typical values between 10 and 35 km s^{-1}.

What are these stars? The presence of high velocity molecular gas, together with the intense far-infrared emission and optical classification, provides strong evidence that they are proto-planetary nebulae in an intermediate stage of evolution between asymptotic giant branch stars and planetary nebulae. During this transition stage, which lasts for not more than a few thousand years, the stellar mass-loss rate decreases whilst the photospheric temperature increases from approximately 3500 to 35000 K. The central star previously enshrouded by the thick circumstellar envelope becomes visible as the density decreases in the inner regions. During this stage it is likely that the star changes from losing mass in a cool, dense envelope to losing mass in a hot, tenuous, fast wind of velocity 1000 to 2000 km s^{-1}.

We interpret the bipolar geometry and the high outflow velocity of the OH and CO molecules in the envelope of HD 101584 as evidence for an interaction between a hot, fast stellar wind and a cool remnant envelope. Hydrodynamic models for planetary nebulae show that their axi-symmetric morphologies probably result from such interactions (Balick, Preston & Icke 1987). The fast stellar wind sweeps up and shocks a circumstellar envelope and expands furthest in the directions of least envelope density resulting in a prolate or bipolar geometry. This mechanism may explain the exceptionally large expansion velocities of the early-type supergiant stars with infrared excesses and may also provide the physical conditions which favour the OH 1667 MHz maser line rather than the OH 1612 MHz line. We suggest that bipolar geometries and large outflow velocities are likely to be a common characteristic of the molecular emission from proto-planetary nebulae.

REFERENCES

Balick, B., Preston, H.L. & Icke, V., 1987, AJ, 94, 1641
Loup, C., Forveille, T., Nyman, L.Å & Omont, A., 1990, A&A 227, L29
te Lintel Hekkert, P., Chapman, J.M. & Zijlstra, A.A., 1992, ApJ in press
Trams, N.R., van der Veen, W.E.C.J., Waelkens, C., Walters, L.B.F.M.
& Lamers, H.J.G.L.M., 1990, A&A 233, 153

OH MASERS IN THE CIRCUMSTELLAR SHELL
OF THE GIa SUPERGIANT AFGL 2343

M. J Claussen
Naval Research Lab
Washington DC 20375-5000

ABSTRACT

High resolution observations of the 1612 and 1667 MHz OH masers in AFGL 2343, a GIa star which is likely a proto-planetary nebula, are presented. Comparison between the OH observations, and recently published SiO and CO observations, as well as modelling of the infrared spectral distribution, suggest that the OH and SiO emission regions may be interior to the cold dust and CO emission regions; a conclusion that may be reached is that the OH and SiO are indicators of a more recent, less massive mass loss epsiode than that which produced the CO and cold dust outflow.

1. Introduction

The late-type star AFGL 2343, which has been identified with HD179821, a GIa star (Bidelman 1981 AJ 86, 553), has been observed in the 1665, 1667 and 1612 MHz OH transitions using the NRAO 43-m telescope. The 1612 and 1667 MHz transitions were detected with very similar profiles, and flux ratios. The peak fluxes are about 50 Jy, and features were detected to about 0.5 Jy. No emission was detected at the 1665 MHz line to an rms noise of about 50 mJy. Thus we confirm the interesting profiles and the non-detection of the 1665 MHz transition to a very low limit as noted by Likkel ((1989, ApJ, 344, 350). CO has been detected by Zuckerman and Dyck (1986, ApJ, 311, 345), and they estimate an expansion velocity from the CO data of 33 km s^{-1}. Bujarrabal et al. (1992, A&A, 257, 701) have mapped the CO (2-1) emission, finding it elongated roughly NE-SW, with a deconvolved size of 18″ by 14″ , and have also observed the v=0 SiO 2-1 and 3-2 lines. Hrivnak et al. (189, ApJ, 346, 265) have modeled the near IR and IRAS data for this source, assuming a low-opacity dust shell at $\lambda > 0.5\mu m$.

2. Data

The NRAO-VLA was used to observe the 1667 and 1612 MHz transitions of OH masers toward AFGL 2343 in the A configuration (beamsize $\approx 1''$) using 1.1 km/s spectral resolution. Sixty-three channels were used to cover the range of the OH emission in AFGL 2343. Continuum emission was also searched toward the source at 6 and 2 cm. No continuum emission was detected to rms noises of 0.8 and 0.2 mJy for the 2 and 6 cm observations respectively. Figure 1 shows spectra for each of the transitions made from the VLA data by summing all the OH emission in each channel. Due to space constraints, each channel map is not shown. Figure 2 shows contour plots of the velocity-integrated maps for both transitions. The cross on each plot represents the stellar position.

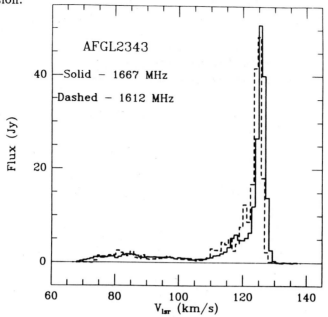

Figure 1. OH Maser Spectra

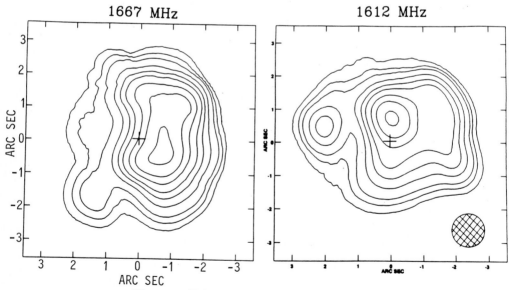

Figure 2. Velocity-averaged OH emission

3. Results and Discussion

I note that the 1667 MHz emission extends to the blueward and to the redward of the 1612 MHz emission. This is not uprecendented, but seems to be rare among sources in which both or all three transitions have been observed (Dickinson and Turner, 1991 ApJS, 75, 1323 and Bowers 1984, ApJ, 279, 350). The velocity extent of the SiO profiles (Bujarrabal et al. 1992) are comparable to that of the OH, and both are smaller than the extent of the CO by about 6 km s^{-1}.

I also note that the maps of the integrated lines have different shapes for the two transitions. For the 1667 MHz line, the overall shape of the emission is roughly circular, while the strongest features show a double morphology oriented roughly N-S. For the 1612 MHz line, the overall distribution is elongated slightly E-W, and the stronger features also share in this elongation. The channel maps of the two transitions for 177.0 km s^{-1} show that the ring-like structure is slightly larger for the 1667 MHz transition. These ring-like structures are reminiscent of the OH maps of the 1612 MHz emission in OH26.5+0.6 and OH127.8-0.0 (Bowers and Johnston 1990, ApJ, 354, 676). However, in both those sources, the extreme red- and blue-shifted emission is more or less coincident; not the case for AFGL 2343. The OH emission from AFGL 2343 is clearly asymmetric in both transitions, and the asymmetry is different for each of the lines. A plot of the distribution of all the maser features in the 1667 MHz line (Figure 3) shows a slight extension in the NE-SW direction. The extent of the OH emission has a maximum radius of about 3″ , about 20% of the CO size.

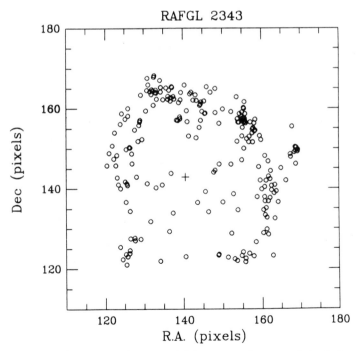

Figure 3. Distribution of the 1667 MHz maser features (pixel size=0.1″)

We don't know the distance to AFGL 2343. Bujarrabal et al. (1992) argue that a distance of 1.5 kpc allows the CO extent and the size of the dust shell as deduced by Hrivnak et al. to be compatible. If the 1.5 kpc distance is adopted, the maximum OH shell radius is about 7×10^{16} cm, comparable to the size of the inner dust shell radius. The luminosity of this object would then be about 10^4 $L_\odot$.

Based on the comparable velocity extent of the v=0 SiO emission and the OH maser emission, and 'spike' features in the SiO at the same red-shifted velocity as the strong OH maser feature, it seems possible that the OH and SiO emission regions are nearly co-spatial. If so, this inner part of the circumstellar shell may be the result of a more recent and much less massive mass loss process than the one which resulted in the large dust and CO shell. This more recent mass loss episode has a small enough mass-loss rate so that the central star can now be seen. A small mass-loss rate may account for maser pumping which suppresses the inversion for the 1665 MHz transition, and makes the 1667 and 1612 MHz emission nearly co-spatial with similar strengths (see Collison's poster at this meeting).

4. Conclusions

• AFGL 2343 has strong, spectrally asymmetric OH emission at 1612 and 1667 MHz. The ratio of the fluxes is about unity, and there is no 1665 MHz emission.

• The 1612 MHz emission has a somewhat smaller velocity and spatial extent than the 1667 MHz emission. The strongest emission, at the red-shifted velocities, is elongated N-S at 1667 MHz, but E-W at 1612 MHz.

• Combined with models of the infrared emission and observations of CO and SiO, the size and velocity extent of the OH emission supports the suggestion that the OH and SiO emission are manifestations of a different mass loss episode than that of the CO and infrared emission, as first suggested by Bujarrabal et al. (1992).

This is part of a more extensive work soon to be submitted to the *Astrophysical Journal*.

CIRCUMSTELLAR OH MASERS

R.J. Cohen
Nuffield Radio Astronomy Laboratories, Jodrell Bank
Macclesfield, Cheshire SK11 9DL, UK

ABSTRACT

The use of OH masers as probes of circumstellar envelopes is briefly discussed, with special emphasis on the mainlines at 1665 and 1667 MHz. In general these are inverted nearer to the star than the 1612 MHz masers, and the envelope structure they reveal is more irregular and asymmetric. The recent detection of proper motions in the OH envelope of U Orionis gives new insights into the nature of the mass-loss from Mira variables.

1. INTRODUCTION

There are three OH maser lines commonly seen in circumstellar envelopes: the ground state satellite line at 1612 MHz, and the ground state mainlines at 1665 and 1667 MHz. All three were discovered in circumstellar envelopes at the same time (Wilson & Barrett 1968), but since then the greatest progress has been made in understanding the 1612 MHz masers. The infrared pumping scheme already hinted at in the discovery paper has been clearly established (Elitzur *et al.* 1976), to the extent that given the infrared flux we can predict the 1612 MHz maser flux. The 1612 MHz masers are usually saturated, and so vary linearly with changes in the pump radiation, following the stellar variations in a predictable way. The 1612 MHz masers nearly always have the twin-peaked spectrum characteristic of a uniformly expanding thin shell. Shell structure has been confirmed for numerous sources by long-baseline interferometry (Booth *et al.* 1981; Bowers *et al.* 1983). The OH shell is produced by photodissociation of outflowing H_2O by the external UV radiation. Given the mass loss rate, the expansion velocity, and the external UV radiation field the OH shell radius can be predicted (eg. Netzer & Knapp 1987, and references therein).

This doesn't seem to leave much for radio astronomers to discover! To first order we know what is going on in these 1612 MHz sources. This makes them useful tools for astronomical investigations, for example distance measurements. The OH mainline masers on the other hand offer more of a challenge, and it is these I want to concentrate on now.

357

2. OH MAINLINE MASERS

You can tell that the OH mainline masers are interesting from their variety of spectral profiles. The profiles fall into two main classes: twin-peaked (as for 1612 MHz), or else plateau-like, often with multiple irregular peaks. There is no way to predict which type of profile a given source will have, or to predict the 1665 MHz profile type given the 1667 MHz profile. There is a simple explanation for the class division, however, in terms of the thin shell model (eg. Chapman & Cohen 1985). If the OH is located far from the star, in a region where the outflow velocity changes very little with radius, then the maximum path length for maser amplification will occur at the front and back of the shell, leading to a twin-peaked spectrum (as for 1612 MHz masers). The maser radiation is beamed radially outwards from the star in this case. If the OH lies nearer the star, in a region where there is an appreciable velocity gradient, then the maximum path length for maser gain occurs in an annular region perpendicular to the line-of-sight. The maser radiation is beamed tangentially rather than radially. This leads to a line profile concentrated nearer the stellar velocity, depending on the velocity gradient.

The variations of OH mainline masers are also interesting. Whereas the 1612 MHz masers vary linearly with the infrared pump, the mainlines are more capricious. For example Sivagnanam (1989) monitored a sample of Miras simultaneously in the three OH lines and found that the mainlines varied over a range 2-3 times wider than that of the 1612 MHz line, with large amplitude variations from cycle to cycle. Furthermore the shape of the spectrum often changed on a timescale of one year. These facts strongly suggest unsaturated masers, which respond exponentially to changes in the maser pump.

One of the few regularities that is observed is that the ratio of the OH mainlines to the 1612 MHz line decreases systematically with increasing mass-loss rate (Kirrane 1987). This may reflect an increase in 1612 MHz pump efficiency at high mass-loss rates (Dickinson 1987). However there are interesting counter-examples to this general pattern. Some of the sources with the highest mass-loss rates have broad OH maser profiles some 100-200 kms^{-1} in velocity width, often with the 1667 MHz line dominant (te Lintel Hekkert *et al.* 1988). The prototype OH 231.8+ 4.2 has a well developed bipolar outflow (Morris *et al.* 1982). The others are suspected to be bipolar also.

Theory suggests that the OH mainlines should be inverted in warmer regions of the circumstellar envelope than the 1612 MHz line (Elitzur 1978). This is confirmed by observation, as shown in Figure 1. Here the OH radius for 1667 and 1612 MHz masers is plotted as a function of mass loss rate. In general the 1667 MHz radius is smaller at a given mass loss rate, although IRC +10420 provides a spectacular counter example. The use of an OH radius is not meant to imply that shell structure is the rule for the mainlines. A diversity of structures is found, ranging from thick and thin shells and discs to irregular filaments. Different parts of the envelope may emit in only one mainline, and even in only one circular polarization (eg. Sivagnanam 1989; Chapman *et al.* 1991). The polarization selectivity indicates that magnetic fields are important. The mainline selectivity may be due to line overlap in the maser pump, whereby kinematic effects could preferentially favour one mainline in one region, and the other mainline elsewhere.

A recent surprise was the detection of proper motions in the OH

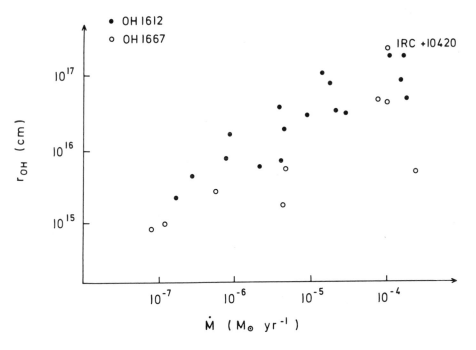

Figure 1. OH radius as a function of mass loss rate for circumstellar 1612 MHz and 1667 MHz OH masers. Data have been taken from Cohen (1989), Bowers *et al.* (1982, 1989), Shepherd *et al.* (1990), and references therein.

envelope of U Orionis (Chapman *et al.* 1991). Most of the mainline emission comes from a ring (tangentially beamed shell) which is expanding at 5 mas yr^{-1}, which is equivalent to 7 km s^{-1} at the distance of the star. The MERLIN data suggest that the mass loss rate has recently increased, and that the material is in filaments. Elongated filamentary structures are also seen in other Mira envelopes (Sivagnanam 1989). Detailed studies of the OH mainline polarization will be important in establishing the possible role of the stellar magnetic field in shaping the circumstellar envelope.

REFERENCES

Booth, R.S., Kus, A.J., Norris, R.P. and Porter, N.D., 1981. *Nature,* **290,** 382.

Bowers, P.F., Johnston, K.J. and Spencer, J.H., 1983. *Astrophys. J.,* **274,** 733.

Bowers, P.F., Johnston, K.J. and de Vegt, C., 1989. *Astrophys. J.,* **340,** 479.

Chapman, J.M., Cohen, R.J., 1985. *Mon. Not. R. astr. Soc.,* **212,** 375.

Chapman, J.M. Cohen, R.J. and Saikia, D.J., 1991. *Mon. Not. R. astr. Soc.,* **249,** 227.

Cohen, R.J., 1989. *Rep. Prog. Phys.,* **52,** 881

Dickinson, D.F., 1987. *Astrophys. J.,* **313,** 408.

Elitzur, M., 1978. *Astr. Astrophys.,* **62,** 305.

Elitzur, M., Goldreich, P. and Scoville, N., 1976. *Astrophys. J.,* **205,** 384.

Kirrane, T.M., 1987. *MSc thesis,* University of Manchester.

Morris, M., Bowers, P.F. and Turner, B.E., 1982. *Astrophys. J.,* **259,** 625.

Netzer, N. and Knapp, G.R., 1987. *Astrophys. J.,* **323,** 734.

Shepherd, M.C., Cohen, R.J., Gaylard, M.J. and West, M.E., 1990. *Nature,* **344,** 522.

Sivagnanam, P., 1989. *PhD. thesis,* Universite de Paris.

te Lintel Hekkert, P., Habing, H.J., Caswell, J.L., Norris, R.P. and Haynes, R.F., 1988. *Astr. Astrophys.,* **202,** L19.

Wilson, W.J. and Barrett, A.H., 1968. *Science,* **161,** 778.

PUMPING OF MAIN LINE MASERS IN TYPE I MIRAS

Alan J. Collison and Gerald E. Nedoluha
Naval Research Laboratory
Washington, DC 20375-5000

ABSTRACT

The conditions for main line maser emission in the circumstellar envelopes of Miras are studied. The radiative transfer is treated using the escape probability formalism and a modified Sobolev approximation. The effects of line overlap are included. Strong inversion of the 1612 MHz satellite line due to overlaps at 120μm dominates the models if all overlapping lines are included. Main line emission would be favored by low outflow velocities, thin maser regions and very asymmetrical or clumpy outflows.

1. INTRODUCTION

The pumping of OH main line masers has previously been studied by Bujarrabal et al. (1980) for circumstellar envelopes and by Cesaroni & Walmsley (1991) for interstellar masers. Bujarrabal et al. model a spherically symmetric, uniform shell with a constant radial velocity gradient. Only far infrared (FIR) overlaps to the ground state are considered and collisions were found not to be important. Also, only overlapping lines with relative velocities below a chosen limit were considered and emission from a central star was not included. Cesaroni & Walmsley model a spherically symmetric, uniform cloud with a given radius and expansion velocity, thus limiting the maximum velocity at which overlaps can occur (which for most of their models is 2 km s^{-1}). They are concerned primarily with collisional pumping by H$_2$ molecules and by line overlap and include the effects of "local" (i.e., thermal) overlaps. Although both conclude that the effects of line overlap can explain the observed emission, no justifications are given for the exclusion of overlaps above a seemingly arbitrary velocity. Our calculations show that the choice of this limit can have significant effects on the population inversions, including those relevant to ground state masers. If *all* possible overlaps are included in the calculations, with contributions limited *only by the chosen geometry and velocity field*, we find that the 1612 MHz line dominates the main lines in virtually all of the model calculations.

2. MODELS

The models presented here are similar to those of Bujarrabal et al. The radiative transfer is treated using escape probabilities and the Sobolev approximation is used with modifications to take into account (approximately) the finite size of the masing region (see Collison & Fix 1992). The effects of line overlap are treated using a local approximation similar to the treatment of Bujarrabal et al. Local overlaps are not included in the calculations.

Our model differs from that of Bujarrabal et al. in several important respects. The outflow and production of OH is calculated in the hope of producing an OH shell with reasonable variation of parameters (number density, gradients, temperature, etc.) through the masing region. The outflow is determined by assuming acceleration due to dust-gas collisions as in, e.g., Kwok (1975) except that a high initial velocity is assumed (Bowers 1992, Bowen 1988). The OH distribution is calculated assuming photodissociation of H_2O due to background UV. No distribution is calculated inside a radius for which chemical reactions are assumed to dominate. This distribution is then scaled to produce various total column densities of OH. We include emission from the central star, collisional cross sections for OH with H_2 (Flower 1989, Corey and Alexander 1988), and overlaps between excited rotational state transitions. In addition, the maximum velocity for overlaps is *not* limited to some arbitrary value in our "standard model." Model calculations are presented to illustrate the effects of artificially limiting the velocity for overlaps. The following parameters specify the standard model: stellar mass = 1 $M_\odot$; stellar temperature = 3000°K; stellar radius = 2.5×10^{13} cm; gas temperature = $300°K(r/10^{14}cm)^{-0.5}$; inner dust shell boundary = 6.2 times the stellar radius which gives a maximum dust temperature of $\approx$ 1000°K; dust grain radius = 0.035μm; optical thickness of dust at 10μm = 0.1; mass loss rate = $3.5 \times 10^{-7}M_\odot$/yr; dust absorption efficiency varies as ν^2.

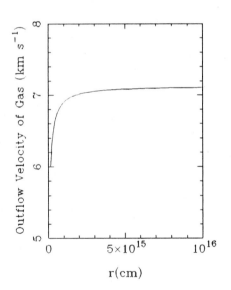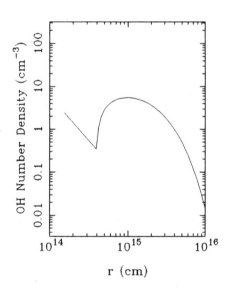

Figure 1. Velocity structure and OH distribution for the standard model.

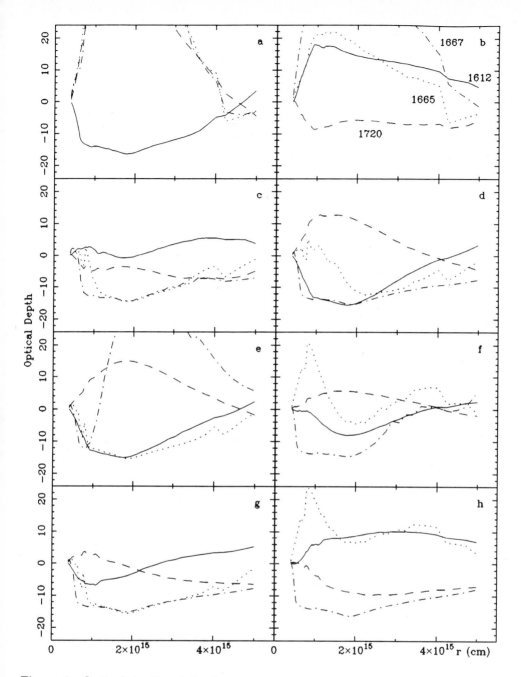

Figure 2. Optical depths of the four ground state maser lines. The following models are shown: (a) standard model; (b) standard model without the 120μm overlaps. Also shown are models with the optical depth of dust multiplied by 5 and: (c) without 120μm overlaps, (d) overlaps limited to 3.0 km s^{-1}, (e) 2.2 km s^{-1}, (f) 2.0 km s^{-1}, (g) overlap intensity reduced by a factor of 10 for overlaps > 2.2 km s^{-1}, (h) > 2.0 km s^{-1}.

3. SUMMARY OF RESULTS

The thrust of these calculations has been to model maser emission for a seemingly reasonable range of conditions thought to be appropriate for Mira variables. The most robust feature of the models is the presence of strong 1612 MHz inversion. This is due principally to line overlaps at 120μm which are in *emission* with respect to the background infrared, thus resulting in the removal of population from the $^2\Pi_{3/2}, J = 3/2, F = 2$ levels and hence favors inversion of the 1612 MHz transition but not 1665 MHz or 1667 MHz. This process also inhibits inversion of the 1720 MHz transition. In *all* the models produced so far in which all overlaps were included, the 1612 MHz transition is strongly inverted unless the column density of OH is too small for significant emission in any of the maser lines. In some cases (e.g., when the dust shell is somewhat thicker than in the standard model and for low column densities of OH) the 1612 MHz and 1667 MHz transitions can both be strongly inverted with little or no 1665 MHz emission. When the dust emission is increased sufficiently, pumping up the $\Pi_{1/2}$ ladder dominates the line overlaps at 120μm and the main lines are more easily inverted. However, the shells then resemble Type II Miras. The only models which resemble Type I Miras are those which neglect overlaps at 120μm altogether (these show no 1612 MHz emission at all) and those which reduce the intensity in overlapping lines at "high" velocity. *However*, we also note that it is significant at what point one makes the distinction between high and low velocity overlaps, as the results clearly show.

While some Mira variables might have dominant 1612 MHz emission due to the mechanism just described, clearly something must inhibit it in Type I Miras. It may be that the intensity of overlaps at high velocities ($\gtrsim 2$ km s^{-1}) is reduced because the masing region is much thinner than in our models. Alternatively, a highly asymmetric geometry may help to reduce the importance of high velocity overlaps by localizing the masing region in a "disk" or similar structure. Such asymmetry is in fact strongly indicated (e.g., Bowers 1989, Johnson and Jones 1991, Collison and Fix 1992). A very clumpy outflow with a small number of large clumps might also have the same effect. Alternatively, near infrared pumping due to stellar radiation, which can have a significant contribution in the small, optically thin shells of Miras, might be important. This would require much larger column densities of OH or some other mechanism to preferentially populate the upper lambda states (e.g., Cimerman and Scoville 1980).

Bowen, G. H. 1988, ApJ, 329, 299
Bowers, P. F., Johnston, K. J. and de Vegt, C. 1989, ApJ, 340, 479
Bowers, P. F. 1991, ApJS, 76, 1099
Bowers, P. F. 1992, ApJL, in press
Bujarrabal, V., Guibert, J., Nguyen-Q-Rieu, and Omont, A. 1980, A&A, 84, 311
Cesaroni, R. and Walmsley, C. M. 1991, A&A, 241, 537
Cimerman, M. and Scoville, N. 1980, ApJ, 239, 526
Collison, A. J. and Fix, J. D. 1992, ApJ, 390, in press
Corey, G. C. and Alexander, M. H. 1988, J. Chem. Phys., 88, 6931
Flower, D. R. 1989, Phys. Rep., 174, 1
Kwok, S. 1975, ApJ, 198, 583

UNUSUAL OH RADIO EMISSION IN MIRA (o CET)

E. Gérard and G. Bourgois
Observatoire de Paris-Meudon
92195 Meudon Pal Cedex

France

ABSTRACT

The prototype Mira (o Cet) has been extensively studied at radio wavelengths in HI, H_2O, CO and SiO. It was only detected once in OH, at 1665 MHz, by Dickinson (1975) and never since. We also found emission in Mira at 1665 MHz, using the Nançay radio telescope in November 1990 and have monitored it regularly. The line profile is narrow and limited to the velocity range 45.3-46.8 kms^{-1} LSR (the stellar velocity is 46.8 kms^{-1}). The total intensity appears to vary periodically but the peak occurs near light minimum. The signal exhibits strong circular polarization.

1. THE 1665 MHZ OBSERVATIONS

The prototype Mira variable O Cet was easily detected in OH at 1665 MHz with the Nançay radio telescope in November 1990 (Gérard and Bourgois, 1992).

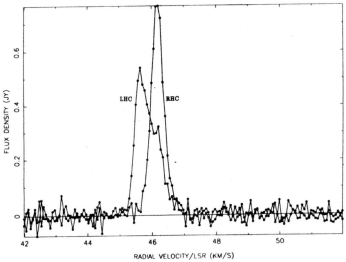

Figure 1: 1665 MHz profile of o Cet, averaged over the period Nov. 1990 to Dec. 1991. Left hand circular (LHC) polarization: filled circles. Right hand circular (RHC) polarization: open circles.

We have monitored the signal since then, in left handed (LHC) and right handed (RHC) circular polarization. The total system noise is typically 45 K and the observations were made regularly, using the autocorrelator receiver in the frequency switching mode, with a resolution of 390 Hz (0.07 km/s at 1665 MHz). The monitoring program includes three calibrators, 3C161 (continuum), W12 and 3C123 (line) which have not been fully processed yet; thus the flux density data presented here are only accurate to about 10%.

Mira was not detected in the 3 other hyperfine transitions at 1667 MHz, 1612 MHz and 1720 MHz and only the 1665 MHz transition was extensively studied during 15 months. The 1665 MHz profile, averaged over the period November 1990-December 1991, is shown on Fig.1 in both polarizations.

The OH emission is strictly limited to the velocity range 45.3-46.8 km/s LSR i.e. blueshifted with respect to the stellar velocity of 46.8 km/s.

The 1665 MHz profile mostly consists of two velocity components. The first one, near 45.6 km/s, is strongly circularly polarized, up to 100% at the low velocity end. The second one, near 46.2 km/s, exhibits both circular and linear polarization.

2. THE RADIO LIGHT CURVE

The total power flux density is clearly variable, as evidenced by Fig.2. The 1665 MHz emission appears to repeat itself with a period of about 340 days although the time interval spanned by our data is admittedly insufficient to provide a better period than the 332 days obtained from the visual light curve analysis over many years (Mattei et al., 1990; Menessier, 1991).

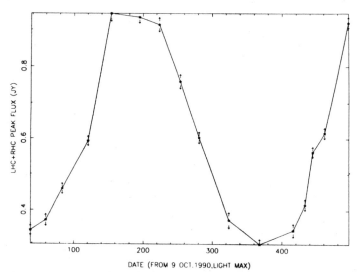

Figure 2: The total power flux density of the 1665 MHz profile of o Cet, as a function of time. The origin is the light maximum (phase 0.0) on 9 Oct. 1990. The 40 individual profiles have been averaged in 16 separate groups in order to improve the signal-to-noise. ratio.

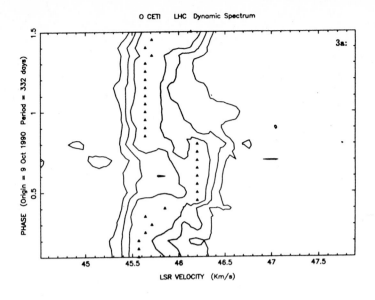

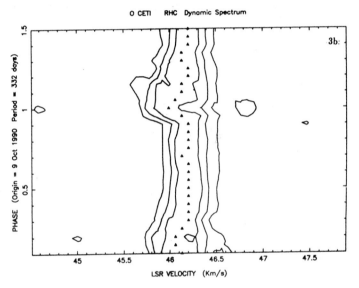

Figure 3a: 1665 MHz LHC polarization dynamic spectrum of o Cet. The individual observations are binned over phase intervals of 0.05 (17 days), using the 332 days optical
period, and then normed. The velocity position of the peaks are indicated with triangles
and the 3 full lines on either sides yield the positions of the 75%, 50% and 25% levels,
respectively.

Figure 3b: Same as a) for the RHC polarization.

The most surprising result is that the radio light curve appears to be anticorrelated with the visual light curve and, in particular, the "radio maximum" occurs near the "light minimum" at about 200 days i.e. phase 0.6 (the origin of the time scale in Fig.2 is the visual maximum on 9 October 1990).

If one extrapolates the radio light curve, the next maximum is predicted between early February and early May 1992.

Taken separately, the LHC and RHC flux density curves (not shown here) are radically different and thus it is very unlikely that the two velocity components, at 45.6 and 46.2 km/s, could be Zeeman counterparts.

3. DYNAMIC SPECTRA

Using the optical period of 332 days (Mattei et al., 1990), we have constructed a dynamic spectrum of o Cet in LHC and RHC polarization, as illustrated in Fig.3 a,b.

Several points are noteworthy:

(i) the velocity of the LHC peak shifts from 45.6 to 46.2 km/s near light minimum (phase 0.5-0.8).

(ii) the velocity of the RHC peak is fairly constant except for a slight redshift about light maximum.

(iii) globally, the radio emission shifts towards higher radial velocity near the light minimum.

(iv) apart from those variations, the velocities of the LHC and RHC peaks are remarkably constant and show little drift over a 15 month period.

A detailed study of the 1665 MHz emission of o Cet is being prepared for publication.

REFERENCES

Dickinson,D.: 1975, Ap.J. 199, 131, 1975
Mattei, J.A., Mayall, M.W., Waagen, E.O.: 1990, "The American Association of Variable Stars Observers": Maxima and Minima of Long Period Variables.
Mennessier, M.O.: 1991, private communication.
Gérard,E., Bourgois,G: 1992, IAU Circular No. 5427

VLBI SPECTRAL LINE POLARIZATION OBSERVATIONS OF THE OH MASERS

IN THE SUPERGIANT IRC+10420

A. J. Kemball and P. J. Diamond
National Radio Astronomy Observatory
Socorro, NM 87801

ABSTRACT

We present spectral line polarization observations of the 1612MHz OH masers in the supergiant IRC+10420. Our observations indicate that the individual maser components are highly circularly polarized in contrast to the polarization observed in total power spectra. Linear polarization of some features was also detected. We estimate that, at the distance of the 1612MHz OH masers from the star, the magnetic field has a strength of 1-2 mG.

1. INTRODUCTION

The circumstellar 1612 MHz OH masers typically show little circular polarization (Wilson and Barrett 1972), although the supergiant sources such as IRC+10420 (Reid et al. 1979) appear to be an exception. Highly circularly polarized features of narrow linewidth are found to be prevalent in supergiant spectra if observed with sufficient velocity resolution (Cohen et al. 1987). A simple Zeeman interpretation implies a magnetic field of the order of $\sim$ mG, which may significantly influence the circumstellar outflow (Cohen et al. 1987). The OH/IR stars are distinguished by much weaker circular polarization at comparable velocity resolution (Zell and Fix 1991).

We report here the results of VLBI observations of IRC+10420 to determine the distribution of the polarized OH maser emission with high spatial resolution.

2. OBSERVATIONS and DATA REDUCTION

The observations were carried out in March 1987 using the antennas at Jodrell Bank, Effelsberg, Onsala, Medicina and Westerbork, operating as part of the EVN. Both circular polarizations were recorded at each site using the MK3 VLBI recording system, with the exception of Medicina where only LCP was recorded. Each polarization was recorded within a 250 kHz bandwidth. All antenna polarization pairs were subsequently correlated in spectral line mode at the Haystack correlator.

All post-correlation data reduction including fringe-fitting was carried out within AIPS. The complex bandpass response of each antenna was removed, the spectra were shifted to the local standard of rest and amplitude calibration of each sense of polarization was performed using an adaptation of standard standard spectral line VLBI techniques (Reid et al. 1980). In addition, residual group delays and fringe rates were removed while taking account of all known phase and delay offsets between the two recorded polarizations. These included ionospheric Faraday rotation and instrumental offsets in the receivers and electronics at each antenna. Residual antenna phases were determined from the self-calibration of a reference velocity channel in one sense of polarization. A linear model of the polarization response of each feed was used (Hjellming 1983),

$$V_k^R = E^R e^{-j\alpha_k} + D_k^L E^L e^{j\alpha_k}$$
$$V_k^L = E^L e^{j\alpha_k} + D_k^R E^R e^{-j\alpha_k}$$

where $E^{(R,L)}$ are the components of the incident electric field, α_k is the time-variable parallactic angle at antenna k, and $V^{(R,L)}$ are the recorded signals in each sense of polarization.

Autocorrelation Spectra (I,V) of IRC+10420 (1 Mar 1987)

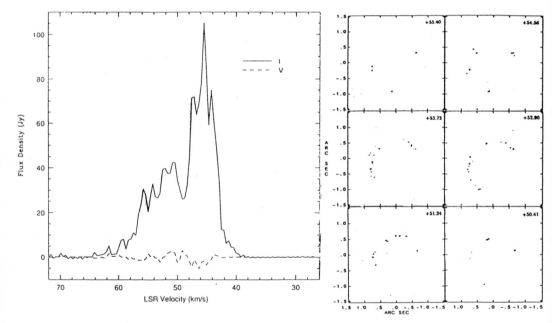

Fig. 1.— Total power spectrum of the 1612 MHz OH emission. Formal Stokes I and Stokes V are shown.

Fig. 2.— Total intensity VLBI maps of the 1612 MHz OH emission The LSR velocity of each channel is indicated in the upper right hand corner in units of km/s. The contour levels are (-30, -15, 15, 30, 45, 60, 75, 90, 100) % of the maximum brightness in each map. These maps show the formal Stokes I emission.

370

The cross-talk terms $D_j^{(R,L)}$ were determined by a simultaneous fit of a range of independent velocity channels within the program source IRC+10420 assuming net linear polarisation ~ 0, but taking into account the total intensity structure in each channel. This is a justifiable approximation as linear polarization is known to be small within the source. Polarization calibration using the continuum calibrators proved unreliable due to the narrow recorded bandwith and limited parallactic angle coverage.

A set of maps in each Stokes parameter $(I, V, Q, U)(\zeta, \eta, v)$ were obtained where (ζ, η) are angular coordinates and v is the velocity with respect to the LSR. The Q and U maps were derived from the complex imaging of $P = Q + jU$ (Conway and Kronberg 1969). The absolute overall orientation χ_0 of the angle of linear polarisation $\chi = \frac{1}{2}\arctan(\frac{U}{Q}) + \chi_0$ is not known due to nature of the polarization calibration.

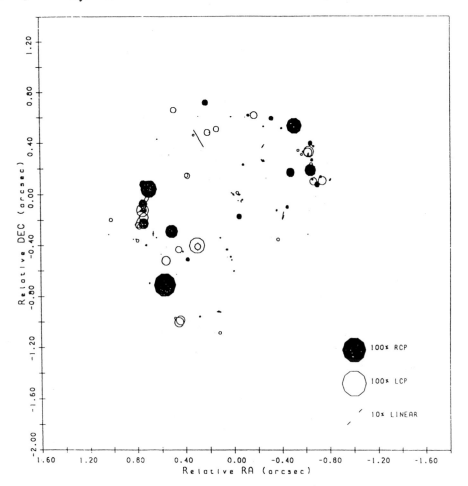

Fig. 3. The spatial distribution of the polarized OH maser components in IRC+10420. See the text for a full explanation of this figure.

3. RESULTS

The total power Stokes I and V spectra of the source are shown in Fig. 1. The total intensity VLBI maps are plotted in Fig. 2, for a selection of velocity channels. The E-W disk-like structure at a position angle of $\sim$ 70 deg reported by Diamond, Norris and Booth (1983) and Bowers (1984) is confirmed, but these observations reveal a more complete shell structure over a comparable velocity range. Preliminary proper motion results derived from an earlier VLBI observation in 1982 imply that the shell is expanding, but the overall kinematics of the source are complex (Bowers 1984).

The individual maser components were found to be highly circularly polarized in contrast to the total power spectrum shown in Fig. 1. The spatial distribution of the polarized components is shown in Fig. 3, where the degree of circular polarization, $m_c = \frac{V}{I}$, is indicated by the radius of the plotted circle and the sign of m_c by the shading of the circle. The degree of linear polarization $m_l = \frac{\sqrt{Q^2+U^2}}{I}$ is indicated by the length of the associated diagonal line, which is drawn at the polarization angle χ. Velocity information is omitted from this plot to limit confusion. Components were excluded from this plot with a peak I or $m_l I$ less than 0.5 or 0.1 mJy/beam respectively. Preliminary estimates of the magnetic field from the polarization data indicate a field of 1-2 mG.

REFERENCES

Bowers, P.F.,1984, *Ap. J.*, **279**, 350.

Cohen *et al.*, 1987, *M.N.R.A.S.*, **225**, 491.

Conway, R.G., and Kronberg, P.P.,1969, *M.N.R.A.S.*, **142** 11.

Diamond, P.J., Norris, R.P., and Booth, R.S.,1983 *Astr. Ap.*, **124**, L4.

Hjellming,, 1983, *"An Introduction to the VLA"*, NRAO Internal Publication.

Reid *et al.*, 1979, *Ap. J.*, **227**, L89.

Reid *et al.*, 1980, *Ap. J.*, **239**, 89.

Wilson, W.J., and Barrett, A.H., 1972, *Astr. Ap.*, **17**, 385.

Zell, P.J., and Fix, J.D.,1991,*Ap. J.*, **369** ,506.

To Mase or Not to Mase? Polarimetry of OH/IR Stars

Geoffrey F. Lawrence
Astronomy Department
University of Minnesota
Minneapolis, MN 55455

Optical and infrared polarimetry has been performed on a sample of OH/IR stars selected by their IRAS colors and a second sample of stars sharing the same IRAS colors but lacking a measurable 1612 MHz OH maser. Infrared photometry indicates these nonmasing stars are from the same population of stars as the masers: old disk dusty Miras (Lawrence et al. 1990). Although photometry is unable to discern a difference between masing and nonmasing stars, polarimetry may provide important clues. Polarimetry can provide information about the geometry of the circumstellar shell.

Histograms of the maximum intrinsic polarization (i.e., ISM removed) of the two samples are presented in Figure 1. On average the masers show a higher degree of polarization than the nonmasers. A two-tailed Kolmogorov-Smirnov test, with a binning of 1% polarization, indicates that the two samples are different with a 95% confidence level.

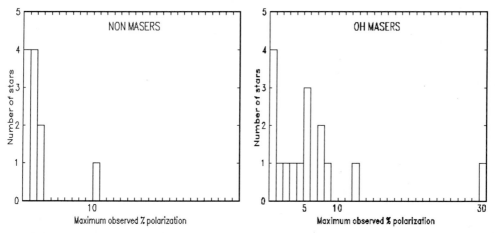

Figure 1.   Histograms of observed maximum intrinsic polarization for nonmasing stars and masing OH/IR stars.

The wavelength dependence of the two samples show a difference between the masers and nonmasers. The nonmasers, as illustrated by the polarization curve of IRAS 16582+0212 (Fig. 2, *left*), generally have a wavelength dependence similar to the polarization found in normal red giants with silicate grains in their circumstellar shells (Shawl 1975a,b). The nonmasers are typified by a polarization that drops continually toward longer wavelengths. The behavior for masers however can be strikingly different as illustrated by the polarization curve of IRAS 19495+0835 (Fig. 2, *right*). A peak is reached in the polarization at wavelengths much redder (~i to H) than in nonmasers.

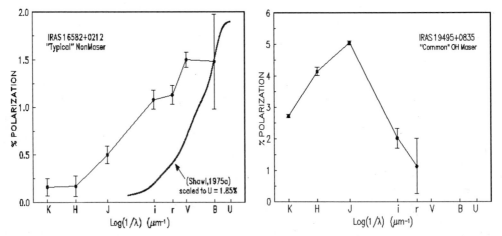

Figure 2.   Typical polarization curves for nonmasing stars (*left*) and masing OH/IR stars (*right*). Also included in the nonmasing figure is a theoretical polarization curve of a normal red giant with a circumstellar silicate shell (Shawl, 1975a).

The final graph (Fig. 3) plots the K-L color of the stars versus the maximum observed polarization.   The K-L color is a tracer of circumstellar shell thickness (Bedijn 1987, Lawrence et al. 1990).   No clear trend is seen, indicating that dust shell optical depth alone is not sufficient to determine whether or not a star masers.

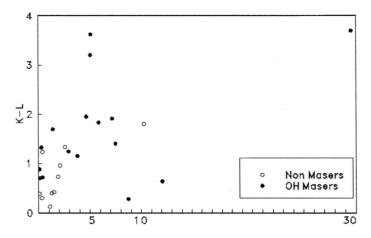

Figure 3.   K-L color versus the observed maximum intrinsic polarization for both masing and nonmasing stars.

Although the number of stars in each sample is small, initial results suggest that dust shell geometry is somehow tied to the ability to mase. Future work will include polarimetry of more stars and extending these observations to 3.6μm. Modelling the observed polarizations will then be undertaken.

References:

Bedijn, P.J. (1987). Astron. Astrophys. 186, 136.

Lawrence, G., Jones, T.J. and Gehrz, R.D. (1990). Astron. J. 99, 1232.

Shawl, S.J. (1975a). Astron. J. 80, 595.

Shawl, S.J. (1975b). Astron. J. 80, 602.

EVIDENCE FOR COMPETITIVE GAIN IN OH TRANSITIONS

OBSERVED IN CIRCUMSTELLAR SHELLS

B. M. Lewis
Arecibo Observatory
PO Box 995, Arecibo PR00612

ABSTRACT

The intensity of circumstellar mainline masers decreases as a shell reddens, and the dominance of the 1612 MHz maser increases. This trend is attributable to competitive gain between ground-state transitions of the OH molecule. Surveys show that (i) the 1665 MHz maser is exhibited by red giants, from the bluest mira shells where the line is comparable with the 1667 MHz maser, through ever thicker shells as these initiate 1612 MHz masers, but it tends to disappear in the thickest shells; (ii) a 1665 MHz maser is suppressed by a strong 1612 MHz maser at the same velocity; (iii) the 1667 MHz peaks are generally at different velocities from the 1612 MHz peaks.

1. INTRODUCTION

Mira variables and OH / IR stars form a seamless sequence in an IR two color diagram. The miras occupy its blue end, as their low mass-loss rates provide them with thin, translucent circumstellar shells, while the heavier mass-loss rates of OH / IR stars result in thicker and more opaque shells. There are other differences too, as OH / IR stars have strong 1612 MHz masers, whereas only the reddest miras generally have any 1612 MHz emission. On the other hand many miras exhibit the 1665 and 1667 MHz mainlines of OH, which usually have similar intensities and integrated fluxes; the mainlines are detected from a mere 53% of OH / IR stars with $(25-12)$ μm > -0.6, though the 1665 MHz line is often weak or absent. These distinctive patterns need explanation. Since the mainlines appear to flourish when the 1612 MHz line is weak, it is natural to suspect that competitive gain between the ground states of the OH molecule is responsible for the changing relative intensities (Lewis 1989). These effects resemble those found in the semi classical modelling of OH masers, which are due to competitive gain (Field 1985), when the OH column density increases in a slab.

A simplified way of looking at competitive gain in OH is to just consider its ground states, where the 1665 and 1612 MHz transitions share the same excited level. So whenever the 1665 MHz radiation field promotes a molecule to the excited state, in the presence of a strong 1612 MHz flux, a propagating 1612 MHz photon is likely to induce its reemission as a 1612 MHz photon. There is then a net transfer of energy into the 1612 MHz flux, with a concomitant weakening of the 1665 MHz flux. This is the simplest and most direct effect of competition between levels. In the case of the 1612 and 1667 MHz lines, there is a shared ground state, so the effects of competition are more indirect. Nevertheless, any interaction between a propagating 1667 MHz photon that raises a molecule to an excited state, increases the inversion of the 1612 MHz line, making it easier for an extra 1612 MHz photon to be added to the 1612 MHz flux. When the column density of OH molecules is suitably large, therefore, competitive gain weakens both main-lines propagating in the same gas column with the same velocity.

There is some supporting evidence for competitive gain. Thus Zylstra *et al* (1989) find a 24 Jy 1612 MHz feature in OH0.9+1.3, which coincides with a distinct hole in the 1665 MHz emission at the same (122-125 km s^{-1}) velocity, even though there is 1665 MHz emission otherwise over a 108 to 131 km s^{-1} range. Similarly Merlin maps of OH127.8 +0.0 (Diamond *et al* 1985) find 1612 MHz emission in a horseshoe shaped distribution, while 1667 MHz emission at the same velocity appears to lie just at the position of its gap. But this interpretation of these observations is not beyond question: it is currently impossible to locate the absolute positions of aperture synthesis maps on the sky to this precision when made with different lines, so we cannot be entirely sure that the outcome is due to conditions in exactly the same gas column. And in the case of the proto planetary nebula OH0.9+1.3 there is such a wide velocity distribution in the 1665 MHz line, that it is difficult to appeal to spherical symmetry, or to be sure that we currently understand its excitation environment. While both of these instances are consistent with competitive gain, we are interested in finding more observational confirmation for it.

2. OBSERVATIONAL EVIDENCE

Perhaps the most clearcut evidence will come from comparing changes in the intensity of the mainlines with changes in the 1612 MHz intensity round a complete pulsation cycle in classical two peaked spectra. We would then expect the occasions of maximum 1612 MHz intensity to correspond with minima in the other lines. I am conducting this experiment in collaboration with Ann-Marie Le Squeren and Pedro David at Nancay, but the results are not yet in. A different approach to this notion though is to intercompare the incidence of largest peaks in the 1612 and 1667 MHz lines in a sample of sources, to see whether they are typically anti-synchronized or random; with the assumption of competitive gain we would anticipate that the velocity of the largest 1667 MHz peak is opposite (redder or bluer) than that of the largest 1612 MHz peak near the same velocity. Yet another approach is to look more closely at the statistics on the detection of the mainlines with IR color from OH / IR stars. These experiments are discussed sequentially next.

The first few IR sources detected in 1612 and 1667 MHz from the *IRAS Preliminary Circulars* exhibit the property that the low (or high) velocity peak which is strongest at 1612 MHz is weakest at 1667 MHz, and viceversa. Examples are 18069+0911, 19440+ 2251 and 19288+2923. But these initial observations were from a small number of sources and were necessarily made at different epochs: Arecibo can only reach one of these lines at a time, and needs a maintenance period to switch feed sensitivity between lines. I therefore sought to test the applicability of competitive gain by making observations of a sample of OH / IR stars with 1667 MHz masers, that had distinct blue and red peaks at both frequencies, and arranged to measure their intensities in both lines, with the same 0.1 km s^{-1} velocity resolution, within 2-3 days of each other.

Basically most of these observations agree with expectations from the application of competitive gain, but the initial scenario was too simple to be literally applicable. I found that the largest 1667 MHz peak is indeed near the (red / blue) velocity of the smaller 1612 MHz peak for 24 of 45 stars. But this on its own would imply a random distribution of peaks. In practise a more assertive result occurs, in that the exact velocities of the peak features are never the same when their intensities are more than a factor of ~4 different (e.g. 19190+1128). Figure 1 shows an example where the 133 Jy 1612 MHz feature coincides with a deep hole in the 1667 MHz feature, which comes to local maxima with intensities of up to 510 mJy at velocities where the 1612 MHz intensity is weaker than the 1667; but the strongest peak of both lines is in the same red velocity range. I find that the 1667 MHz feature almost always arises inside or outside the velocity of the 1612 MHz feature (whether blue or red shifted). The small ($\leq$ 1 km s^{-1}) shifts are readily explained as a 1667 MHz line arising from the extreme doppler cohorts of molecules on the wings of the Boltzmann distribution, where the number of molecules does not provide a saturated OH column density for masing at 1612 MHz. But at velocities where the 1612 MHz line is clearly saturated, competitive gain all but eliminates the 1667 line at the same velocity. It is less clear at present that it is necessary to resort to this explanation for explaining the discrete 1667 MHz features with ~2 km s^{-1} velocity separations.

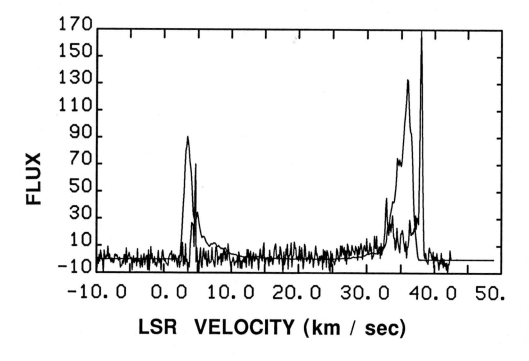

Fig. 1: Superposed 1612 MHz and 1667 MHz spectra for OH39.7+1.5. The flux scale for the 1612 MHz spectrum is in Jy, and shows little noise; that for 1667 is in mJy after multiplying by 3.

The intensity of the mainlines decreases vis a vis the 1612 MHz intensity, as (25-12) μm color reddens, and the OH column density increases. This is nicely illustrated in Fig. 1 of Cohen (1990), which shows a sharp decrease by a factor ~400 in the ratio of peak intensities with color, as sources change from mainline dominance to parity at (25-12) μm ≈ -0.4, which is followed by an average factor of ~40 difference in the ratio for redder type II sources where the 1612 MHz line dominates. Cohen rightly points out in his talk that the mainline pump is an order of magnitude weaker than that for the 1612 MHz line, which could perhaps explain these changes, though it does not explain why the 1665 MHz line is differentially weaker. The ratio of 1665 to 1667 MHz flux for sources with no detected 1612 MHz emission, and so no competition, is plotted in Fig. 2; these have a mean ratio circa one from the bluest miras to the reddest proto planetary nebulae. A new feature of this diagram though is the existence of some type I sources in the color range -0.5 to 0.0, which is usually dominated by OH / IR stars with saturated 1612 MHz masers. Since these have the usual mainline flux ratio, the mainline excitation mechanism is clearly independent of color, and of subtleties in the IR spectrum. But 1665 MHz masers are detected in just 36% of Arecibo OH / IR stars with (25-12) μm > -0.6, and in ~67% of those with 1667 MHz detections. These frequencies are consistent with the suppression of the 1665 MHz maser by competition with the 1612 MHz line.

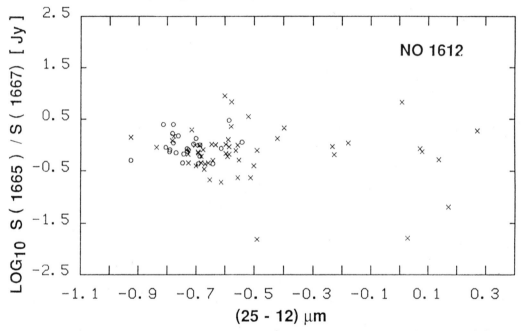

Fig. 2: Mainline integrated flux ratio as a function of color, for sources with no 1612 MHz emission. Arecibo survey sources are (x); miras as (o).

This work is supported by NAIC, which is operated by Cornell University under a management agreement with the National Science Foundation.

Cohen, R. J. 1990 *"From Miras to Planetary Nebulae; Which Path for Stellar Evolution"* 239
Diamond, P. J., Norris, R. P., and Booth, R. S. 1985, *MN,* **216**, 1P
Field, D. 1985, *MN,* **217**, 1
Lewis, B. M. 1989, *Ap.J.,* **338**, 234
Zylstra, A. A. *et al.,* 1989, *A&A,* **217**, 157

DISTANCE MEASUREMENTS OF OH-IR SOURCES

M.C. Shepherd[1], M.E. West[2], R.J. Cohen[1] and M.J. Gaylard[2]

[1]Nuffield Radio Astronomy Laboratories, Jodrell Bank,
Macclesfield, Cheshire SK11 9DL U.K.

[2]Hartebeesthoek Radio Astronomy Observatory, FRD,
P.O. Box 443, Krugersdorp 1740, South Africa.

ABSTRACT

Phase-lag techniques for determining the light travel time diameters of OH 1612 MHz maser shells are discussed. The measurement of angular diameters from radio interferometer maps is also discussed. Systematic errors appear to limit the accuracy of distance measurement to 10%.

1. INTRODUCTION

The technique for measuring distances of OH–IR sources is well known. The OH 1612 MHz maser emission comes from a thin spherical shell, the angular diameter of which can be measured using a long–baseline interferometer such as MERLIN (Booth *et al.* 1981). The masers are pumped by infrared radiation from the central dust envelope. They are saturated masers and so vary coherently with the (radiative) pump. The light travel time across the shell then leads to measurable phase-lags between different features across the 1612 MHz spectrum (Schultz *et al.* 1978; Jewell *et al.* 1980; van Langevelde *et al.* 1990, and references therein). Comparison between the angular size and the light travel time then yields the distance to the source (Herman & Habing 1985). In principle distances should be very accurately determined by this method. In practice systematic errors can limit the accuracy. In this paper we discuss recent work on reducing these effects and refining the distance measurement technique.

2. ANGULAR SIZE MEASUREMENT

For a large OH shell the angular size can be measured very accurately. Figure 1 shows MERLIN data on the source OH357.3-1.4, which has one of the largest shells known (Shepherd 1991). The source is strong and well resolved, and the OH emission can be mapped right across the spectrum. Complete rings of emission are seen in many spectral channels, so that any reasonable technique will give an accurate shell size. Indeed the variation of angular size with velocity can be studied so closely that small departures from the standard model of a uniformly expanding thin shell become apparent. These effects appear in Figure 1 as systematic departures from constant ring radius in the individual channel maps, and as systematic errors of 5-10% in a plot of ring radius against velocity (Shepherd 1991). This sets a natural limit to the accuracy of the angular size measurement.

A more serious problem arises when the interferometer does not resolve the shell structure. A good example is the source OH 0.9-1.3, which has an OH shell comparable in size with the MERLIN beam (Shepherd *et al.* 1990).

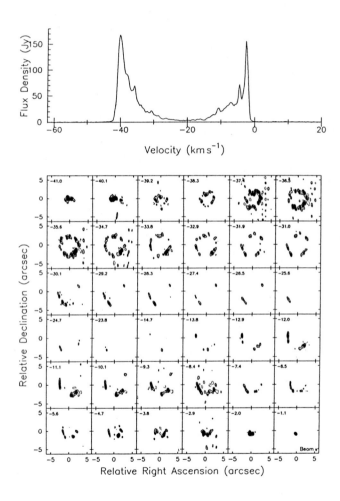

Figure 1. MERLIN maps of the OH 1612 MHz maser emission from OH357.3-1.4, showing the emission in each channel across the spectrum. The radial velocity is indicated at the upper left of each map. The beam is 0.3x0.6 arcsec^2 (Shepherd 1991).

The technique we have developed to recover the angular size of the shell in such cases is based on an analysis of the first and second moments of the channel maps. The first moment is the weighted mean emission centroid. For each channel map we calculate the second moment about this position. If the map showed a fully resolved ring (a slice through a thin spherical shell centred on the star) the second moment would be proportional to the square of the ring radius. The first and second moments of the channels are then used in a least squares fitting process to determine the radius of the maser shell and its central position (making use of the parallel axis theorem to recalculate second moments about various possible centres). Effects of noise and beam-smearing are reduced by using clean-components rather than the clean map. It is also possible to remove the contribution to the second moment from maser spot size, including effects of interstellar scattering, by examining the second moments of the extreme (zero radius) channels. By careful use of this technique it is possible to make useful angular size measurements of shells which are smaller than the interferometer beam (Shepherd 1991). The error analysis is complicated by the fact that the moments are calculated from the clean components of each map. This introduces pixel quantization errors which we are still investigating.

3. PHASE-LAG MEASUREMENTS

A number of techniques have been devised for measuring the phase-lag diameter of an OH-IR source. All the methods published to date assume that the maser emission is saturated, and that there is a one-to-one mapping between velocity and phase-lag, so that each spectral channel follows the same "light curve", but with a different phase-lag. In the standard approach the relative phase lags between different channels are obtained by interpolation and correlation, and then combined in a suitable way to yield the light travel diameter of the OH maser shell (e.g. van Langevelde *et al.* 1990). According to the thin shell model the phase-lag should vary linearly with velocity across the spectrum, and so in principle it is possible to check for departures from the model. In practice it is difficult to do this check because of the poor signal-to-noise ratio in the much weaker central channels of the spectrum.

An alternative approach which we have developed involves modelling the observed spectra. The method is based on the fact that the OH spectrum changes its shape during the stellar cycle. Using the OH "light curve" of the brightest feature, the average OH spectrum and the thin shell model, we

382

calculate sets of model spectra for each epoch of observation, with a series of assumed shell diameters. The optimum phase lag can then be found by least squares techniques. Since the shape is of primary importance, and since there are calibration errors caused by GLONASS transmissions, we normalize the observed and model spectra to have the same area before subtracting them in the least squares analysis. This makes the process much more robust.

The technique has been tested on data from the Hartebeestheok monitoring programme (Cohen *et al.* 1989). With data of this quality the statistical error which should be achieved is at the level of only a few percent. However a number of systematic errors are present, as evidenced by high χ^2 values, and these are still under investigation. For example the data cover several stellar periods in some cases, and the light curves obtained are clearly not periodic. Attempts to fit the data with a periodic light curve not only force up the χ^2 values, but also systematically bias the resulting phase-lag diameter. Best results are obtained by fitting independently to individual stellar cycles and then combining the results in a weighted mean. Another problem is that the OH spectra may change slightly with time. Again fitting to subsets of the data individually is better than attempting to model the whole data set at once.

Efforts to identify and eliminate the systematic errors are continuing, and we are confident that the techniques will yield reliable distances to OH–IR sources with errors of less than 10% in the best cases.

REFERENCES

Booth, R.S., Kus,A.J., Norris, R.P. and Porter, N.D., 1981. *Nature,* 290, 382

Cohen, R.J., Shepherd, M., Gaylard, M.J. and West, M.E., 1989. *IAU Symp. No.* 136, 51-4, ed. M.Morris, Kluwer Academic Publishers, Dordrecht.

Herman, J. and Habing, H.J., 1985. *Astron. Astrophys. Suppl.,* 59, 523.

Jewell, P.R., Webber, J.C. and Snyder, L.E., 1980. *Astrophys. J.,* 242, L29.

Schultz, G.V., Sherwood, W.A. and Winnberg, A., 1978. *Astron. Astrophys.,* 63, L5.

Shepherd, M.C., 1991. *Ph.D. Thesis,* University of Manchester.

Shepherd, M.C., Cohen, R.J., Gaylard, M.J. and West, M.E., 1990. *Nature,* 344, 522.

van Langevelde, H.J., van der Heiden, R. and van Schooneveld, C., 1990. *Astron. Astrophys.,* 239, 193.

FURTHER EVIDENCE FOR THE PHOTOPRODUCTION MECHANISM OF THE OH MASER FORMATION

M. Szymczak
Toruń Radio Astronomy Observatory
Nicolaus Copernicus University
PL–87100 Toruń

ABSTRACT

The interstellar ultraviolet field, G was computed at locations of Mira variables in homogeneous subsamples selected on their optical and infrared properties. Miras with non–OH masers are illuminated slightly less by UV photons than those with OH masers. The OH absolute luminosity increases steeply with increasing G. This finding strongly supports the photoproduction mechanism of OH shell formation.

1. INTRODUCTION

Although it is established that OH maser emission occurs in the gas–dust envelopes of oxygen–rich, long–period Mira variables we still do not know distinctive criteria that separate maser stars from non-maser stars. Some properties of these objects such as the shape of visual light curve, period, spectral type, infrared colors, strength of the $9.7\mu m$ feature have been found to correlate well with OH maser probability (Bowers & Kerr 1977; Sivagnanam et al. 1988). However, there are OH and non–OH Miras with apparently the same characteristics. We consider the possibility that the detection rate for OH maser emission from Mira variables in a homogeneous sample is affected by the ambient ultraviolet radiation. The photoproduction of OH from H_2O induced by the interstellar UV photons has been proposed as a main mechanism of OH molecule formation in the outer parts of the circumstellar envelopes (Huggins & Glassgold 1982; Netzer & Knapp 1987). Here we present the results of a study of the interstellar UV radiation field in the solar neighborhood at locations of Mira variables. These results give strong confirmation to the OH photoproduction model.

2. THE MIRA SAMPLE

We crosscorrelated all Mira variables M5.0e and later in the *General Catalog of Variable Stars* (Kholopov et al. 1985) with the *IRAS Low Resolution Spectrometer* (LRS) observations (IRAS Science Team, 1986) and selected those with the *IRAS* colors $-0.8 < (25-12) < -0.5$ and $-1.3 < (60-25) < -1.0$. The infrared colors are defined as $(\lambda_2 - \lambda_1) = log(\nu S_\nu(\lambda_2)/\nu S_\nu(\lambda_1))$, where S_ν is the flux density. From these we extracted the sample of stars within ~ 800 pc of the Sun with the visual periods 300–500^d which have been searched for OH maser emission. Distances were computed by using the Foy et al. (1975) method. We further refined the sample using the automatic classification of LRS objects (Cheeseman et al. 1989) and estimating the strength of the 9.7μm silicate feature defined by $E_{9.7} = 2.5log(S_{9.7}/S_{cont})$. The continuum flux density, S_{cont} was found from the interpolation between 8μm and 15μm. On the basis of $E_{9.7}$ value and OH maser detection our sample (Table 1) was divided into four homogeneous groups with very narrow ranges of optical and infrared properties. The OH absolute luminosity at maximum light integrated for the main lines was taken from Sivagnanam et al. (1988) and Slootmaker et al. (1985).

TABLE 1

THE SAMPLE OF M5.0e AND LATER MIRAS

Star*	$E_{9.7}$	G/G_0	Star	$E_{9.7}$	G/G_0	Star	$E_{9.7}$	G/G_0	Star	$E_{9.7}$	G/G_0
S Scl	0.05	0.9	RR Per	0.05	1.3	W Eri	0.40	3.8	Z Pup	1.00	8.5
T Cas	0.05	1.1	S Col	0.07	3.7	U Ori	0.61	2.0	V Ant	0.85	3.8
RX Tau	0.05	2.7	R Cnc	0.05	1.8	U Lyn	0.45	1.5	SY Aql	0.91	0.8
R Aur	0.12	1.1	X Hya	0.10	3.0	R Com	0.45	0.8	UX Cyg	1.10	5.9
S CMi	0.05	4.0	X Oph	0.05	0.8	RU Hya	0.60	1.4	RV Peg	1.33	2.0
S Vir	0.06	1.0	RR Sgr	0.12	1.0	RS Vir	0.40	0.9			
R CVn	0.06	0.7				S CrB	0.40	0.9			
S UMi	0.07	0.9				RW Sco	0.65	1.6			
BG Cyg	0.05	1.3				V Lyr	0.60	1.2			
T Cep	0.06	1.3				RT Aql	0.39	1.0			

* without OH emission

3. INTERSTELLAR ULTRAVIOLET RADIATION FIELD

The interstellar radiation field, G at 156.5 nm wavelength was computed by using procedure described elsewhere (Szymczak 1987). G exhibits considerable variations at different points in the solar neighborhood by a factor of 2–15 as compared with its mean value $G_0 = 8\times10^{-18} erg cm^{-3} nm^{-1}$. A comparison of our distance estimates to those from the literature indicates agreement to within about $\pm25\%$. This uncertainty causes for some Miras a change in the G/G_0 ratio by a factor of 4–5.

A plot of the asymmetry factor defined as the fraction of the period between minimum and maximum light versus the G/G_0 ratio for Miras with weak 9.7μm feature is shown in Figure 1. The average value of this ratio for stars without OH maser is slightly less than for stars detected in OH. However, more precise distance determinations and more sensitive OH observations are needed to confirm this tendency. On the other hand this plot supports the notion that OH masers are more frequently associated with Miras with asymmetric light curves than those with symmetric light curves.

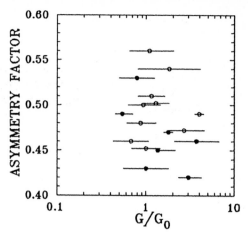

FIG. 1 Asymmetry factor of the visual light curve versus the interstellar UV field. Open circles: Miras without OH masers. Filled circles: OH Miras. The error bars indicate the G/G_0 uncertainties due to $\pm 25\%$ inaccuracy of distance estimates.

In Figure 2 the OH absolute luminosity is plotted as a function of the interstellar UV field. In spite of the scatter and the uncertainty in the G/G_0 ratio there is a clear trend of increasing OH luminosity with increasing interstellar UV field. Since the all three subsamples of OH Miras are strictly homogeneous then this correlation suggests that OH molecules can be effectively produced in these parts of the circumstellar envelopes where conditions for the main line excitation occur. Thus, the photoproduction model proposed to explain the 1612 MHz OH emission can be also justified for the main line maser emission.

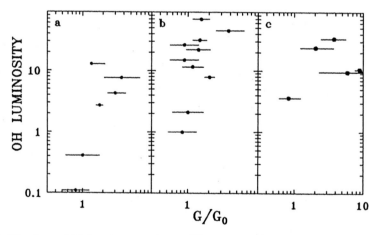

FIG. 2 Absolute OH luminosity (in 10^{17} W) as a function of the interstellar UV field for OH Miras with weak (a), moderate (b), and strong (c) 9.7μm silicate feature listed in Table 1. The bars mean the same as in Figure 1.

4. CONCLUDING REMARKS

The foregoing analysis suggests that anisotropy of the ambient galactic UV field can affect the OH maser probability in the Mira stars. Future observations, in conjuction with better distance estimates, should provide definitive conclusion. A correlation between the OH maser luminosity and the interstellar UV field found for limited but homogeneous samples of Miras indicates that the ambient UV photons are involved in the production of OH molecules in the masing regions. Despite the complexities of the maser emission processes a similar relationship should be also observed for OH sources identified only with the infrared stars.

REFERENCES

Bowers, P.F., and Kerr, .J. 1977, *Astr.Ap.*, **57**, 115.

Cheeseman, P., *et al.* 1989, *Automatic Classification of Spectra from the Infrared Astronomical Satellite (IRAS)*, NASA Ref.Pub. 1217.

Foy, R., Heck, A., and Mennessier, M.O. 1975, *Astr.Ap.*, **43**, 175.

Huggins, P.J., and Glassgold, A.E. 1982, *Astron.J.*, **87**, 1828.

IRAS Science Team 1986, *Astr.Ap.Suppl.*, **65**, 607.

Kholopov, P.N., *et al.* 1985, *General Catalog of Variables Stars* (4th ed; Moscow)

Netzer, N., and Knapp, G.R. 1987, *Ap.J.*, **323**, 734.

Sivagnanam, P., Le Squeren, A.M., Foy, F., and Tran Minh, F. 1989, *Astr.Ap.*, **211**, 341.

Slootmaker, A., Herman, J., and Habing, H.J. 1985, *Astr.Ap.Suppl.*, **59**, 465.

Szymczak, M. 1987, *Ap.Space Sci.*, **139**, 63.

MONTE CARLO MODELS OF 1612 MHz OH MASERS IN CIRCUMSTELLAR SHELLS

Huib Jan van Langevelde & Marco Spaans
Sterrewacht Leiden
Postbus 9513, NL–2300 RA Leiden
the Netherlands

ABSTRACT

We present models of the radiative transfer in circumstellar OH masers. In particular we model the 1612 MHz maser line in OH/IR stars. The coupled radiative transfer in expanding circumstellar shells can be simulated by means of a Monte Carlo technique (Spaans & Van Langevelde 1992). We show that such an approach is relatively advantageous for the radiative transport in masers. The combined effects of saturation and beaming can be observed in such models. In the geometry of expanding shells spectra can be obtained that resemble the average spectra in OH/IR stars quite well.

1. INTRODUCTION

Qualitatively the shape of the spectra of circumstellar OH 1612 MHz are well understood (see reviews by Herman & Habing 1985a and Cohen 1989). The maser spectrum shows on average two equally strong peaks separated by 20–50 km/s with a steep decline on the outer edges and a more tapered decline in the inner part of the spectrum. Also the variability of these masers is relatively regular. Due to saturation the maser process is self-regulating in these objects, so that a well-behaved radiation pattern emerges. Here we attempt to model the radiation field in these expanding shells, with the goal to obtain a quantitative description of the spectrum and brightness distribution of the 1612 MHz masers in OH/IR stars. We will assume that the pumping mechanism for the 1612 MHz is known; no attempt is made to solve the excitation of many states of the OH molecules (e.g. Gray et al., these proceedings). Instead we will concentrate on the complicated radiative transfer problem.

2. SIMULATING SATURATION AND BEAMING WITH A MONTE CARLO MODEL

Consider the problem of radiative transfer in an inverted medium (Elitzur, 1982). The transport equation for radiative transfer is:

$$\frac{dI_\nu}{ds} = -\alpha_\nu(\vec{x})I_\nu + \sigma_\nu(\vec{x}),$$

where α_ν is the absorption coefficient and σ_ν describes the spontaneous emission. Here α_ν depends on the population of the upper and lower levels, n_2 and n_1 respectively. The population of these levels is set by the pumprate that inverts the molecules and the intensity integrated over all directions of the radiation field $J_\nu = \frac{1}{4\pi} \int I_\nu(\Omega)\, d\Omega$.

Since we are dealing with a maser medium ($\alpha_\nu < 0$), a beam of radiation traveling in a particular direction will at first be amplified exponentially, while the inversion in the medium is maintained by the pump process. As long as the maser intensity is small, the inversion in the medium will hardly be affected by the growing intensity. This is the unsaturated regime, where the transfer problem is relatively easy, although already the geometry of the problem is important. Small changes in pathlenght will lead to large changes in intensity, thus beaming the maser. When the intensity is large it will deplete the ambient inversion and the maser wil saturate. Then beams of radiation propagating in different directions will compete for the same inverted molecules. This will lead to the effect that strong beams propagating through the medium may quench the growth of the weak ones by "eating away" the available inverted molecules. This will further enhance the beaming of the maser; the geometry of the radiating medium will have a very strong impact on the maser radiation. The effect of this "beam crossing" has been discussed by several authors. Especially Elitzur (1990) and Alcock & Ross (1985a,b) give a clear description of the differences between one- and two-dimensional transfer and the essential role played by saturation.

We solve the radiative transfer problem numerically using a Monte Carlo method. The basic idea is that we use a limited number of "photon packages" to represent the radiation field. Such a Monte Carlo approach is limited by statistics: a finite number of packages results in a noisy model. Also one has to demand that the resolution in, for instance, time and space, is sufficient to follow the physics.

We divide the medium up in cells which have a density, population distribution, pump rate and a velocity. The size of these cells should be such that we can resolve the changes of the physical conditions with position in the medium. By using a random number generator we initialize the radiation field with spontaneous emission. The packages travel through the medium and we calculate their amplification by the process of stimulated emission. This can be done without changing the total number of photon packages in the computer for a maser, which makes the Monte Carlo approach particularly suited for this type of problem.

To test if our code can reproduce the effect of beam crossing we have modeled homogeneous, static rectangular slabs with different axis ratios. Figure 1 shows the results of these calculations. The curves show the intensity ratio as a function of axis

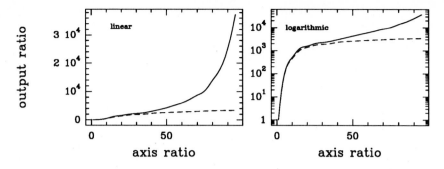

Figure 1. The ratio of flux that emerges from the long axis of a grid to that of the short axis as a function of axis ratio, plotted linearly on the left, logarithmic on the right. Dashed line shows calculation in which the effect of beam crossing is artificially switched off.

ratio. The effect of beam crossing is evident, as we move towards saturated models, the stronger beams will quench the growth of the weaker ones more and more.

3. RESULTS FOR CIRCUMSTELLAR SHELLS

Next we apply the same approach to the geometry of expanding shells. In the model the expansion velocity is taken to be 20 km/s and is assumed to be constant throughout the shell. The Doppler profile of the OH molecules is represented by a Gaussian profile with a FWHM of 1.0 km/s. The simple model presented here uses a constant pump rate, which is estimated from the 35 μm FIR pump cycle described by Elitzur et al. (1976). The OH density is taken to be constant at $n_0 = 2$ cm^{-3} between an outer and inner radius (Goldreich & Scoville, 1976, Netzer & Knapp, 1987). The region where the maser operates lies between $2.5 \cdot 10^{16}$ cm and $3.5 \cdot 10^{16}$ cm. The radius of the maser region is known from phase lag measurements (Herman & Habing, 1985b, van Langevelde et al., 1990). The thickness of the shell is in accordance with theoretical work by Netzer & Knapp (1987).

The maser profile in Figure 2 shows a nice resemblance with the average profile observed in OH/IR stars. Obviously observed OH/IR spectra will show irregularities. However it should be realised that a small amount of inhomogeneity suffices to account for this. Relatively small changes in optical depth (either due to velocity turbulence, density irregularities or non-uniform pumping) will be enhanced by the maser efficiently. We emphasize here that the spectral shape produced by our method is not a strong function of input parameters. Physical parameters such as thickness, pump rate and density can all be varied within reasonable values without changing the result in Figure 2 much.

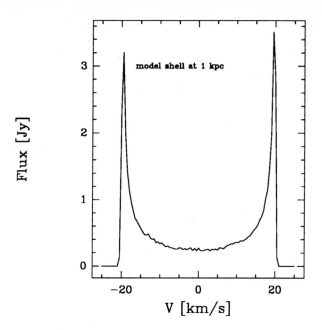

Figure 2. The result of model calculations for an expanding shell.

4. CONCLUSIONS

We have shown that a Monte Carlo simulation offers a very transparent way to model the complicated radiative transfer in masers. It is capable of reproducing important effects such as saturation and beam crossing. A big advantage is that everything is coded in local physics. This, in principle, makes it straightforward to simulate complicated situations, such as velocity fields, detailed pump mechanisms and density fluctuations. When we use the code to simulate an OH/IR star in a circumstellar shell, we find that a rough estimate of the parameters readily reproduces a spectrum that closely resembles the average spectra of OH/IR stars.

REFERENCES

Alcock, C., Ross, R.R., 1985a, ApJ **290**, 433
Alcock, C., Ross, R.R., 1985b, ApJ **299**, 763
Cohen, R.J., 1989, *Rep. Prog. Phys.*, **52**, 881
Elitzur, M., 1982, *Reviews of Modern Physics*, **54**, 1225
Elitzur, M., 1990, ApJ **363**, 628
Elitzur, M., Goldreich, P., Scoville, N., 1976, , ApJ **205**, 384
Goldreich, P., Scoville, N., 1976, , ApJ **205**, 144
Herman, J., Habing H.J. 1985a, *Physics Report* **124**, 255
Herman, J., Habing H.J. 1985b, A&A, Suppl. Ser. **59**, 523
Netzer, N., Knapp, G.R., 1987, ApJ **323**, 734
Spaans, M., Van Langevelde, H.J., 1992, MNRAS, in press
Van Langevelde, H.J., Van der Heiden, R., Van Schooneveld, C., 1990, A&A **239**, 193

PROPERTIES OF OH/IR STARS IN THE GALACTIC CENTRE

Anders Winnberg, Michael Lindqvist
Onsala Space Observatory
S-43900 Onsala
Sweden

Harm J. Habing
Sterrewacht Leiden
P.O. Box 9513
NL – 2300 RA Leiden
The Netherlands

1. INTRODUCTION

We have used the Very Large Array at 1612 MHz to search for OH/IR stars close to the Galactic Centre (GC) and we have found 134 such stars (Lindqvist et al. 1992a). We have used these stars as gravitational probes to derive the mass density distribution within the range $5 - 100$ pc from the GC (Lindqvist et al. 1992b). However the present paper deals with the physical properties of the OH/IR stars analyzed in a statistical sense (Lindqvist et al. 1992b; Lindqvist et al. 1992c).

This project has lead to several other projects. Most of them are based on the fact that all the OH/IR stars are at the same distance from us, i.e. at the distance to the GC. In this sense they constitute an ideal sample of OH/IR stars for which one can study properties which are dependent on distance (Lindqvist et al. 1990; Lindqvist et al. 1991; Winnberg et al. 1991). van Langevelde et al. (1992) have monitored 37 of the stars and determined periods and sizes and Blommaert et al. (1992), have observed 34 of the stars in the infrared and measured their bolometric luminosities.

2. CORRELATION BETWEEN ENVELOPE EXPANSION VELOCITY AND AGE

The 134 OH/IR stars plus 15 more from Habing et al. (1983) form an elongated system with the major axis parallel to the galactic equator and they show a concentration toward the Sgr A West radio source (Lindqvist et al. 1992b).

There is a positive correlation between the radial velocity of the stars and the galactic longitude which we interpret as galactic rotation but the dispersion from the regression line is large (Lindqvist et al. 1992b). Thus many of the stars have elongated orbits around the centre.

Baud et al. (1981) found an anticorrelation between the envelope expansion velocity (v_{exp}) and the dispersions in radial velocity (σ_{v_c}) and in galactic latitude (σ_b) for OH/IR stars in the galactic disk. In order to investigate this for the OH/IR stars close to the GC we divided our sample into two parts with roughly equal number of stars according to the value of v_{exp}. In Group I we include stars with $v_{exp} < 18\,\mathrm{km\,s^{-1}}$ and in Group II stars with $v_{exp} > 18\,\mathrm{km\,s^{-1}}$. When plotting the $(l - b)$ and the $(l - v)$ diagrams for the two groups we see the same effects for the OH/IR stars in the GC as Baud et al. (1981) did for OH/IR stars in the galactic disk (Lindqvist et al. 1992b).

Group I stars are much more spread out in galactic latitude ($\sigma_b \approx 44 \pm 4\,\mathrm{pc}$) than Group II stars ($\sigma_b \approx 20 \pm 2\,\mathrm{pc}$). Moreover Group II stars are more concentrated to the GC itself than Group I stars. A similar effect can be seen in the $(l - v)$ diagrams. Linear regressions to the two data sets give slopes which are both about $1.2\,\mathrm{km\,s^{-1}\,pc^{-1}}$. The dispersion from this regression is much higher for Group I stars ($\sigma_{v_c} \approx 82 \pm 7\,\mathrm{km\,s^{-1}}$) than for Group II stars ($\sigma_{v_c} \approx 65 \pm 6\,\mathrm{km\,s^{-1}}$). Errors in the dispersions are $\sigma/\sqrt{2N}$ (Trumpler&Weaver 1953)

Thus there is a difference in average age (or in main-sequence mass) between Group I and Group II. However, there is no sharp border between the two groups but rather a continuous correlation between v_{exp} and σ_b and between v_{exp} and σ_{v_c}. We have split the sample along the median value of v_{exp} ($18\,\mathrm{km\,s^{-1}}$) just to get maximum statistical significance.

There are thus strong evidence for Group I stars being older than Group II stars. However, the *absolute* age of these stars is much more difficult to estimate because there is no independent and well established age determination of stars in this part of the Galaxy. In addition there is the complication of higher metal abundance of these stars compared to stars further out in the disk. A higher metal abundance is expected to lead to a higher dust density in the circumstellar shell and therefore to a more efficient radiation pressure mechanism to drive the shell. Therefore a star with a higher metal abundance is expected to show a higher value of v_{exp} than a star with the same age but with a lower metal abundance. Furthermore there might be a larger gradient in the metal abundance at these short distances to the GC ($< 100\,\mathrm{pc}$) than farther out in the disk. Recently evidence has emerged that v_{exp} and metallicity are indeed correlated in the sense indicated above from studies of OH/IR stars in the LMC by Wood et al. (private comm.) and in the galactic anti-centre direction by Blommaert et al. (private comm.). Metallicity effects make correlations between statistical properties of stars at the GC more difficult to understand than elsewhere in the Galaxy.

3. COMPARISON WITH THE $2-\mu\mathrm{m}$ DATA OF Catchpole et al. (1990)

We have calculated the surface density of OH/IR stars on the sky averaged over elliptical annuli (N) as a function of the elliptical radius,

$$R = \left[(l/\rho)^2 + (b)^2\right]^{1/2},$$

where ρ is the axial ratio and l and b are the galactic coordinates. The distribution of all the OH/IR stars (Groups I and II) fit very well to the distribution of IR stars with K magnitudes in the range $6 - 7$ taken from Figure 10b of Catchpole et al. (1990) (Fig. 1). The slopes $[d(\log N)/d(\log R)]$ are -1.09 ± 0.09 and -1.19 ± 0.06 for the OH/IR stars and the $6 - 7^{\mathrm{m}}$ stars, respectively. The average factor between the two surface densities is 26 ± 2. The Group II stars compare very well with the IR stars with K magnitudes in the range $5 - 6$ (Fig. 1). The slopes are the same within their errors, -1.4 ± 0.1, and the average factor between the surface densities is 8 ± 1. This tells us first of all that only the complete sample (Groups I plus II) forms a space distribution with an equilibrium configuration, i.e. $n \propto r^{-2}$. The Group I stars show a much slower falloff ($n \propto r^{-1.6}$) and the Group II stars a much faster one ($n \propto r^{-2.4}$). Secondly, it tells us that the Group II OH/IR stars (the younger ones) are associated with the stronger K-band stars and, thirdly, that there is a constant ratio between the surface density of the K-band stars and that of the OH/IR stars.

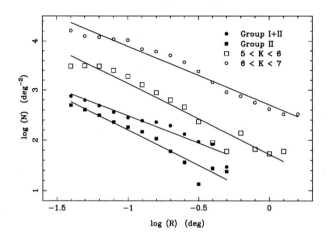

Fig. 1. The surface densities averaged over elliptical annuli as a function of the elliptical radius defined in the text. The filled symbols represent the OH/IR stars and the open symbols the stars observed in the K band by Catchpole et al. (1990).

The nature of the K-band stars is not established. It is believed that most of them are red giants. If this hypothesis could be confirmed and if a complete survey for OH/IR stars could be made, then an estimate of the fraction of the red giant evolution occupied by the OH/IR stage could be made.

4. THE OH LUMINOSITY FUNCTION

This leads us into the question of the OH luminosity function. Since the OH/IR stars in this sample are all at the same distance from us (within a few percent) the distribution of flux densities has the same shape as the luminosity function. We therefore believe that the distribution shown in Fig. 2a is the most correct representation of the 1612-MHz OH luminosity function of OH/IR stars published. It shows the distribution of the peak flux density of the strongest OH line component. The cutoff at the low-flux side is in accordance with the probability of detection, i.e. we have not reached the maximum of the luminosity function. The high-flux side ($\log(S_{max}) > -0.5$) corresponds to $N(S_{max}) \propto S_{max}^{-2.0\pm0.1}$. For stars of Group I (Fig. 2b) the exponent is (-1.7 ± 0.2) and for stars of Group II (Fig. 2c) it is (-2.7 ± 0.6). This implies that the strongest OH emitters are found among Group I whereas there are more Group II stars with weaker OH flux (≤ 0.4 Jy) than Group I stars. This is contrary to Baud&Habing (1983) who argue that the strong OH masers show high expansion velocities. Again, metallicity effects may make it more complex to interpret the results than elsewhere in the Galaxy.

The comparison with the K-band stars and the OH luminosity function will be discussed further in Lindqvist et al. (1992c).

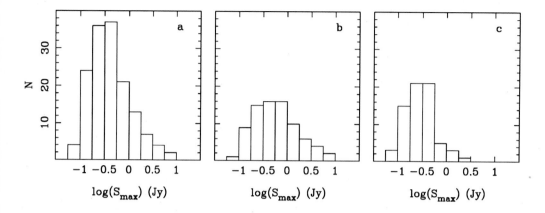

Fig. 2. The distributions of peak flux density of the strongest OH line component for (a) Group I+II, (b) Group I, and (c) Group II.

Acknowledgements. This project is supported by the Swedish Natural Science Research Council (NFR).

REFERENCES

Baud B., Habing H.J., Matthews H.E., Winnberg A., 1981, A&A, 95, 156

Baud B., Habing H.J, A&A, 127, 73

Blommaert J.A.D.L., van Langevelde H.J., Habing H.J., van der Veen W.E. C.J., Epchtein N., 1992, A&A, in preparation

Catchpole R.M., Whitelock P.A., Glass I.S., 1990, MNRAS, 247, 479

Habing H.J., Olnon F.M., Winnberg A., Matthews H.E., Baud B., 1983, A&A, 128, 230

Lindqvist M., Winnberg A., Forster R., 1990, A&A, 229, 165

Lindqvist M., Ukita N., Winnberg A., Johansson L.E.B., 1991, A&A, 250, 431

Lindqvist M., Winnberg A., Habing H.J., Matthews H.E., 1992a, A&AS, 92, 43

Lindqvist M., Habing H.J., Winnberg A., 1992b, A&A, 259, 118

Lindqvist M., Winnberg A., Habing H.J., 1992c, A&A, in preparation

Trumpler R.J., Weaver, H.F., 1953, *Statistical Astronomy*, Dover Publications, New York

van Langevelde H.J., Janssens M., Goss W.M., Habing H.J., 1992, A&AS, in preparation

Winnberg A., Lindqvist M., Olofsson H., Henkel C., 1991, A&A, 245, 195

14. CIRCUMSTELLAR WATER MASERS

THE EXCITATION OF VIBRATIONALLY EXCITED H$_2$O MASERS

Javier Alcolea

Harvard-Smithsonian Center for Astrophysics
60 Garden St., Cambridge MA 02138, USA
and Centro Astronómico de Yebes
Apartado 148, E-19080 Guadalajara, Spain

and

Karl M. Menten
Harvard-Smithsonian Center for Astrophysics

ABSTRACT

We describe a pumping mechanism that is capable of inverting a number of rotational lines within the first vibrationally excited state of the bending mode ($\nu_2=1$) of H$_2$O under the physical conditions present in the inner layers of circumstellar envelopes around oxygen-rich evolved stars. All the predicted maser lines result from a systematic overpopulation of the "transposed backbone" levels, which have quantum numbers $J_{k_a=J,k_c}$, with respect to their neighboring levels. We find that this systematic inversion can be explained by means of differential trapping in the radiative vibrational decays of the levels involved in those maser transitions. Although this mechanism leads to the inversion of the maser lines detected so far, numerical calculations we have performed do not quantitatively agree with the (still scarce) observational data.

1. INTRODUCTION

In recent years, about half a dozen new water maser transitions have been detected at (sub)millimeter wavelengths. Three of these lines are rotational transitions within the lowest vibrationally excited state of H$_2$O, the $\nu_2=1$ bending mode, namely the $4_{4,0}-5_{3,3}$ at 96.3 GHz, the $5_{5,0}-6_{4,3}$ at 232.7 GHz and the $7_{7,0}-8_{6,3}$ at 263.5 GHz (Menten & Melnick 1989; Menten 1992). These $\nu_2=1$ masers have been observed only toward oxygen-rich evolved stars with circumstellar envelopes (VY CMa and W Hya). Their observational characteristics, line shape and velocity range, together with the high temperatures necessary for their excitation, suggest that they arise from the hot and dense innermost layers of these envelopes, which also give rise to the vibrationally excited SiO masers (Menten & Melnick 1989; 1991).

One of the most remarkable characteristics of the $\nu_2=1$ maser transitions is that the upper level is always a $J_{k_a=J,k_c=0}$ level. This regularity resembles that of the masers within the vibrational ground state, which have their upper levels in or near the so-called backbone. Backbone levels have $k_c=J$. In the $\nu_2=1$ state, the upper levels of all the maser transitions detected so far lie in the part of the energy diagram that is opposite of the backbone and which we will refer to as the "transposed backbone" ($k_a=J$, see Fig. 1). In analogy to the ground state masers one might suspect that a systematic overpopulation of these "transposed backbone" levels might be achievable by a very general pumping mechanism.

2. NUMERICAL MODEL OF THE EXCITATION OF H_2O

In order to study the excitation of H_2O and in particular the nature of the $\nu_2=1$ masers we have developed a numerical code that, using the escape probability formalism for spherical symmetry under the LVG approximation, computes the population of the H_2O levels and the approximate intensity of the corresponding radiative transitions. In our calculations we consider excitation of water by collisions with molecular hydrogen and by radiation from the central star (assumed to be a perfect black body), and radiative coupling with the background radiation. We ignore the possible effects due to the circumstellar dust, line overlapping and, therefore, interaction between the para and the ortho species. We treat ortho and para water as independent species.

We include all levels in the ground and first vibrational excited ($\nu_2=1$) states of H_2O with energies up to ~ 5500 cm^{-1}. The line frequencies and strengths and the level energies and multiplicities are from the AFGL HITRAN database (Rothman et al. 1987). For the transitions within the ground state of H_2O the collisional rates have been computed following Neufeld & Melnick (1991). Collisional rates for pure rotational transitions within the $\nu_2=1$ state have been assumed to be identical to the rates of the corresponding transitions in the ground state. Finally, for the rotational-vibrational transitions between the ground and $\nu_2=1$ states we have assumed that the collisional rates are proportional to those for the pure rotational transitions. The a priori unknown proportionality factor has been varied between several runs of the program.

From the outset, our calculations are not expected to accurately reproduce the observed intensities, but we expect to obtain information on the physical conditions under which the $\nu_2=1$ masers are likely to occur. In particular, we only solve the local problem, i.e., we do not integrate along the line of sight through the circumstellar envelope to obtain the intensities and profiles of the different lines. The only dependence of our model on the geometry is introduced by a certain logarithmic gradient of the velocity field $d\log V/d\log r$, for which we have adopted different values to test its influence on the results.

3. THE PUMPING OF THE $\nu_2=1$ H_2O MASERS

For physical conditions typical of the inner shells of circumstellar envelopes (kinetic temperatures about 1500 K and densities of 10^9–10^{10} cm^{-3}) we find inversion in all the detected $\nu_2=1$ masers for H_2O column densities about 10^{18}–10^{21} cm^{-2} (coherence lengths of 10^{14}–10^{15} cm for a $[H_2O]/[H_2]$ relative abundance of 10^{-4}). These conditions are very similar to those required for the inversion of the high excitation $v=2,3$ SiO masers (Alcolea et al. 1989). We also obtain inversions in all the other "transposed

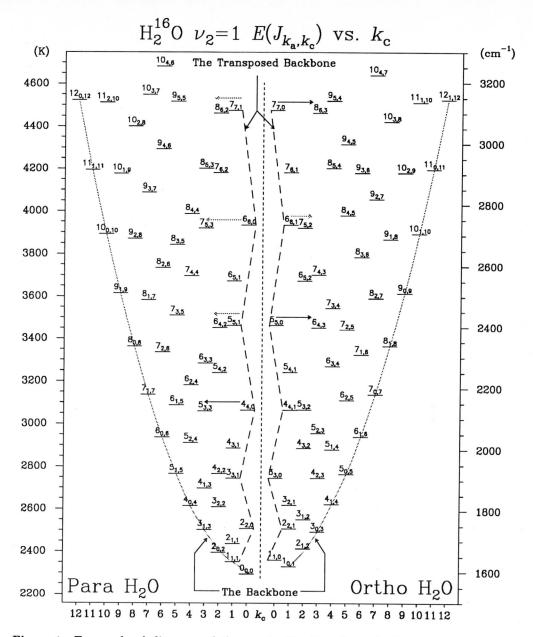

Figure 1. Energy level diagram of the $\nu_2=1$ vibrationally excited state of H_2O. The dotted line shows the levels which form the "backbone" while the dashed line shows those which form what we have called the "transposed backbone". The solid arrows represent the lines that have been found to be masing in the $\nu_2=1$ state; note that in all cases the upper level belongs to the "transposed backbone". The doted arrows represent other lines, also with a "transposed backbone" upper level, predicted to be maser by our model.

backbone" transitions (see Fig. 1) under the same physical conditions. At kinetic temperatures below 1000 K the inversion of all these lines is very weak, if present at all; their intensity also drops when either the amount of stellar radiation or the value of the logarithmic velocity gradient are increased. In our model the "transposed backbone" $\nu_2=1$ masers are collisionally pumped, their intensity being strongly dependent on the assumed values for the rotational-vibrational collisional coefficients. For densities larger than 10^{10} cm^{-3} these masers are quenched by the collisions within the $\nu_2=1$ state.

The systematic inversion of the $\nu_2=1$ "transposed backbone" transitions is produced by the different radiative trapping in the vibrational de-excitation of the upper and lower levels of the masers. The molecules are collisionally pumped to the $\nu_2=1$ state. Provided that the pump rate is the same for the upper and lower levels of the $\nu_2=1$ maser transitions, and neglecting the collisional de-excitation of these levels and the radiative transitions within the $\nu_2=1$ state, the level populations are determined by the rates of the radiative decay to the ground state. For the column densities quoted above, some of the decay routes of the "transposed backbone" levels become optically thick, while those for the lower states of the masers do not. This reduces the radiative decay rate of the "transposed backbone" levels, leading to their systematic overpopulation and to the observed $\nu_2=1$ masers.

This inversion mechanism has however a very low efficiency because of the relative low line strength of the radiative vibrational transitions, which, in spite of their larger frequencies, have Einstein's A coefficients similar to those of the $\nu_2=1$ rotational lines, which therefore cannot be neglected. Moreover, at the physical conditions necessary for this mechanism to work, the collisional de-excitation to the ground state and the collisions within the $\nu_2=1$ state are also important. Both collisions and $\nu_2=1$ radiative rotational transitions tend to reduce the overpopulation of the "transposed backbone".

Although the model calculations predict the inversion of the detected maser lines, we cannot reproduce their intensities. Because of the low efficiency of the pumping, in order to obtain for the detected lines intensities similar to those observed, we must assume that the rotational-vibrational collisions are as strong as the pure rotational ones, which seems to be very unlikely. Moreover, the 96.3 GHz line ($4_{4,0}$–$5_{3,3}$), which has been observed to be the strongest $\nu_2=1$ maser, is the one that is most difficult to invert in our calculations. It is of course possible that other phenomena that we have not taken into account so far, like line overlapping (in particular between infrared lines of the para and the ortho species) or excitation via higher vibrational states, could also play an important role in the pumping of the $\nu_2=1$ H$_2$O masers.

REFERENCES

Alcolea J., Bujarrabal V., Gallego J.D., 1989, A&A 211,187
Menten K.M., Melnick G.J., 1989, ApJ 341, L91
Menten K.M., Melnick G.J., 1991, ApJ 377, 647
Menten K.M., 1992, in preparation
Neufeld D.A., Melnick G.J., 1991, ApJ 368, 215
Rothman et al., 1987, Appl. Opt. 26, 4058

J.A. is partially supported by the Spanish Ministerio de Educación y Ciencia.

THE 22 GHZ WATER MASER OF RX BOOTIS

D. Engels, Sternwarte der Universität Hamburg, Germany
A. Winnberg, Onsala Space Observatory, Sweden
J. Brand, Istituto di Radioastronomia, Bologna, Italy
C.M. Walmsley, Max-Planck-Institut für Radioastronomie Bonn, Germany

RX Bootis is an SRb-type semiregular variable star of spectral type M6.5e-M8e, varying optically between mpg=8.6 and 11.3 mag (Khopolov et al., 1985). The star loses mass at a rate of $4 \cdot 10^{-7}$ $M_\odot$/yr creating a circumstellar shell expanding with a final velocity of ~11.5 km s^{-1} (Gehrz and Woolf, 1971; Knapp and Morris, 1985). The shell contains a strong 22 GHz H_2O maser, which occasionally reaches peak flux densities of several hundred Jansky (Engels et al. 1988 and ref. therein), making RX Boo an outstanding case to study the water maser properties in semiregular variables. Additionally the shell sustains an SiO maser, but no OH maser emission (Benson et al., 1990). The stellar radial velocity is +0.5-1.0 km s^{-1} (Knapp and Morris, 1985; Bujarrabal et al. 1986), and the distance is 225 pc (Morris et al. 1979).

We have included RX Boo in an ongoing monitoring program of water maser emission of selected late-type stars. Single-dish monitoring is performed with the radiotelescopes in Effelsberg and Medicina measuring approximately every two months. The program was started in 1990. Interspersed with the single-dish observations, mapping is done with the VLA in its A-configuration. Until now 3 maps at epochs February and June 1990 and October 1991 were obtained. Typical rms-sensitivities achieved are 2500 mJy (Medicina), 300 mJy (Effelsberg) and 30 mJy (VLA). Preliminary results from the 1990 data are presented here.

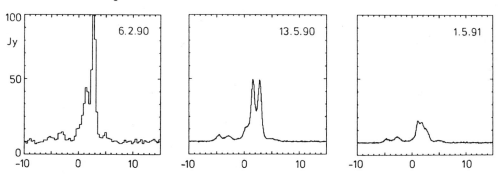

Fig.1: Water maser spectra of RX Boo. Velocities are in km s^{-1}.

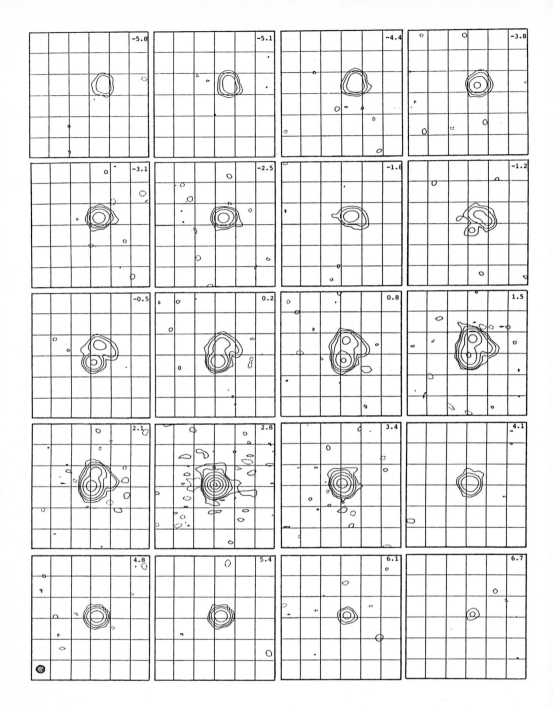

Fig.2: VLA maps of 1990, February of RX Boo. The corresponding velocities are given in the upper right corners. The grid is centered at the strongest component at 2.8 km s^{-1} and has a spacing of 0.2x0.2 arc sec. The contour levels are at -0.04, 0.04, 0.12, 1.2, 2, 10, 30 and 60 Jy/beam. The beam size was 0.08 arc sec.

Some representative spectra from the Effelsberg/Medicina monitoring program are shown in Fig. 1. We found that the intensities of the different maser features vary by a factor 2-3 relative to each other on the scale of months. They vary on an absolute scale by a factor of 10 and more on the scale of a year. During the two years of observations the integrated flux declined by a factor of ~3. The corresponding photon luminosities are $1.4-4.1 \cdot 10^{42}$ s^{-1}.

The maps obtained at the VLA in February and June 1990 are almost identical in structure but the intensity of the features varied. We therefore display only the first map (Fig.2). The comparison of single-dish spectra and maps allows a spatial identification of all single-dish maser features and their development with time. In the Feb. 90 map we identified at least 10 maser spots (Fig.3). The position of the star is not directly evident, so that we assumed that the components at 1.5 and 2.8 km s^{-1} lie in a ring around the star. The adopted stellar position is indicated by a cross. This geometry has the advantage that the v=0.3-component, which is close to the systemic velocity, lies at the tangent point of the shell. The radius of the shell is then r=0"13, i.e. $4.3 \cdot 10^{14}$ cm. Assuming $R_* \sim 10^{13}$ cm the radius is <45 R_*. The H_2O outflow velocity is 6 km s^{-1} as estimated from the total H_2O velocity interval. This is only 50% of the final expansion velocity.

The water maser profiles of SR-variables vary strongly within several weeks (Engels et al. 1988). The structural similarity of the two 1990 maps of RX Boo shows that these variations are pure intensity fluctuations. To cross the H_2O masing shell a mass element needs about 20 years pointing to time scales of several years for structural changes. Such a change has probably taken place in 1982/1983. Before 1983 a component at ~6 km s^{-1} dominated the spectrum. VLA-observations in 1982.2 by Johnston et al. (1985) show that the velocity range extended from -11 to +9 km s^{-1}, almost the full range over which thermal emission of CO and SiO is observed, whereas in 1990.2 the velocity range was considerably smaller (Fig.2). In particular the blue-shifted emission between -7 and -11 km s^{-1} was completely absent in 1990. Furthermore in 1982 the H_2O maser had a clear-cut double-peaked profile and the size of the shell was larger by a factor of 1.5. At the end of 1983 the 6 km s^{-1} component had completely vanished (Lane et al. 1987) and since then the strongest components of the maser were always observed within 3 km s^{-1} of the radial velocity of the star (Engels et al. 1988, Comoretto et al. 1990).

The observations are consistent with the assumption that before 1983 RX Boo could sustain H_2O maser emission at larger radial distances than later on. At this larger distance the gradient of the outflow velocity is smaller allowing

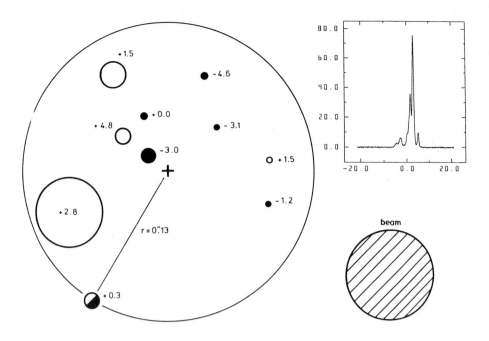

Fig.3: Distribution of the maser components in the shell. The size of the spots indicates the strength of the emission. Filled circles give the blueshifted components with respect to the radial velocity of the star. The VLA-beam is indicated. The single-dish spectrum shown was taken in Effelsberg near the VLA observation.

radial beaming to become prominent, which leads to the well-known double-peaked profile. After 1983 the maser region was completely confined to the accelerating part of the shell, giving preferential raise of tangentially beaming components (Chapman and Cohen, 1985). The profile of the maser is then dominated by features close to the stellar velocity. Such changes in the maser shell size may be caused by changes in the mass loss rate, which controls the dissociation rate of the H_2O molecules, or by changes of the luminosity which controls the excitation temperature.

References

Benson P.J., Little-Marenin I.R., Woods T.C. et al., 1990, ApJS 74, 911
Bujarrabal V., Planesas P., Gomez-Gonzalez J. et al, 1986, A&A 162, 157
Chapman J.M., Cohen R.J., 1985, MNRAS 212, 375
Comoretto G., Palagi F., Cesaroni R. et al., 1990, A&AS 84, 179
Engels D., Schmid-Burgk J., Walmsley C.M., 1988, A&A 191, 283
Gehrz R.D., Woolf N.J., 1971, ApJ 165, 285
Johnston K.J., Spencer J.H., Bowers P.F., 1985, ApJ 290, 660
Khopolov P.N., Samus N.N., Frolov M.S. et al., 1985, GCVS, Moscow
Knapp G.R., Morris M., 1985, ApJ 292, 640
Lane A.P., Johnston K.J., Bowers P.F. et al., 1987, ApJ 323, 756
Morris M., Redman R., Reid M.J., Dickinson D.F., 1979, ApJ 229, 257

THE WATER MASER PROPERTIES OF THE ARECIBO SAMPLE OF OH/IR STARS

D. Engels, Sternwarte der Universität Hamburg, Germany
B.M. Lewis, Arecibo Observatory, Puerto Rico

1. ABSTRACT

We discuss some initial findings from our survey for H_2O masers among the Arecibo sample of 392 OH/IR stars, including a tentative luminosity function.

2. INTRODUCTION

Previous studies of the properties of water maser emission in circumstellar shells were hindered by variability and selection effects. The IRAS survey provided a suitable database for selecting a flux limited sample of AGB stars and therefore allows past restrictions to be bypassed. We have searched the Arecibo sample of OH/IR stars (Lewis et al. 1990), which is complete for $S(25\mu m) > 2$ Jy and $-0.65 < [25-12] < +0.2$, for 22 GHz water maser emission. This search used the Effelsberg 100m-radiotelescope at several epochs between 1987 and 1990. Some preliminary results were reported by Lewis and Engels (1988, 1991) and Engels and Lewis (1990). A total of 205 OH/IR stars were detected giving an overall detection rate of 52%.

3. THE EVOLUTION OF THE MASER PROFILE WITH MASS LOSS RATE

As reported by Engels and Lewis (1990) the most obvious result from the profile analysis is the strong evolution of the maser profiles with mass loss rates. At low mass loss rates (blue IRAS colors) the H_2O maser emission is located in the vicinity of the stellar radial velocity, i.e. in the middle of the OH velocity interval. At larger mass loss rates the profile changes to a double-peaked one, as it is usual for OH maser profiles. This was interpreted as a change in the location of the H_2O masers, the masers in the denser shells being radially further away from the star. A change from more tangential gain paths to radial ones might accompany it (Lewis and Engels, 1991). The increase of the H_2O maser shell size is confirmed by a recent VLA measurement of the OH/IR star 39.7+1.5 (Engels, Winnberg, Brand,

Walmsley, in prep.). The radius of the shell is ~10^{15} cm, about twice the radius of H_2O maser shells of Mira variables (Johnston et al. 1985).

4. THE DECREASE OF THE DETECTION RATE WITH IRAS COLOR

The water maser detection rate of OH/IR stars in the Arecibo sample decreases strongly with increasing IRAS color. Among the bluer sources the detection rate is about 75%, decreasing to ~30% for the reddest OH/IR stars (Engels and Lewis, 1990). This decrease cannot be understood if the increase of the maser luminosity with mass loss rate is as strong as claimed by Bowers and Hagen (1984) and Engels et al. (1986). Indeed, Lewis and Engels (1991) found evidence that the H_2O maser emission in high mass loss rate stars is weakened or even fully suppressed. Thus the detection rate of sources with redder IRAS colors should drop as they are statistically at larger distances.

We can understand the IR color dependence of the H_2O maser luminosity and its time-variability as resulting from the interplay of three parameters. The inner boundary of the maser shell is given by the *quenching radius* R_q, inside which the maser is suppressed due to deexcitation by collisions. The outer boundary is set by either the *dissociation radius* R_d, outside which the H_2O molecules are destroyed by the interstellar UV-field, or by the *excitation radius* R_e, outside which the temperature is too low to excite the maser. While R_q and R_d remain stable (unless the mass loss rate changes), R_e is varying with the luminosity variation of the central star. At high mass loss rates even at maximum light $R_e < R_q$ might occur and the maser is fully quenched. There will be other cases where the actual zone between R_q and R_e is so small even at maximum light, that the maser can achieve only low luminosity. In still another cases the maser will be luminous, if $R_e >> R_q$ at maximum light, and weak during minimum light. The net effect might be that in the mean there is no increase of maser luminosity with mass loss rate.

5. THE LUMINOSITY FUNCTION OF CIRCUMSTELLAR H_2O MASERS

To determine maser luminosities distances to the sources have to be known. We derived distances for the Arecibo sample from the 12μm-flux, by assuming a total luminosity of 5000 $L_{\odot}$ and applying the bolometric correction given by van der Veen and Breukers (1989). Based on these distances, which are typically between 1 and 3 kpc, water maser photon luminosities were determined (Fig. 1a). Nearly all sources have $10^{42} < L(H_2O) < 10^{44}$ s^{-1}, with the lower limit being only partially an observational selection effect. This is more obvious in Fig. 1b showing the distribution of the photon luminosity limits L^{lim} of the full sample. L^{lim} is the minimum luminosity needed to be detected in this survey. Out of 36 sources with $L^{lim} < 10^{42}$ s^{-1} only 10 have actually a luminosity below 10^{42} s^{-1}. The inset in Fig. 1a shows the distribution of photon luminosities of a volume limited subset (see below) for two different IRAS color ranges. No significant difference in the distributions is found, corroborating our earlier suggestion that there is no or only a very weak increase of the maser luminosity with mass loss rate for OH/IR stars. However, there is an important limitation. If the initial assumption of a uniform mean total luminosity for all OH/IR stars is too strong and redder OH/IR stars are in the mean more luminous (Likkel, 1989), their maser luminosities will be greater as well. Thus, some evolution of the maser luminosity with mass loss rate cannot be excluded.

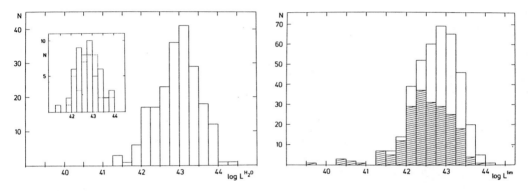

Fig. 1: a) Distribution of the photon luminosities $L(H_2O)$ of the N=205 detected sources. The inset shows the distributions of subsets distinguished by their IRAS colors: -0.7 < [25-12] < -0.5 (solid line) and -0.5 < [25-12] < -0.1 (dashed line). b) Distribution of the photon luminosity limits L^{lim} of the full sample. The hatched area shows the detected sources.

Fig. 1b also shows the importance of sufficient sensitivity to achieve a high detection rate. The detection rate for $L^{lim} < 10^{42}$ s^{-1} is 90% (N=36), and is still 80% for $L^{lim} < 3 \times 10^{42}$ s^{-1} (N=138). This strong increase of the detection rates with sensitivity agrees with an earlier result of Bowers and Hagen (1984), that the detection rate rises to 75% among nearby Mira variables, when these are searched to a level of 10^{41} s^{-1}.

The well established selection criteria of the sample allow for the first time the determination of a luminosity function of water maser emission in circumstellar shells. We constructed a volume limited sample, with the volume placed symmetrically to the galactic plane. The volume is a wedge with l ~ 35 - 75 deg, z < 1000 pc and d < 3.0 kpc. The wedge contains 150 OH/IR stars implying a density of 23 kpc^{-2} projected on the galactic plane.

For comparison, Jura and Kleinmann (1989) derived a density of 13 kpc^{-2} of dust-enshrouded oxygen-rich AGB-stars within 1 kpc of the sun. We computed then the number of OH/IR stars having a water maser luminosity above a given level, correcting for the number of stars, which were not observed down to this level. The resulting photon luminosity function normalized to the mean density in the wedge is shown in Fig. 2. The slope of the function is -1.0 for $10^{43} < L(H_2O) < 10^{44}$ and -0.4 below. Note, however, that the high luminosity tail of the function might be underestimated because of the assignment of a uniform total luminosity to all OH/IR stars, as it was discussed before.

6. CONCLUSIONS

Water masers are commonly present in the circumstellar shells of OH/IR stars. With increasing mass loss rate the masers are located at larger radial distances from the star, where the excitation conditions are usually less favorable. A decrease of the detection rate among redder OH/IR stars follows and the maser luminosity range of the whole sample is restricted to a rather narrow interval covering two orders of magnitude and centered on $L(H_2O) \sim 10^{43}$ s^{-1}.

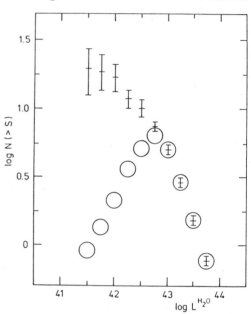

Fig. 2: Luminosity function of water masers in circumstellar shells. Given are the number of stars/unit volume with S > L(H$_2$O) and L^{lim} < S (o) and the corrected values (+) (see text).

REFERENCES

Bowers P.F., Hagen W., 1984, ApJ 285, 637
Engels D., Lewis B.M., 1990, From Miras to Plan. Neb., Eds.
    M.O. Mennessier, A. Omont, Gif sur Yvette
Engels D., Schmid-Burgk J., Walmsley C.M., 1986, A&A 167, 129
Johnston K.J., Spencer J.H., Bowers P.F., 1985, ApJ 290, 660
Jura M., Kleinmann S.G., 1989, ApJ 341, 359
Lewis B.M., Eder J., Terzian Y., 1990, ApJ 362, 634
Lewis B.M., Engels D., 1988, Nat 332, 49
Lewis B.M., Engels D., 1991, MNRAS 251, 391
Likkel L., 1989, ApJ 344, 350
van der Veen W.E.C.J., Breukers R.J.L.H., 1989, A&A 213, 133

THE EVOLUTION OF WATER MASER SPECTRA IN M TYPE STARS

Hiroshi Takaba
Communications Research Laboratory
893-1 Hirai Kashima Ibaraki, 314 Japan

Makoto Miyoshi, Nobuharu Ukita
Nobeyama Radio Observatory
Nobeyama Minamisaku Nagano, 384-13 Japan

ABSTRACT

The results of an almost simultaneous observation of the 22 GHz H_2O maser and the 43 GHz SiO masers in M type stars are presented. A systematic change of H_2O maser spectra which shows the expansion of maser emitting region was found. Except for massive stars, the expansion velocity of the H_2O maser correlates with the IRAS 12/25 micron intensity ratio that is thought to be correlated with the mass loss rate.

These results are interpreted by the collisional excitation model calculated by Cooke and Elitzur and our H_2O maser's beaming model. Also the blocking effects of red shifted H_2O maser emission are discussed.

1. Introduction

From recent works based on the IRAS, it becomes clear that the oxygen-rich M type stars evolves in a sequence of visible Mira variables, obscured IRC/AFGL objects, and highly obscured OH/IR stars (e.g. van der Veen and Habing 1988). In this sequence, the mass loss rate increases from $10^{-7} M\theta/yr$ for a Mira variable to $10^{-4} M\theta/yr$ for an OH/IR star and at the last stage of the OH/IR phase, the mass loss stops and the star evolves to a proto-planetary nebula and finally to a planetary nebula.

The 22 GHz H_2O maser and the 43 GHz SiO masers are commonly observed in Mira variables, IRC/AFGL objects, and OH/IR stars. SiO masers are thought to be emitted from closest part of the stellar gas envelopes whereas the H_2O masers from outer envelopes (e.g. Reid and Moran 1981). Then we made almost simultaneous observations of those two masers for most of the known stellar sources if there exist some differences of maser spectra in objects of different evolutional stage.

2. Observations

A snapshot 22 GHz H_2O and 43 GHz SiO (J=1-0,v=1) maser survey was made between 1991 April 26 and May 11 using the CRL's 34 m radio telescope at Kashima (Takaba et al. 1991). About 200 stars were selected from the catalog of stellar masers by Benson et al. (1990) and about 160 late type stars were observed almost simultaneously with both two maser transitions.

411

3. Results

Thirty-six objects exhibited both H_2O and SiO maser emission except for supergiants and some massive semi-regular variables. The detection of SiO maser means that the source was likely at the maximum phase because SiO maser is being pumped by infrared emission of the star. Also the detection of SiO maser excludes interstellar objects because only Ori-KL, W51, and Sgr-B2 are known to be interstellar SiO maser sources (there exists some confusion of interstellar sources in their catalogue as written in the notification).

By comparing those spectra, we found a systematic change of H_2O maser spectra in M type stars. In figures' 1-3, H_2O and SiO (J=1-0,v=1) maser spectra toward three stars (a visible Mira variable R LMI, a near infrared source IRC+60169, and a far infrared source OH12.8-1.9) are shown. In Mira variables, peak velocities of SiO and H_2O masers were coincided within 1 km/s (rms). Since the SiO maser is thought to be a good indicator of the stellar velocity (Jewell et al. 1991), we can say from our observations that the H_2O maser of Mira variables is also a good tracer of the stellar velocity. Also we can say from Figures 2 and 3 that the H_2O maser comes from expanding shell and the maser should be beamed toward the line of sight in IRC/AFGL objects and OH/IR stars. It was found that the H_2O maser expansion velocity is correlated with the IRAS 12/25 micron intemsity ratio (Figure 4).

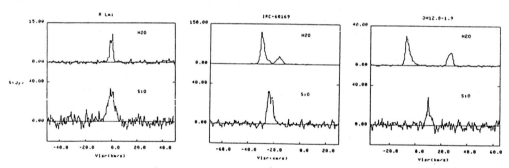

H_2O (Top) and SiO (Bottom) maser spectra of R LMI (Fig.1), IRC+60169 (Fig.2), and OH12.8-1.9 (Fig.3).

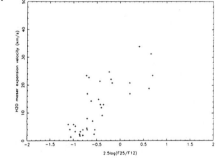

Figure 4. Plot of the H_2O maser velocity (half intensity width for single peaked sources and differences between two peaks for double peaked sources) versus the IRAS two color ratios.

4. Discussion

Cooke and Elitzur (1985) calculated that the location of 22 GHz H_2O maser emitting region moves out with the mass-loss rate increase based on the collisional excitation model. According to the model, the H_2O (6_{16}) decays in dense region by the radiative process $(6_{16} - 5_{05})$, then 22 GHz H_2O maser $(6_{16} - 5_{23})$ does not observe.

Furthermore, we need three types of beaming models to explain observed maser spectra.

In Mira variables, SiO and H_2O masers are excited in the closest region of the star. The mass loss occurs from the surface of the star, the gas motion around the star should be highly scrambled because infall motion also occurs with the shock wave propagation. In such the velocity field, the maser should be beamed toward the limb direction (Reid and Menten 1990). Then the SiO and H_2O maser velocities are coincided with the stellar velocity.

In AFGL/IRC objects and OH/IR stars, SiO maser still comes from closest region of the stellar photosphere because it needs high temeperature to excite SiO maser. Then the velocity of the SiO is also coincided with the stellar velocity.

Whereas H_2O maser in IRC/AFGL objects and OH/IR stars is emitted from outer region and the radius is a function of the mass loss rate. Where the temperature becomes lower than 1500K, the gas start expands by the radiative pressure of the central star, and the acceleration terminates after the density becomes low enough.

In IRC/AFGL objects, the H_2O maser comes from expanding shell, the maser should be conically beamed and shows double peaked spectra.

In OH/IR stars, the H_2O maser comes from the outermost shell of the almost constant velocity, the maser should be beamed toward cylindrical direction like the OH maser, then the separated double peak spectra are detected.

From the observation, we found that the blue shifted emission were stronger than the red shifted emission in about 70 % objects in IRC/AFGL objects and OH/IR stars. The same result was also obtained from our recent H_2O maser survey. These results are explained by the blocking of red shifted emission by the central star because the H_2O maser should be highly beamed toward the line of sight.

To check if the blocking plays an important role in H_2O maser, we had started single dish monitoring and VLBI observations. Figure 5 shows a long time variablity of IRC-10414. The red shifted emission varied but the blue shifted emission was almost unchange. If we think that the red shifted emission comes very close region from the star, some components must be masked according to the stellar pulsation.

Figure 6 shows a short time variability of RW Lyr. The high velocity red shifted emission seems to change in 10 minutes. If we think also that the high velocity red shifted components are highly beamed towards the line of sight, the emission must be passing very closest region of the stellar photosphere with highly mass losing. The maser might be difracted or scintilated and was observed short time variation.

The most important evidence if the red shifted component are blocked by the star or not will be obtained by the VLBI observations. Figure 7 illustrated the result of a VLBI between Kashima and Nobeyama (about 200km baseline) of an H_2O maser source W Hya. The red shifted emssion was highly resolved out and this result may be interpreted by the blocking hypothesis.

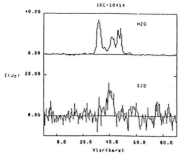

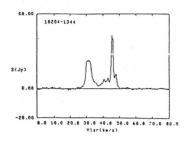

Figure 5. Time variation of H_2O maser spectra toward IRC-10414 in a 6 month interval.

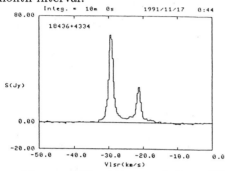

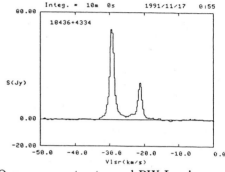

Figure. 6 Short time variation of H_2O maser spectra toward RW Lyr in a 10 minute interval.

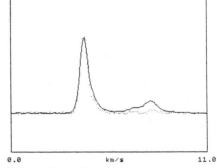

Figure 7. Auto correlation (Solid) and cross correlation (Dots) spectra of H_2O maser W Hya. The data were taken between Kashima and Nobeyama.

References

Benson, P.J. et al., 1990, Ap.J.(Supple), 74,911

Cooke, B. and Elitzur, M., 1985, Ap.J., 295, 175

Jewell, P.R. et al., 1991, Astr. Ap., 242, 211

Reid, M.J. and Menten, K.M., 1990, Ap.J.(Letters), 360, L51

Reid, K.M. and Moran, J.M., 1981, Ann. Rev. Astr. Ap. 19, 231

Takaba, H. et al., 1991, 'Frontiers of VLBI', 265, Universal Academy Press, Inc. Tokyo

Van der Veen, W.E.C.J. and Habing, H.J., 1988, Astr. Ap., 195, 125

CIRCUMSTELLAR WATER MASERS

J. A. Yates, University of Manchester, Nuffield Radio Astronomy Laboratories, Jodrell Bank, Macclesfield, Cheshire SK11 9DL.

ABSTRACT

The circumstellar envelopes of a sample red giants and supergiants have been investigated using MERLIN 22GHz H_2O maser observations and single dish observations of the 22, 321, and 325GHz water maser lines. The MERLIN data shows that inner circumstellar envelopes are generally asymmetrical with evidence for discs and bipolar outflows among the supergiants. The single dish data shows that the 22 and 325GHz masers occur in the same regions. The 321GHz masers seem to occur further in towards the star. The appearance of 'wings' on all three spectra may point to quasi-thermal emission or may be further evidence for the bipolar outflows seen on some MERLIN maps of these stars.

1. INTRODUCTION

The study of the structure and kinematics of the inner regions of circumstellar envelopes around late type, long period variable, stars has mostly been done with interferometric observations of the SiO, H_2O, and OH mainline masers found in this region. In particular the 22GHz H_2O water masers have proved an excellent probe. This maser is collisionally excited in high density regions near the star, with characteristic densities of 10^9-10^7 cm^{-3} at characteristic radii of 5-20 stellar radii. The pump mechanism of this maser has been modelled (Cooke & Elitzur 1985) and predicts that the quenching radius R_q of the maser emission depends on the mass loss rate of the star, $\dot{M} \propto R_q^{3/2} v_q^{3/2}$, where v_q is the expansion velocity of the inner quenching radius. The model assumes spherically symmetric mass loss. This paper discusses recent MERLIN data which were used to test the model of Cooke & Elitzur (1985).

Recently, with the advent of submm telescopes on high mountain sites, there has been much theoretical consideration of the higher frequency water maser lines. Neufeld & Melnick (1991) proposed that simultaneous multifrequency observations of these lines could yield temperature and density information, as well as clues to the physical structure, of the inner circumstellar envelopes. This has been attempted by Menten & Melnick (1991) who observed the 22 and 321GHz lines seen towards a sample of late type variable stars. This paper outlines the preliminary results from recent observations of the 22, 321, and 325 GHz water maser lines.

2. MERLIN 22GHz WATER MASER RESULTS

Recently MERLIN maps of circumstellar 22GHz H_2O masers around three miras (IK Tau, RT Vir, and U Her) and three supergiant sources (VY Cma, NML Cyg, and S Per) have been made to test the model of Cooke & Elitzur (1985) and to investigate the physical, and kinematic, structure of their inner circumstellar envelopes (Yates & Cohen 1992). These maps show a large degree of asymmetry on a scale of $\sim 10^{15}$cm, which is strongest in the more massive stars. Structural asymmetries seem to be the rule according to the results of our sample. This is contrary to the assumption of spherically symmetric mass loss implicit in models of the mass loss mechanism of long period, variable, stars. The sizes of the H_2O masing regions are $\sim 5 \times 10^{14}$cm for miras and $\sim 5 \times 10^{15}$cm for supergiants. The overall size correlates with the mass loss rate of the star. An inner zone of avoidance (quenching zone) is clearly seen in velocity space about the stellar velocity. The quenching radii that we have measured are from 3–10 stellar radii. The measured expansion velocities of the inner and outer radii show that the gas and dust are radially accelerated past the stellar escape velocity in this region. The velocity fields in this region of our six sources are similar to that of VX Sgr (Chapman & Cohen 1986). Preliminary results show that the 'thick shell' model of Chapman & Cohen (1986) is valid and that radiation pressure is the cause of the radial acceleration we see in the inner circumstellar envelope. From the quenching radii the mass loss rates for the sources in our sample have been calculated using the model by Cooke & Elitzur (1985). These are found to be a factor of 5-10 times higher than mass loss rates found by other methods. It is thought that this difference is due to the non-spherical structure and mass loss in the inner part of the circumstellar envelope. Two supergiants show clear evidence for a disc plus polar jet structure, with blobs of material ejected along the polar axis of the disc. Data on VY Cma are shown in figure 1. The change from circumstellar disc to ellipsoid as radial distance from the star increases, and the unexpected discovery of bipolar outflows, shows that non-spherical large scale structure, such as the asymmetrical OH 1612 MHz maser envelope of NML Cyg, can be produced internally. This could be due to the action of rotation and/or the stellar magnetic field.

3. SUBMILLIMETRE WATER MASER OBSERVATIONS

In March 1992 we made simultaneous single dish observations of three H_2O maser lines using the JCMT and the new 32m telescope at Cambridge. The lines observed were at 321GHz (10_{29}–9_{36}) , 325GHz (5_{15}–4_{22} para), and 22GHz (6_{16}–5_{23} ortho). The last two transitions have very similar excitation temperatures, ~ 600K above the ground state, and are both backbone transitions (Neufeld & Melnick 1991). The 321GHz transition's excitation temperature is much higher at 1841K. The 325GHz line had been seen previously in only two circumstellar envelopes (Menten and Melnick, unpublished). The fact that we detected it in 4 out of 5 sources suggests that it is a widespread maser species. The spectra for VY Cma are shown in figure 2. From figure 2 it is possible to compare the velocity structure in the spectra directly, although detailed flux comparisons will have to await calibration. The 22 and 325GHz spectra have very similar spectral profiles. This is to be expected because of their very similar excitation conditions. (The 'wings' are more obvious on the 325GHz spectrum). The brightest emission is close to, but not at, the stellar velocity. The 321GHz emission covers half the velocity range of the other lines, which is to be expected because of its higher excitation temperature. From these simple comparisons we can say that the 22 and 325GHz maser emission arises in the same part of the circumstellar envelope. Because of its different profile the 321GHz emission probably arises in a different part of the circumstellar envelope in warmer and denser conditions than the others. Three more sets of spectra were obtained for mira variables R Crt, IK Tau, and R Leo. These show similar results. The flux ratios of the different lines for each individual star will be calculated. Using these and the inversion model proposed by

Neufeld & Melnick (1991) limits may be placed on the temperature and density conditions in the inner circumstellar envelopes of these stars (Yates, Cohen & Hills 1992). Further joint JCMT-Cambridge 32m sessions are planned in order build on the success of our first session. It is hoped that long term MERLIN and single dish monitoring of these lines can be arranged. This will allow us to see changes in the physical conditions in circumstellar envelopes and relate them to structural changes.

References

Bowers, P. F., Johnston, K. J. & Spencer, J. H., 1983. *Astrophys. J.*, **274**, 733.

Chapman, J. M. & Cohen, R. J., 1986. *Mon. Not. astr. Soc.*, **220**, 513.

Cooke, B. & Elitzur, M., 1985. *Astrophys. J.*, **295**, 175.

Menten, K. & Melnick, G., 1991. *Astrophys. J.*, **377**, 647.

Neufeld, D. A. & Melnick, G. J., 1991. *Astrophys. J.*, **368**, 215.

Yates, J. A. & Cohen, R. J. 1992. In preparation.

Yates, J. A., Cohen, R. J. & Hills, R. E. 1992. In preparation.

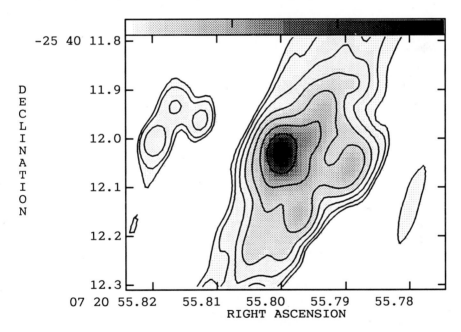

Figure 1: An integrated line emission map of the 22GHz H_2O maser emission around VY Cma (Yates & Cohen 1992). The disc and bipolar flow is clearly seen. The disc is at the same position angle as the OH 1612 MHz maser shell (Bowers, Johnston & Spencer 1983) and the two blobs are at the same position angle as the polar axis of the OH 1612 MHz maser shell (Bowers, Johnston & Spencer 1983).

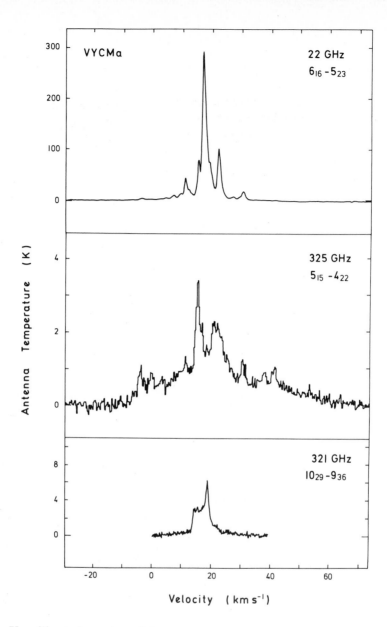

Figure 2: Uncalibrated spectra of 3 water maser transitions, observed simultaneously towards VY Cma, using the JCMT and the 32m MERLIN telescope at Cambridge U.K. The 325 and 22GHz transitions have similar excitation temperatures ($\sim$ 600K), which is reflected in their similar spectral profiles. The 321GHz transition has a much higher excitation temperature (1841K), which explains its smaller velocity coverage.

15. CIRCUMSTELLAR SiO MASERS

RECENT WORK ON SiO MASERS AT THE CENTRO ASTRONÓMICO DE YEBES

V. Bujarrabal, J. Alcolea, A. Martínez, A. Barcia,
J.D. Gallego, J. Gómez-González, A. del Pino, P. Planesas,
R. Bachiller, A. Rodríguez, A. del Romero, M. Tafalla, P. de Vicente
Centro Astronómico de Yebes
Apartado 148, 19080 Guadalajara, SPAIN

ABSTRACT

We have been observing SiO masers in evolved stars at the Centro Astronómico de Yebes since 1984, using the Yebes 14m dish. With this telescope it is possible to observe the 7mm lines, i.e. the rotational transitions J=1-0 in different vibrational states. Observations at higher frequencies were also carried out with the IRAM 30m telescope at Pico de Veleta (Spain). The data include monitoring and systematic observations, as well as search for new maser transitions or exotic emitters. The theoretical interpretation of the results is being undertaken, with emphasis on the discussion of pumping mechanisms.

The main recent results will be presented, concerning two main topics : Variability of maser emission (v=1,2 J=1-0) and statistical properties of the SiO maser emission in evolved stars.

1. MONITORING OF SiO MASER EMISSION

We are monitoring since (in the best cases) 1984 the v=1,2 J=1-0 SiO maser emission with a periodicity of about once a month. The observed objects are :

M-type Miras : R Aql, **R Aqr, TX Cam,** R Cnc, **R Cas,** *o* **Cet (Mira),** U Her, X Hya, **W Hya, R Leo, R LMi,** IK Tau, IRC+10011, OH26.5

S-type Mira : χ **Cyg**

M-type supergiants : **VY CMa,** μ Cep, **VX Sgr,** AH Sco

M-type semiregulars : **GY Aql, RT Vir**

Star-forming region : **Ori A**

(Boldface characters indicate the best studied objects.) The relative calibration has been carefully treated, observing each star always at the same sidereal time (same parallactic angle and elevation), observing only under good weather conditions, correcting possible pointing errors by doing small, fast maps around the star position, and checking the calibration by observing a *constant* continuum source (the HII region W51). We expect that the relative calibration errors are not larger than $\sim$ 10%. The first part of this monitoring is already published (Martínez, Bujarrabal and Alcolea, 1988).

The variations in intensity (Fig. 1) present two main components. The first one corresponds to a periodic variation with the stellar cycle, with a typical phase-lag with respect to the optical maxima of about 0.1-0.2. This variation is therefore in phase with the near- and middle-IR stellar variability. A second component corresponds to long-term variations, that affect both the amplitude and the average intensity; it is not clear whether some kind of *superperiod* exists.

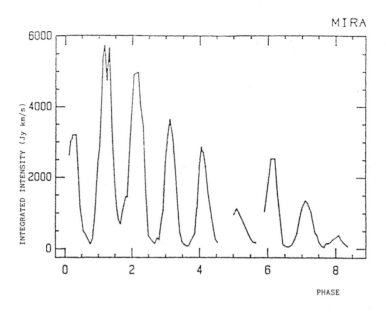

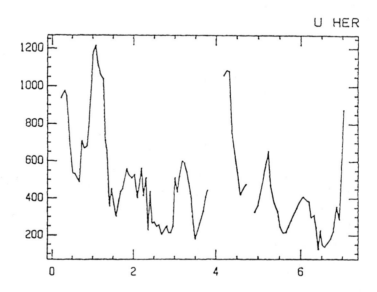

Fig. 1.- Variations in velocity integrated intensity (Jy km/s) of the SiO emission from Mira and from U Her as a function of the stellar phase.

In addition to these variations in intensity, we also find an important variability with time of the profile shape. The strongest profile variations seem to mainly occur during epochs of relatively low emission and to be associated with the long-term changes in the intensity. We think that the strong variations in the profile are related to changes in the structure of the inner envelope, from which the SiO emission probably comes.

2. SYSTEMATIC OBSERVATIONS OF SiO MASER EMISSION

We have observed a large number of objects of different types in the SiO maser lines (mainly in v=1,2,3 J=1-0, v=1 J=2-1, and ^{29}SiO v=0 J=1-0). One of our main results is the good correlation found by Bujarrabal, Planesas and Del Romero (1987) between the maser and stellar $8\mu m$ luminosities in M-type Miras (8 μm is the wavelength of the vibrational transition expected to pump the masers according to the radiative models), Fig 2; we often present other results with respect to this relation. In this diagram both SiO and IR fluxes are observed at maximum and corrected for the distance. The relation is not only useful by itself (in particular supporting the radiative pumping of SiO masers), it has also allowed a meaningful comparison of the maser intensity in different kinds of stars. Note that the representation of the SiO emission on such a diagram eliminates two major sources of the SiO intensity variations from source to source in M-type Miras, namely the distance and the IR flux availability.

The relative intensity of different maser transitions, including new maser lines, can also be quantitatively studied by means of this relation. As an example, in Fig 3 we represent (Alcolea and Bujarrabal 1992) the comparison between the high v=1 J=1-0 intensity of the main isotope (^{28}SiO) with that of the ^{29}SiO v=0 J=1-0 maser. The statistical analysis of ^{29}SiO v=0 J=1-0 emission, that takes into account the measured upper limits, indicates that the peak intensity of this line is more than fifty times weaker than for the ^{28}SiO v=1 J=1-0 line.

These observational data are also useful to be compared with theoretical results. In Fig. 2 we show the results of different models. In relatively recent models (Alcolea, Bujarrabal and Gallego, 1989; Bujarrabal, unpublished recent calculations using a non-local treatment of transfer for a stationary SiO layer) the agreement is reasonable. We also note, as an example, the case of the calculations by Langer and Watson (1984). These authors concluded that their model was unable to explain the SiO emission, since the predicted intensities are too low. However, the continuum emission of the model star is also relatively weak, and the precise comparison of the calculations with the observations is not so bad, being both results almost compatible.

REFERENCES

Alcolea J., Bujarrabal V. 1992, Astr. Ap., 253, 475

Alcolea J., Bujarrabal V., Gallego J.D., 1989, Astr. Ap., 211, 187

Bujarrabal V., Planesas P, Del Romero A. 1987, Astr. Ap., 175, 164

Langer S.H., Watson W.D. 1984, Ap. J., 284, 751

Martínez A., Bujarrabal V., Alcolea J., 1988, Astr. Ap. Suppl. 74, 273

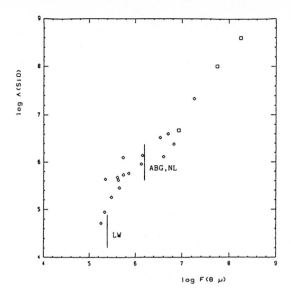

LW : Langer and Watson, 1983

ABG : Alcolea et al. 1989

NL : Non-local rad. model

Fig. 2.- Relation between the SiO (v=0 J=1-0) profile area and stellar emission at 8 μm for O-rich Miras and some supergiants. Some theoretical predictions are represented (see text).

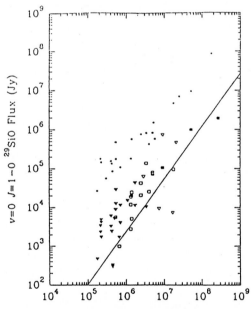

Fig. 3.- Relation between the peak intensity of the ^{28}SiO v=1 J=1-0 (small circles) and ^{29}SiO v=0 J=1-0 masers vs. the stellar emission at 8 μm. Downward triangles represent ^{29}SiO upper limits.

SiO MASERS IN VY CMa : DETECTION OF MASER EMISSION IN THE v=4 STATE

J. Cernicharo, V. Bujarrabal
Centro Astronómico de Yebes
Apartado 148, 19080 Guadalajara, SPAIN

J.L. Santarén
IRAM
Av Divina Pastora 7, 18012 Granada, SPAIN

ABSTRACT

We present radio observations of ^{28}SiO (abundant isotope) and of ^{29}SiO masers in the red supergiant VY CMa and other O-rich giants. We report the detection for the first time of ^{28}SiO v=4 (J=5-4) maser emission. Note that this v=4 line is placed at 7000 K from the ground, being by far the highest microwave transition ever detected. We also present the detection of ^{29}SiO v=1 rotational lines in a number of objects. The observations were made with the 30m IRAM radiotelescope at Pico de Veleta (Spain).

1. OBSERVATIONS AND OBSERVATIONAL RESULTS

We have observed in the red supergiant VY CMa a high number of rotational transitions of ^{28}SiO in the vibrational states v=0 to 4; we have also observed in this object and other O-rich giants the rare isotopic substitution ^{29}SiO. We present the profiles of the observed J=2-1, 3-2, 4-3, 5-4, 6-5 transitions of ^{28}SiO (v=1,2,3,4) in Fig. 1, and of ^{29}SiO (v=0) in Fig. 2. In most cases the spectral resolution is 1 MHz, higher resolution spectra (100 Khz) are also shown for some ^{29}SiO transitions. The first detected ^{29}SiO v=1 masers are presented in another drawing (Fig. 3) for some stars (see Cernicharo, Bujarrabal and Lucas, 1991). In order to point out the variations of the profiles from transition to transition, we show in Fig. 4 the profiles of some masers of ^{28}SiO in VY CMa.

The observations of the ^{29}SiO transitions have been already presented (Cernicharo, Bujarrabal and Lucas, 1991). The observations of the ^{28}SiO v=1 to 4 J=2-1 to J=6-5 transitions were carried out in September 1990 and January 1991 with the IRAM radiotelescope. Three SIS receivers at 1mm, 2mm and 3mm were tuned at the frequency of the different maser lines and used simultaneously. The pointing was monitored every 30 minutes using a Schottky receiver tuned at the frequency of the strong J=2-1 v=1 line of SiO. The relative pointing of the four receivers was better than two arcseconds. The 2mm receiver is sensitive to a linear polarization perpendicular to that of the other two (1mm and 3mm) receivers. The observations were made with a wobbler system that provides particularly good baselines. See for more details Cernicharo, Bujarrabal and Santarén (1992).

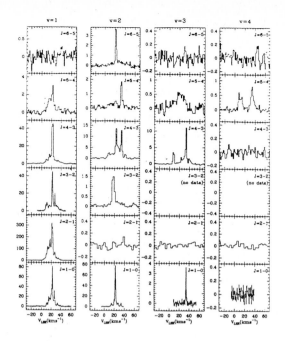

Fig. 1.- ^{28}SiO masers observed towards VY CMa.

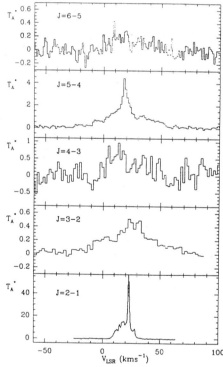

Fig. 2.- ^{29}SiO v=0 transitions observed towards VY CMa.

2. DISCUSSION

We have observed SiO lines in the red supergiant VY CMa. It is remarkable the large number of SiO masers detected in this object, that is the most intense emitter in these lines. We note that masing and not masing ^{29}SiO transitions alternate in the same vibrational ladder (thermal emission is detected in some ^{29}SiO v=0 lines). We suggest to explain such a striking behaviour the presence to line overlaps (probably between IR vibrational lines of ^{29}SiO and of the main isotope, ^{28}SiO), the effect of which can strongly enhance or depress the ^{29}SiO masers. This effect is necessarily very selective and only changes the excitation of certain lines, it could therefore explain the strong differences measured between the intensities of adjacent lines. For example, the overlap J=9-10 v=1-0 of ^{28}SiO with J=4-5 v=1-0 of ^{29}SiO can be responsible for the pumping of the ^{29}SiO v=1 J=4-3 maser (the velocity shift between these lines is -4.8 km/s).

It is remarkable in the data on ^{28}SiO masers presented here the detection for the first time of a v=4 transition (J=5-4), involving levels at more than 7000 K from the ground. We also note that, as the energy of the observed ^{28}SiO transition increases (Fig. 4), only some of the features present in the v=1 J=2-1 (very strong) maser are detected. In particular, for the v=3 J=4-3 and v=4 J=5-4 lines only the clumps of relatively high velocity are detected. This result could be interpreted as confirming the rotating dish model presented by Zhen-pu and Kaifu (1984), since the highest rotational velocity is expected in the inner (more excited) regions of the rotating structures. (In any case, the model by Zhen-pu and Kaifu introduces an expansion that leads to some features than are not agreement with the observations.) As already mentioned, we interpret our results as showing that the conditions for the excitation of high energy transitions become more restrictive and are only present in some clumps of peculiar properties. Numerical calculations (using a standard LVG approximation) indicate that the clumps responsible for this high-energy emission must have high volume densities ranging between 1 and 5 10^{10} cm^{-3} and be placed at less than 5 stellar radii from the central object. Though such conditions are probably quite restrictive, they do not seem extraordinary for the envelopes of evolved stars, and we believe that the previous nondetection of v=4 lines is probably related to the fact that only low-J transitions were searched for. The observation of high-frequency SiO masers is surely a very promising observational field. (For more details on the conditions and models to explain the high-energy SiO emission in VY Cma, see Cernicharo, Bujarrabal and Santarén, 1992.)

REFERENCES

Cernicharo J., Bujarrabal V., Lucas R., 1991, Astron. Astrophys. Letters, 249, L27
Cernicharo J., Bujarrabal V., Santarén J.L., 1992, Astrophys. J. Letters, submitted
Zhen-pu Z., Kaifu N., 1984, Astron. Astrophys. 138, 359

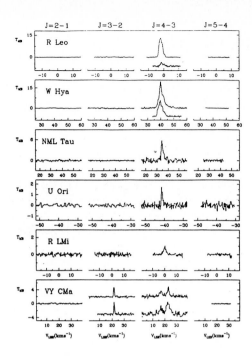

Fig. 3.- ^{29}SiO v=1 masers detected in O-rich red stars.

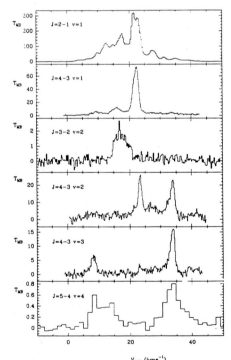

Fig. 4.- Selected ^{28}SiO profiles in VY CMa.

RECENT VLBI OBSERVATIONS OF SiO MASERS

Francisco Colomer
Onsala Space Observatory
S - 439 92 Onsala, Sweden

ABSTRACT

Very long baseline interferometric observations of the SiO maser emission in evolved stars and in the star-forming region Orion A have been performed in Europe since 1990. We report on the progress in the analysis of the data and present some preliminary results. We find evidence for the existence of complex structures at milliarcsecond scales in the form of multiple maser spots distributed in the region. This is important for the understanding of the physical conditions close to the stellar photosphere and the SiO pumping mechanism. Repeated observations at several epochs should allow the modelling of the circumstellar envelope. We also discuss the suitability of the circumstellar SiO masers as a link between the optical and radio reference frames as they are bright radio sources and their parent stars are sometimes optically bright.

1. INTRODUCTION

Very little is known about the physical environment of the stellar SiO masers. Because of the excitation energy of the upper level of the maser transition (1750 K for the $v = 1$, $J = 1 \rightarrow 0$ transition at 43 GHz), it is proposed that they are much closer to the stellar photosphere than are the associated OH or H_2O masers. Thus we may assume that detailed observations of the SiO masers may provide some information on the physical conditions in this region. Indeed the distribution of the maser spots as a function of their velocity should provide some dynamical detail of the circumstellar material at distances of a few stellar radii from the star. Interferometric observations of the 1612 OH masers in evolved stars showed that the OH lies in an expanding shell some 10^{17} cm from the central star (Booth et al. 1981).

VLBI observations in both the $v = 1$ and $v = 2$, $J = 1 \rightarrow 0$ lines near 43 GHz, and in the $v = 1$, $J = 2 \rightarrow 1$ transition near 86 GHz are also of importance for the understanding of the SiO maser theory. Elitzur (1982) has discussed the importance of measuring the distance of the masing material from the exciting star in order to understand the pumping mechanism. Theories that rely on shocks require the masing material to be within one stellar radius of the stellar surface. At this location, gas densities and collision rates are large enough to sustain the maser action. Theories which rely on infrared pumping are unable to account for the maser intensities under the assumption of a spherical isotropic stellar wind. Some form of anisotropy and possible maser clumping are under investigation. VLBI measurements will help to differentiate between these theories.

An exciting aspect of this work is the probable association of the SiO masers with turbulent cells in the stellar photosphere. Schwarzschild (1975) suggested that convection in red giant envelopes might be dominated by a few giant cells and Salpeter (1974) has proposed a scenario in which grains form in local, temporarily cool patches on the surface of the star. Perhaps, as has been discussed by Alcock and Ross (1986), the masers are associated with these cool dense knots which are being ejected from the star, although some probably fall back in. Indeed, it has even been suggested by Muchmore et al. (1987) that the formation of SiO may cause a cooling instability. VLBI observations should yield the sizes of these knots and repeated observations will have sufficient relative positional accuracy that the motion of these cells may be determined giving for the first time a really close look at the physics of these evolved stars.

2. RECENT VLBI OBSERVATIONS OF SiO MASERS IN EUROPE

As a detection test, we performed a spectral line VLBI experiment at 43 GHz in June 1990 towards selected galactic late-type stars and in Orion. With 15 hours of VLBI observations using three stations (Effelsberg (MPIfR, FRG) – Onsala (OSO, Sweden) – Yebes (CAY, Spain)), we detected 11 out of 19 late-type stars and Orion. Since the baseline range was 330-1740 km, the observations already indicated that the detected maser spots were much more compact than previously expected. The maser emission appeared in distinct regions separated by several tens of milliarcseconds. We found typical (Gaussian) sizes for the maser spots of ~ 1.0 mas. The data were analyzed using the NRAO-SAO spectral line VLBI calibration package (Reid et al. 1980). The results of this experiment are published and further details can be found in Colomer et al. (1991 and 1992).

Table 1: List of collaborators

Institute	Collaborators
Onsala Space Observatory	F.Colomer, B.O.Rönnäng, R.S.Booth
Max-Planck-Institut für Radioastronomie	D.A.Graham, T.P.Krichbaum, A. Witzel
Centro Astronómico de Yebes	P.de Vicente, A.Barcia, J.E.Garrido
	J.Gómez-González, J.Alcolea
Observatoire de Bordeaux	A.Baudry, N.Brouillet, G.Daigne

The experiment of June 1990 was intended as a search for detectable sources. A second experiment was performed in April 9 to 11, 1991. The institutions and scientists involved are listed in Table 1. We observed 10 stars (the best detections in June 1990 and also R Aqr) and Orion using the same VLBI-setup for a longer time (48 hours), in order to obtain visibilities with much better uv-coverage. The data analysis is not yet complete: all the data are correlated, 50% of the data are already calibrated and the first maps are being produced. In Figure 1 we show a preliminary low resolution map of μ Cep produced with the multiple-point fringe-rate mapping technique (Walker, 1981). The position of each maser spot, relative to an unresolved feature at 24.8 km s^{-1}, and the 1σ formal error are shown. The area of each circle is proportional to the correlated flux of that maser spot. From the data it is obvious that the SiO maser sources have complex structures at milliarcsecond scales and that most of the observed objects can

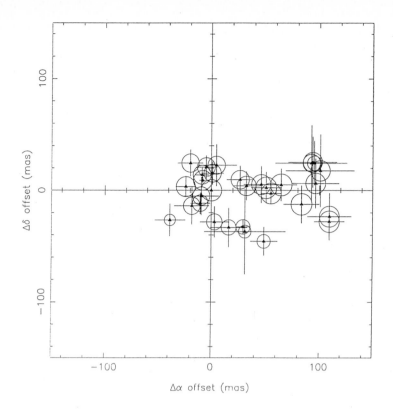

Fig. 1: Spot map of the SiO emission around the supergiant μ Cep.

be mapped. The somewhat low quality of the maps results from the short u-v coverage and the geometry of our interferometer, with mostly N-S baselines. Absolute positions of the SiO maser spots have not been obtained, as very few extragalactic calibration sources are bright enough at 43 GHz to be detected with our present system.

3. CONCLUSIONS AND FUTURE PROSPECTS

Our observations show that the SiO maser emission in the circumstellar envelopes of late-type stars has complex compact structures at milliarcsecond scales and can be mapped with VLBI.

However, we need good u-v coverage to model the relative spatial distribution and proper motions of the maser spots with multi-epoch VLBI monitoring. With better sensitivity, more extragalactic objects will be available. Their observation is important for calibrating the system in order to reduce the systematic errors that dominate the individual maps obtained from our few, widely separated experiments. This might allow as well the determination of the absolute positions of the SiO maser spots by phase-referencing to these extragalactic objects. We have proposed the participation of the telescopes at Pico Veleta (IRAM, Spain) and Metsähovi (Finland). Once front-end receivers are available, Cambridge (U.K.) and also Medicina (Italy) could participate, resulting in a very good array in Europe for the study of these masers.

Regarding absolute position measurements, it has been recognized for some time that the stellar masers may be suitable as a link between the extragalactic radio reference frame (EGRF) and the galactic optical reference frame because they are bright radio sources and their parent stars are sometimes optically bright (e.g. Baudry et al. 1984; Rönnäng, 1989). The stellar positions will be measured with the Hipparcos satellite and the radio positions can be linked to those of the quasars by VLBI phase-referencing.

The still difficult step in this application is to estimate the position of the central star from the observed distribution and kinematics of the SiO maser emission, as it strongly depends on the actual distribution of the maser spots. We need good maps of the SiO emission in order to study if we can locate the star with reasonable accuracy. Another interesting approach would be to try to detect the radio continuum emission from the stellar photosphere itself using the SiO masers as a phase-reference as done for W Hya using the H_2O masers at 22 GHz (Reid and Menten, 1990).

Acknowledgements. We are grateful to the staff operating the Effelsberg, Onsala, and Yebes observatories. Special thanks to F.Bosq, R.Ganesan, M.Hagström, A.Kristensson, C.O.Lindström, G.Montignac, B.Nilsson, K.Standke, J.M.Torre, and N.Whyborn for technical support, and H.Olofsson and V.Bujarrabal for helpful discussions. The 100 m radiotelescope at Effelsberg is operated by the Max-Planck-Institut für Radioastronomie in Bonn. The Onsala Space Observatory at Chalmers University of Technology is the Swedish National Facility for Radio Astronomy. The Yebes radiotelescope is operated by the Centro Astronómico de Yebes (Instituto Geográfico Nacional of Spain), and is partially supported by the Spanish CYCIT under project number PB 88-0453. F.C. acknowledges financial support from the Spanish Ministerio de Educación y Ciencia.

References.
Alcock,C., Ross,R.R., 1986, ApJ 310, 838
Baudry A.,Mazurier J.M.,Requième Y., 1984, Proc. IAU Symp.110, VLBI and Compact Radio Sources. Reidel, Dordrecht, p.355
Booth,R.S., Kus,A.J., Norris,R.P., Porter,N.D., 1981, Nature 290, 382
Colomer,F., Graham,D.A., Krichbaum,T.P., Rönnäng,B.O., de Vicente,P., Barcia,A., Booth,R.S., Witzel,A., Gómez-González,J., Baudry,A., 1991, IAU Colloquium No. 131, Radio Interferometry. Theory, Techniques and Applications, pp. 338
Colomer,F., Graham,D.A., Krichbaum,T.P., Rönnäng,B.O., de Vicente,P., Witzel,A., Barcia,A., Baudry,A., Booth,R.S., Gómez-González,J., Alcolea,J., Daigne,G., 1992, A&A 254, L17
Elitzur, 1982, ApJ 231, 124
Krichbaum,T.P., Witzel,A., 1991, Frontiers of VLBI, Universal Academy Press, p.297
Muchmore,D.O., Nuth III,J.A., Stencel,R.E., 1987, ApJ 315, L141
Reid,M.J., Haschick,A.D., Burke,B.F., Moran,J.M., Johnston,K.J., Swenson Jr.,G.W., 1980, ApJ 239,89
Reid,M.J., Menten,K.M., 1990, ApJ 360, L51
Rönnäng,B.O., 1989, Proc. 7th Working Meeting on European VLBI for Geodesy and Astrometry, Instituto de Astronomia y Geodesia, Madrid, p.113
Salpeter,E.E., 1974, ApJ 193, 585
Schwarzschild,M., 1975, ApJ 195, 137
Walker,R.C., 1981, ApJ 86 (9), 1323

86 GHz SURVEY OF IRAS POINT SOURCES:
156 SiO, v=1, J=2→1 MASER SOURCES*

L. K. Haikala

University of Helsinki Observatory

SF-00130 Helsinki 13, Finland

and

I. Physikälisches Institut, Universität zu Köln

D-5000 Köln 41, Federal Republic of Germany

ABSTRACT

We have used the IRAS Point Source Catalog to select circumstellar SiO maser objects solely according to their IR properties in the catalog and to acquire a large and uniform database. The power of the point source catalog in selecting the maser candidates is demonstrated by the fact that 156 new 86 GHz SiO, v=1, J=2→1 masers were found. The number of known SiO maser sources at the moment is approximately 350 which already allows a statistical study of their properties. The IRAS colours and the low resolution spectra of the SiO maser objects are discussed in this paper.

INTRODUCTION

Searches for SiO masers have been made towards nearby, well known variable stars with circumstellar envelopes (CSEs), including Mira variables, semiregular variables, supergiants and OH/IR objecs. Many such objects have escaped optical detection mainly because of their thick CSE which obscure the stars. The IRAS point source catalog (IPSC) has provided us with an oppoturnity of easily finding stars surrounded by CSEs (e.g. Olnon et al. 1984, van der Veen and Habing, 1988). Searches for SiO maser sources selected from the IPSC have been previously made by e.g. Allen et al. (1989), Deguchi et al. (1990).

The SiO maser candidates can be found from the IPSC eiher by means of their IRAS colours or by using their IRAS Low Resolution Spectral (LRS) types. The colour selection approach was tested in 1986 at the Metsähovi 14m

---

*Based partly on observations collected at the European Southern Observatory at La Silla, Chile

radio telescope at the 86 GHz SiO, v=1, J=2→1 frequency by observing strong northern point sources which had IRAS colours similar to those of the known SiO masers. Two new sources were detected. A southern SiO, v=1, J=2→1 maser survey was conducted with the Swedish-ESO Submillimeter Telescope (SEST) in 1988 where 51 new 86 GHz masers were detected. This survey is described in Haikala (1990). The results of the SEST survey pointed out the usefullness of the IRAS LRS data and subsequently a second southern survey at SEST and a northern survey using the Onsala 20m telescope were done using the LRS spectral type information. Alltogether 156 new 86 GHz SiO, v=1, J=2→1 maser sources have been detected in these three surveys.

THE SELECTION CRITERIA

The IRAS sources observed in the first survey were chosen according to their position in the IRAS [25]-[60], [12]-25] two-colour diagram only The colours are defined by

$$[12] - [25] = 2.5 log \left(\frac{F_{25}}{F_{12}} \right), [25] - [60] = 2.5 log \left(\frac{F_{60}}{F_{25}} \right) \tag{1}$$

where F_i is the non-colour-corrected flux in IPSC in band i. A thorough discussion of the stellar sources in the IRAS [25]-[60], [12]-[25] two-colour diagram is given in van der Veen and Habing (1988) (hereafter VH) where the IRAS two-colour plot is divided into different regions according to the optical and IR properties of the objects.

The sample observed at SEST in 1988 consisted of point sources associated both with oxygen rich (LRS classes 1N and 2N, N=1 to 9) and carbon rich (spectral type 4N) stars. Even though there were five strong point sources (12 μm flux > 500 Jy) associated with the latter spectral types none of them were detected in SiO. Of the remaining ten strong sources eight were detected in SiO.

For the second SEST survey it was decided to observe only stars having LRS classes indicating oxygen rich CSEs. Sources having LRS classes 18 and 17 were excluded because these objects are normal late type stars without CSEs. The northern LRS-based sample was observed in 1991 with the Onsala 20m telescope.

The observations were done in the dual beam switching mode covering ±150 km s^{-1} with respect to the Local Standard of Rest. The average RMS level in the observed spectra was 2.5 Jy. For details and data reduction see Haikala (1990). Spectra oftwo of the detected masers are shown in Figure 1.

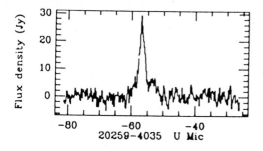
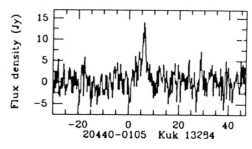

Figure 1.   Sample spectra of the new SiO masers. Only linear baseline correction has been made. The flux density is given in Janskys and the velocity in km s^{-1} with respect to LSR

DISCUSSION

The positions of the presently known SiO masers on the [12]-[25], [25]-[60] two-colour plot are shown in Fig. 2. Only sources having good quality fluxes (IRAS flux flag 3) at 12, 25 and 60 μm in IPSC (altogether 263 sources) are plotted. The different areas as defined in VH (continuous lines) and the region from where the SiO candidates are were chosen in Paper 1 (dashed line) are also shown.

There are 9 sources which have medium quality (2) IRAS fluxes at 60 μm band. In the colour-colour plot these objects lie systematically above those having a good quality flux at 60 μm . There are altogether 77 sources in IPSC which have LRS spectral type 2N and IRAS flux flags 3 3 2 in 12, 25 and 60 μm respectively. Half of these sources have a [25]-[60] colour more positive than -1.5. If the distribution of these sources were similar to those with a good quality flux at 60 μm one would expect only few sources to have [25]-[60] colour higher than -1.5. The 60 μm flux is obviously systematically overestimated for the sources having only medium quality flux at this wavelength.

The histogram of the LRS classes of the SiO masers is shown in Figure 3a. Forty SiO maser sources do not have IRAS LRS spctrum and are plotted in the left of the figure as LRS "class" -1. The hatched part of the histogram is shows the new SiO sources from this paper.

Figure 2.   [12]-[25], [25]-[60] two-colour plot of point sources associated with SiO masers. The division of the two-colour plot by VH is also shown. Only point sources which good quality fluxes at 12, 25 and 60 μm are shown.

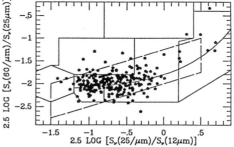

The increase in the number of maser sources is pronounced for LRS classes 26 to 29 (*i.e.* for sources having spectral types typical to dense CSEs). This is to be expected because the sources selected from the IPSC are mostly IRAS sources only (they have no identification in IPSC) or are associated with faint suspected variable stars or near-IR sources.

The histogram of LRS classes of all IRAS sources in the IPSC from classes 0 to 50 is shown in Figure 3b. The Figure shows how well SiO maser sources are favored by certain IRAS LRS classes. The peak at spectral type 18 in Figure 3b is due to the Rayleigh-Jeans tail of the stellar photospheric emission for late type stars. The absence of SiO emission is therefore readily explained. There is a dip in the distribution at LRS class 25 in Figure 3a. The dip is not seen in Figure 3b. The LRS class increases from 21 to 29 with increasing strenght of the 9.7 μm silicate emission feature. It is difficult to imagine any physical reason why the sources having their 9.7 μm SiO emission at the particular level corresponding to LRS class 25 should show no SiO maser emission. The feature is therefore probably due to selection effects. Pre-IRAS observations concentrated more on visible stars (*i.e.* not so dense CSEs, LRS class < 25) and IRAS based surveys concentrate possibly more on the dense CSEs (LRS> 25).

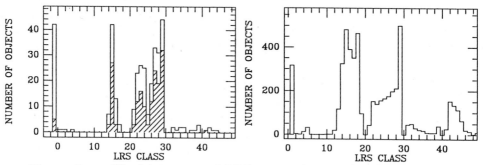

Figure 3   **a** The histogram of LRS spectral types for the presently known SiO masers. The hatched part of the histogram shows the SiO sources from this work. **b** The histogram of IPSC sources for LRS spectral classes 0 to 50.

Alcolea, J., Bujarrabal V., and Gomez-Gonzales J.: 1989, *Astron. Astrophys.* in press

Allen, D. A., Hall, P. J., Norris, R. P., Troup, E. R., Wark, R. M. and Wright, A. W. E.: 1989, *Monthly Notices Roy. Astron. Soc.*, **236**, 363

Deguchi, S., Nakada, Y. and Forster, J. R.: 1989, *Monthly Notices Roy. Astron. Soc.*, **239**, 825

Haikala L.: 1990, *Astron. Astrophys. Suppl.*, **85**, 875

Olnon, F. M., Baud, B., Habing, H. J., De Jong, T., Harris, S. and Pottasch, S. R.: 1984, *Astrophys. J., Letters*, **278**, L41

van der Veen, W. E. C. J. and Habing, H. J.: 1988, *Astron. Astrophys.*, **194**, 125

MODELING SiO MASER EMISSION FROM LATE-TYPE STARS

Philip Lockett
Department of Physics
Centre College
Danville, KY 40422

Moshe Elitzur
Department of Physics and Astronomy
University of Kentucky
Lexington, KY 40506

ABSTRACT

We have performed a thorough study of both radiative and collisional pumping of the SiO masers around late-type stars, carefully considering the combined and separate actions of each type of pump in order to gauge their effectiveness. Our model is based on observations and theoretical arguments that place the SiO masers in high density clumps rather than in the smooth stellar wind. Significantly, general conclusions can be reached which are independent of the pumping mechanism. Most importantly, the overall molecular density is restricted to lie between $\sim 10^9 - 10^{10} \mathrm{cm}^{-3}$. Although all pumps result in the production of a maser chain within each vibrational state, radiative pumps have more of a problem explaining the observed simultaneous production of the same maser transition in different vibrational states. We find that collisional pumping produces the strongest maser emission and, in contrast to radiation, does not require fine tuning of the physical conditions for its operation. Collisional pumping thus appears to be the primary pumping mechanism responsible for the SiO maser phenomenon.

1. INTRODUCTION

Maser emission is commonly observed in the lower rotational transitions of the $v = 1$ and 2 vibrational states, and less frequently in $v = 3$. Recent observations presented at this conference by Cernicharo et al. have detected the $J = 5 \rightarrow 4$ transition of $v = 4$. Interferometry studies place the masers at $2-6R_\star$. The SiO masers are produced not in the stellar wind but rather in a clumpy region having complicated mass motions. The existence of these maser cells is indicated by the complicated and variable line profiles, which consist of many spikes having typical lifetimes of a fraction of a stellar period. This clumpiness has been observed by McIntosh et al. (1989), who combined VLBI and polarization measurements to create a polarization map of R Cas. The photon luminosities of the maser lines in the $v = 1$ and 2 vibrational states are comparable in strength, although there is a significant decrease that occurs at the $J = 6 \rightarrow 5$ transition. The emission from $v = 3$ and 4 maser is much weaker. It appears that the same rotational masers in adjacent vibrational states are produced in the same region.

The most direct evidence for this is obtained using interferometry. Recent observations presented at this conference by Miyoshi show that the $J = 1 \rightarrow 0$ masers in $v = 1$ and 2 usually occur in precisely the same position.

2. MODELING

Out of necessity, numerical calculations are limited to using a finite number of levels. This introduces two significant sources of error. The first problem arises because most of the SiO population will be located in higher rotational levels that are not included in the calculation. Unless corrections are made, all of the population will be artificially forced into only those levels retained in the calculation, resulting in a significant overestimate of the level populations. The major effect is that vibrational optical depths, a primary factor in determining when inversion occurs, will be greatly overestimated. We account for the missing molecules by reducing the total SiO population by the number of molecules that would be found in the excluded rotational levels assuming them to be thermally populated. The significance of this correction may be appreciated by noting that, without it, including as many as 12 rotational levels in each vibrational state still results in an overestimate of the populations by a full order of magnitude!

The second major effect of using a finite number of levels is that the collisional pump rate is severely underestimated. Unlike radiation, collisional coupling between levels is significant for large values of ΔJ, making the collisional pump rate very sensitive to the number of rotational levels included in the calculation. We have developed a corrective procedure that adjusts the collisional pump rate by the ratio of the total to partial pump rates. If a collisional pump rate correction procedure is not used, we recommend that at least 18 rotational levels be retained in each vibrational state (previous studies have typically used 5-12).

3. PUMPING MECHANISMS

The primary factor that affects all pumping mechanisms is the monotonic decrease of vibrational decay rates with J when the vib-rot transitions become optically thick (Kwan and Scoville 1974). This decrease in the loss rate with J results in inversion even for pump rates that are "flat", ie. independent of J. This fundamental inverting mechanism is always present and, as a result, SiO maser emission is easy to produce with either collisional or radiative pumping. We performed extensive investigations of both types of pumps.

Watson et al. (1980) showed that a robust collisional pumping mechanism arises once the vibrational decays become optically thick. These studies were not able to produce detailed results due to the lack of specific cross sectional data. The availability of the Bieniek and Green (1983) cross sections allows a more complete study of collisional pumping to be performed. A significant result of the Bieniek and Green calculations is that the rotational collision cross sections are significantly larger than had been previously assumed, resulting in collisional quenching of the masers at molecular densities above $\sim 10^{10}\,\mathrm{cm}^{-3}$. Calculations show that collisional pumping leads to strong masers and also results in a maser chain occurring in each vibrational state. The relative strengths of the maser lines are controlled by the optical depths of the vibrational transitions. When these become thick, there is a range of column density where many transitions are inverted. As the column density increases, rotational collisions across the maser levels destroy the inversion. This occurs first for the high J masers because they have the largest vibrational opacities and thus suffer more radiative trapping resulting in smaller net radiative decay rates. When the calculated maser intensities reach the observed strengths, the maser chain extends up to about $J = 5$, in agreement with observations. The $J = 2 \rightarrow 1$ and $J = 1 \rightarrow 0$ masers are inverted over a wider range of optical depth and have the largest total emission. The total calculated photon luminosity in these lines is of the order of the typically observed value of $\sim 10^{43}\,s^{-1}$. The relative strengths of maser lines in different vibrational states follow a similar pattern of dependence on optical depth. The higher vibrational state masers become strong at larger values of column density. This behavior is a direct consequence of the vibrational transitions becoming optically thick. It is significant that *there is a large region of overlap in column density where the $J = 1 \rightarrow 0$ masers in both $v = 1$ and 2 are strong*, in agreement with observations. $v = 3$ masers require SiO column densities in excess of $10^{20}\,\mathrm{cm}^{-2}$. These column densities are more rarely achieved and thus detected $v = 3$ maser emission will be rarer and weaker than that from $v = 1$ or 2. The newly discovered $v = 4$ maser in VY CMa requires even larger column densities to produce maser emission. These large column densities will be found only in special sources such as the luminous supergiant VY CMa. It is also not surprising that the first maser to be discovered in $v = 4$ is in a higher rotational transition, $(J = 5 \rightarrow 4)$. A characteristic of all pumping mechanisms is that the higher rotational masers tend to be strongest when the masers in a given vibrational state first turn on. Pumping mechanisms must also be able to explain the time variability of the masers. A significant factor affecting collisional pumping is the increase in the collisional pump rate with temperature. There is a significant increase (by ~ 2 orders of magnitude!) in the maser emission as the temperature increases from 1000 K to 1600 K.

Two different radiative pumping mechanisms have been proposed. The first is an indirect pump proposed by Kwan and Scoville (1974) that inverts the rotational levels of vibrational state v by cycling molecules through the higher vibrational state $v + 1$. This indirect pump relies on small optical depths for the transitions that absorb the pump photons and this property prevents the total pump rates from becoming large. We find that the peak maser emission of the indirect radiative pump is significantly less than that of a collisional pump acting alone and is unable to produce the observed maser emission. A direct pump model proposed by Deguchi and Iguchi (1976) is based on the anisotropic trapping of stellar radiation in a region having a very large velocity gradient $(\epsilon = d\ln v / d\ln r > 1)$. When this occurs, the optical depth is smallest in the radial direction, producing an asymmetry between the absorption of stellar IR photons and the escape of the emitted radiation. However, this mechanism will produce inversion only if $\epsilon > 1$ and then only over a narrow range of optical depth that depends sensitively on the actual value of ϵ. We find that no radiative pump is able to produce enough emission at distances which delineate the outer edge of the maser region. The direct radiative pump is able to work close to the star, but an extremely large velocity gradient

($\epsilon = 5$) is required. In addition, the direct radiative pump is not able to produce strong $J = 1 \rightarrow 0$ masers in both $v = 1$ and 2 at the same position as required by the observations presented by Miyoshi at this conference.

4. SUMMARY

Some conclusions can be reached which are pump independent. First, the total molecular density needed to create the observed maser emission in $v = 1$ and 2 is restricted to lie between $\sim 10^9\,\mathrm{cm}^{-3}$ and $\sim 10^{10}\,\mathrm{cm}^{-3}$. The lower limit is required to create the necessary column density, while the upper limit is set by quenching via rotational collisions. Second, the maser chains observed within the vibrational states are a natural result of both collisional and radiative pumping.

Some important differences between pumping mechanisms do exist. All radiative pumps become less efficient with distance from the star because of the inverse square decrease in pump photons. This prevents even the direct pump from being able to produce the observed emission for masers at the outer boundary of the maser region. Second, direct radiative pumps have a problem in explaining the simultaneous production of the same rotational maser in adjacent vibrational states as discussed above. Third, peak maser luminosities are less for a radiative pump than for a pure collisional pump.

In conclusion, collisional pumping is a robust inversion mechanism that operates under a wide range of physical conditions and, in all likelihood, is the primary pump mechanism responsible for SiO masers. Although this mechanism appears to provide a satisfactory explanation for the ubiquitous presence of SiO masers around late-type stars, detailed comparisons with observations are difficult due to the great complexity of the extended atmospheres. Radiative pumping requires fine tuning of specific physical conditions and thus is less likely to be the dominant mechanism for these masers, whose occurrence is widespread among all types of red giants and supergiants.

REFERENCES

Bieniek, R. J. and Green, S. 1983, ApJLett, 265, L29
Deguchi, S. and Iguchi, T. 1976, PubAstSocJapan, 28, 307
Kwan, J. and Scoville, N. 1974, ApJLett, 194, L97
McIntosh, G., Predmore, C. R., Moran, J. M., Greenhill, L. J., Rogers, A. E. E. and Barvainis, R. 1989, ApJ, 337, 934
Watson, W. D., Elitzur, M. and Bieniek, R. J. 1980, ApJ, 240, 547

THE RESULTS OF KNIFE (JAPANESE MM-VLBI) ON SiO MASERS

MAKOTO MIYOSHI
Nobeyama Radio Observatory
Minamimaki Minamisaku Nagano 384-13 Japan

A N D

KNIFE team
(Communication Research Lab., Inst. of Space Astronaut. Sci. and NRO)

ABSTRACT

We made VLBI simultaneous observations of the two SiO maser lines, and found that the spatial distributions of the v=1and v=2 (J=1-0) of SiO masers were exactly the same around μ Cephei. On the basis of the results, we point out an astrometric use of SiO masers with VLBI.

1. INTRODUCTION

In the SiO maser lines, the vibrational states v=1 and v=2 of J=1-0 often show similar spectral shapes (for example, see figures in Martinez et al. 1988). This fact leads to the idea that the both masers originate in the same physical region. Pioneering VLBI observations of VX Sgr were made by Lane (1984). She showed the SiO maser spots extended to a few stellar radii from the central star and the distributions of these different vibrational states were roughly similar, but there were no detailed correspondence between the same velocity components.

2. KNIFE

KNIFE (Kashima-Nobeyama Interferometer) is a Japanese domestic mm-wave VLBI, using the 34-m radio telescope of Kashima (CRL) and the Nobeyama 45-m radio telescope. These stations provide an east-west 197km baseline with the minimum fringe spacing of 7 milli arcseconds at 43GHz. Following the first fringe at 43 GHz in 1989, several VLBI observations of SiO masers have been done.

3. OBSERVATIONS OF μ CEPHEI

We present here the first results in KNIFE observations. We observed μ Cephei — a red supergiant star (M2 Iab, variability type SRc) in November

10th 1990. Correlation of the VLBI data was made by the K3 correlator at Kashima. The cross power spectra were computed from complex-128 lag cross correlation functions. At the time of observations, the spectral shapes of the both SiO masers were similar to each other. In figure 1 (a) and 2 (a), we present the cross-power spectra of the v=1 and v=2 lines, respectively. Spectra in the v=1 and v=2 are strikingly similar to each other. With keeping this close resemblance between the two v-states, the phase and amplitude of the cross-spectra changed substantially during the 6.5 hours' observations. These indicate that a) most of velocity components are resolved, in other words, the emission is not from single point, and b) the distributions of the two lines are almost exactly the same. Figure 1 (b) and 2 (b) show some examples of 2-points model fitting to the data of the v=1 and v=2 respectively. The phase data are refereed to that of unresolved velocity channel at v_{LSR}=26.78 km/sec . In the figures, Gaussian and contour map presentations are used in order to show the brightness ratio between the spots. The model fitting maps show clearly that the distributions of the v=1 and v=2 are just the same.

3. Discussions

VLBI observations of SiO masers will provide significant informations about the maser mechanism and the physical conditions of circumstellar envelopes. Here, we point out a new application for astrometry. On condition that the v=1 and v=2 of SiO masers emit from the same spots, we can apply the bandwidth synthesis technique to measure the group delay and determine the masing stellar positions. The VLBI technique, coupled with the bandwidth synthesis, have attained 1 milliarcsecond accuracy in constructing the extragalactic radio reference frame. This accuracy is enough to detect the movements of galactic objects. For example, our galactic center moves 7 milli-arcseconds in a year. For lack of compact strong continuum sources in our galaxy, the VLBI have not reached the galactic dynamics researches. There are a few hundred SiO maser sources in our galaxy, and the emissions are strong enough to detect fringes. Only if the measurement of the group delay is possible, SiO maser sources will be suitable for high precision galactic astrometry .

REFERENCES

Lane, A.P. 1984, in VLBI and Compact Radio Sources, ed.R.Fanti (Dordrecht:Reidel) , p.329.
Martinez, A., Bujarrabal, V., and Alcolea, J. 1988, Astron. Astrophys. Suppl., 74, 273.

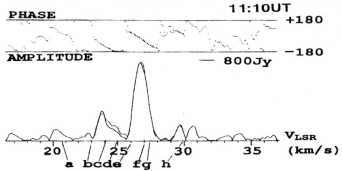

Fig. 1 (a) The cross-power spectra of the v=1 (J=1-0) of SiO masers from μ Cephei. The v=1 main spectral feature extend over two recording channels, the data are overplotted each other.

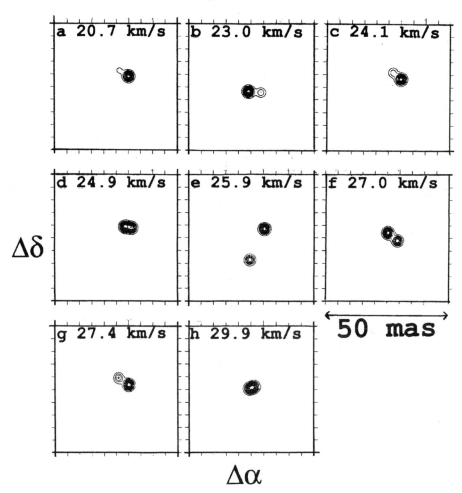

Fig. 1 (b) the 2-points model fittings to the velocity channels marked a to h in the figure (a).

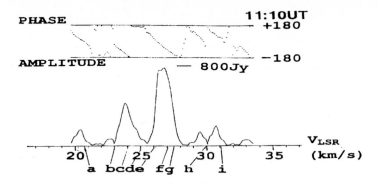

Fig. 2 (a) Same as figure 1 (a) for the v=2 line

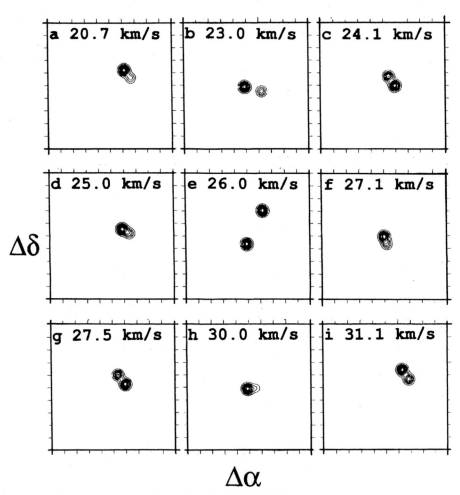

Fig. 2 (b) Same as figure 1 (b) for the v=2 line

SiO Maser Emission and the Intrinsic Properties of Mira Variables

Nimesh A. Patel[1], Antony Joseph, R. Ganesan, C. S. Shukre
Raman Research Institute
Bangalore 560080, India

Introduction

We have made an observational survey of about 160 Mira variables with the aim of investigating the dependence of the SiO masing phenomenon on intrinsic stellar properties. Previous studies to explore this relationship have remained inconclusive due to lack of adequate number of observations and a limited range of spectral-types i.e., M6—M10. The chronological sequence of circumstellar masers proposed by Lewis (1989) does not extend towards the hotter end of spectral-type (<M6) as it is based on available observations which have concentrated on spectral-type >M6. Our survey extends in mean spectral-type from M3 to M10. The results of our survey provide observational constraints for the theories of circumstellar SiO masers, and relate the SiO masing phenomenon to the evolutionary stage of the Mira variables.

Observations

The observations were carried out at 86.243 GHz (J=2→1,v=1), using the Raman Research Institute 10.4m Telescope at Bangalore, India. Some of the characteristics of this instrument, and the observational procedure are described in detail by Patel (1990). 166 Mira variables were selected from the General Catalogue of Variable Stars (Kholopov et al. 1985) and SiO maser emission was searched with a detection limit of 10 to 15 Jy. The observations were made during the winter seasons of 1988 to 1990. Our sample is not biased towards any particular value of the mean spectral-type or period (Patel, Joseph and Ganesan 1992). The sample included a few symbiotic stars, but except the well known R Aqr, none of the other symbiotic sources showed SiO maser emission. Among the few carbon stars in our sample, there appears a tentative detection from T Cnc. This star has been classified as a carbon star with the presence of silicates in its atmosphere, according to the IRAS-LRS (Lloyd-Evans 1990), however, this infra-red signature is not completely unambiguous (Little-Marenin I., private communication) and the presence or absence of SiO maser emission can be used to check whether this and other similar sources classified as J-type stars have an oxygen rich chemistry in their environment as suggested from their infra-red spectra (e.g., Vardya 1990).

To express our results in terms of luminosity, we have calculated distances of the Mira variables in our sample using the observed values of the K-magnitudes and the absolute values obtained from a K-magnitude period-luminosity relation obtained by Feast (1984). This method is discussed along with other available methods by Patel (1990). The small scatter in the period-luminosity relation of 0.16 magnitudes, and the relatively low interstellar extinction at these

---

[1]Present address: Five College Radio Astronomy Observatory, 619 Lederle GRC, University of Massachusetts, Amherst, MA 01003, USA.

wavelengths make this method more reliable than those based on visual magnitudes. From the errors in the measurements of the apparent K-magnitudes, we expect a maximum error in the distances to be about 20%.

Results

Fig. 1 shows the masing and non-masing Mira variables in our sample to be separated on the H-R diagram roughly along a constant radius line corresponding to a value of about $290R_\odot$ (Patel, Joseph and Ganesan, 1992). This minimum radius of the star required for masing may simply reflect a lower limit on the path-length which is related to the gain of the maser. With the currently available optical and IR interferometry techniques, it should be possible to measure the mean value of the diameters of the masing and non-masing Mira variables to confirm this result.

The non-detections could not be attributed to the pulsation phases at the time of observation being close to those corresponding to minimum emission. The masing M-type Mira variables seem to have mean spectral-types later than M6 as seen in Fig. 2. It is important to note that at the maximum brightness, this limit can be as low as M4-M5 (Patel 1990). So far, the time-monitoring observations of SiO maser emisison have been compared only with the variation of visual magnitudes. It will be interesting to also compare the SiO flux with simultaneous measurements of the effective temperature. The effective temperature seems to be an important parameter related to the abundance of SiO as indicated from observations of SiO absorption band at $4\mu m$ by Rinsland and Wing (1982). (See also the theoretical work presented by R. Stencel at this meeting.) Both works indicate a depletion of SiO beyond a temperature of about 3000K. The super-giants showing SiO maser emission do have spectral-types earlier than M6. From the VLBI observations (e.g., Lane 1984), one sees that in these stars the masing seems to occur at a larger distance away from the star, $\sim 5R_*$, compared to a typical value of $\sim 3R_*$ for the regular M-type stars. It is possible that the masing conditions at this greater distance in the super-giants are similar to those closer to the surface of the M-type stars.

The SiO maser luminosity appears to be correlated with the bolometric luminosity of the star as seen from Fig. 3. The scatter in this plot is mainly due to the fact that our observations were made at random values of pulsational phases (a typical expected variation in the luminosity is indicated by the error-bar). This correlation supports a radiative pump for the SiO maser and the 'cut-off' value of -4.8 magnitude suggests a photon-luminosity of a few times 10^{42} photons/sec which is smaller than the lowest observed SiO maser photon-luminosity. Stars fainter than -4.8 magnitudes simply do not generate enough radiation to pump the maser.

The mean value of maser luminosity which lies between 10^{43} to 10^{44} photons/sec, is higher than what is explainable by the current models of SiO masers (Alcolea et al. 1989, Langer and Watson 1985). These models do not include the radiation from the circumstellar dust-shells for pumping the maser. From the dust-shell models of Rowan-Robinson (1984), which are based on IR observations, including the IRAS data, we find that the stars which pose this 'power problem' also have dust with an optical depth at visual and shorter wavelengths, greater than unity. (See Table 1 and Patel and Shukre 1992). If W_ν is the ratio of the dust emission to stellar emission at a given frequency, then, at the location of the masing region, for an optically thin dust shell,

$$W_\nu = \frac{\tau_d I_\nu(T_d)\Delta\Omega_d}{I_\nu(T_*)\Delta\Omega_*}$$

where, I_ν denotes the sum of specific intensities for the two frequencies corresponding to v=0 to v=i and v=0 to v=i+1, for the excitation of v=i level. The solid angle subtended by the dust-shell at the masing region $\Delta\Omega_d = 4\pi - \Delta\Omega_*$. Assuming typical values of $T_* = 2500K$ and $T_d = 1000K$, we obtain the following values of W_ν, for $R = 4R_*$,

v	W_ν ($\tau_d = 0.2$)	W_ν ($\tau_d \gg 1$)
1	1.54	7.7
2	0.71	3.55
3	0.29	1.45

Clearly, the circumstellar dust-shell can contribute significantly to the pumping radiation.

We gratefully acknowledge the insightful guidance received from V. Radhakrishnan and N. V. G. Sarma during the course of this work. It is a pleasure to thank P. G. Ananthsubra-manian, R. Nandakumar, G. Rengerajan, M. Selvamani, K. Sukumaran and other colleagues at RRI for their technical assistance. We thank Janet Mattei and the AAVSO for supplying us the light-curves of many Mira variables in our sample.

REFERENCES

Alcolea J., Bujarrabal V., Gallego J. D., 1989,
    Astron. Astrophys. **211**,187

Feast M., 1984, *Mon. Not. Royal Astr. Soc.* **211**,51p

Kholopov P. N. et al. (Ed), 1985, *General Catalogue of Variable Stars*, Moscow Publishing House

Lane A. P., 1984, *VLBI and compact radio sources* (IAU Symposium No. 110), Ed: Fanti R., Kellermann K., Seti G., Pub: D. Reidel Pub. Co., p329

Langer S. H., Watson W. D., 1984, *Ap. J.* **284**,75

Lewis B. M., 1989, *Ap. J.* **338**,234

Lloyd-Evans T., 1990, *Mon. Not. Royal Astr. Soc.* **243**,336

Patel N. A., 1990, *Ph. D. Thesis*, Indian Institute of Science,

Patel N. A., Shukre C. S., 1992, *in preparation*

Patel N. A., Joseph A., Ganesan R., 1992, *in preparation* Bangalore, India

Rinsland C. P., Wing R. F., 1982, *Ap. J.* **262**,201

Rowan-Robinson M., 1982, *Mon. Not. Royal Astr. Soc.* **202**,767

Vardya M., 1989, *Evolution of peculiar red-giants* (IAU Colloqium No. 106), Ed: Johnson H. R., Zuckerman B., Pub: Cambridge University Press, p359

Source	SiO 10^{44} Photons S^{-1}	τ_d (uv)
R Aqr	0.45	1
RX Boo	0.024	0.2
TX Cam	0.57	5
o Cet	0.81	0.2
R Cnc	0.23	0.2
χ Cyg	5.92	2
U Her	0.49	0.2
R Hor	0.05	0.2
W Hya	0.52	0.5
R Leo	0.25	0.2
R LMi	0.14	0.5
S Per	1.9	2
VX Sgr	4.38	10
IK Tau	0.59	5
NV Aur	2.07	20
WX Psc	2.71	40

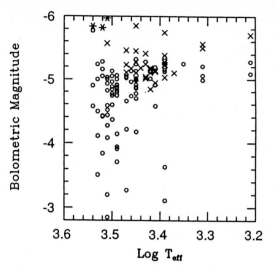

Table 1. Strong SiO sources have optically thick dust shells suggesting a significant contribution to the pumping radiation from dust

Fig. 1 HR diagram. The masing and non-masing (86 GHz) Mira variables seem to separate along a constant radius line corresponding to $\sim 290 R_\odot$

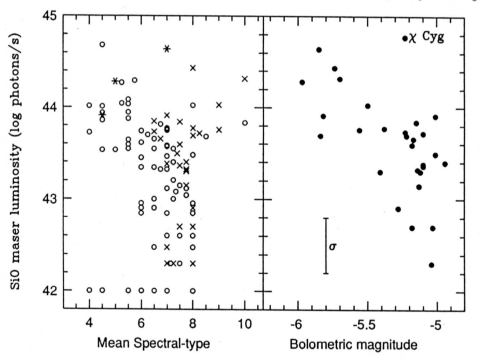

Fig. 2 The masing M-type Mira variables have mean spectral-type $> M6$

Fig. 3 Correlation between SiO luminosity and the Bolometric magnitude implying a radiation pump. σ represents a typical amplitude of SiO flux variation during pulsation

In all figures, O represent non-detections, X represent detections and * denote detection from supergiants

MOLECULAR CATASTROPHES AND CIRCUMSTELLAR SiO MASERS

Robert E. Stencel
CASA and JILA
University of Colorado
Boulder, Colorado 80309-0391 USA

ABSTRACT

Understanding the complex SiO maser regions of highly evolved stars can be improved through multiwavelength studies of "pre-maser" stars, such as M0-M4 giants and semi-regular variables, which can be placed on normal H-R diagrams unlike most of the OH-IR stars. I argue that SiO masers are a key part of the transformation of hot stellar plasma into cold circumstellar silicate dust, in the outflows from evolved, oxygen rich stars. Evidence for this statement rests on the following: (a) red giant mass loss originates in a stochastic, anisotropic manner; (b) SiO maser maps of Miras and red supergiants show numerous maser spots within a few stellar radii; (c) molecules and dust naturally form in a cooling outflow (e.g. SN1987A after 410 days showed strong shortwave infrared bands of CO and SiO molecular emission, and novae show dust production episodes after maximum light); (d) the IRAS Low Resolution Spectrometer provided evidence for diverse and variable 10 micron silicate features in Miras, and these shapes correlate well with the proposed maser chronology, suggesting a formation and annealing sequence. The theory for the occurence of SiO masers involving thermal instability, related "new" physics, recent calculations and a prediction are discussed.

1. SiO Masers Within the Extended Stellar Atmospheres: Evidence for Stochastic Mass Loss from Red Giant Stars

Interferometric studies by Lane et al. (1987), Chapman and Cohen (1986) and McIntosh et al. (1989) suggest that the SiO maser emission at 43 GHz appears in highly inhomogeneous clumps within 1 to 3 stellar radii of the surfaces of selected asymptotic giant branch [AGB] and supergiant stars. Monitoring suggests individual clumps might persist for weeks to months. Analysis of these data indicates that densities of $\sim 10^{10}$ cm^{-3} occur within these condensations (Elitzur 1980; Langer and Watson 1984; Lockett and Elitzur – these Proceedings). New atmospheric models (e.g. Johnson 1991) indicate that densities of 10^8 cm^{-3} persist to 5 stellar radii in the circumstellar envelopes of comparable stars. The enormous scale height of these atmospheres is due to the underlying stellar oscillation – usually a fundamental or first order pulsation among the Asymptotic Giant Branch [AGB] stars, and higher order oscillation among the supergiants and first ascent red giants (see Cuntz and Stencel 1992; Judge and Stencel 1991). Let us examine the hypothesis that the maser "cells" are a consequence of underlying dynamical phenomena.

A number of observational studies suggest that the atmospheres of evolved stars are dynamically active. One of the best studied is the visually bright red supergiant star, α Ori. While α Ori does not exhibit SiO masers itself, it does show thermal SiO emission and is closely related to slightly cooler red supergiants with SiO masers, such as VX Sgr. Smith, Patten and Goldberg (1989) found variations in photospheric radial velocities on timescales as short as a week. Drake et al. (1992) have monitored the radio free-free continuum of α Ori at three different wavelengths, monthly from July 1986 to September 1990 and report variability in the form of flux density decreases which were most pronounced at 2 cm, evident at 3.7 cm and possibly present at 6 cm wavelength, although the changes at different wavelengths do not appear to be correlated. The 25% amplitude of the variations suggest that a large portion of the atmosphere is involved, and the month to month variations are inconsistent with the much slower rotation timescale, and hence not the cause. The origin of these changes, I argue represent transient opacities reducing the chromospheric free-free emission (see below).

Dupree et al. (1987) monitored ultraviolet Mg II emission from α Ori and reported that in addition to a 420 day quasi-periodicity, short term variations in total flux on timescales of days and weeks were found. The 420 day period was also noted by Smith et al. (1989) in low amplitude photspheric radial velocities. Similar variation timescales were reported in optical polarization measurements by Hayes (1984) and Schwarz and Clarke (1984). New high resolution optical imaging has discovered non-uniform and changed disk brightness of α Ori over a 32 month period (Buscher et al. 1990; Wilson et al. 1992). Similarly, Bloemhoef et al. (1984) found variations in the infrared/circumstellar dust output of α Ori.

In each case, variations on timescales of $\sim 10^5$ to 10^7 seconds are reported in adjacent layers of the stellar atmosphere of α Ori. The stellar wind speed of α Ori is approximately 10 to 20 km/sec, measured using optical or UV features or radio CO lines. At such speeds, variation scale sizes of 10^{11} to 10^{13} cm are implied, which are small to moderate fractions of the stellar radius of α Ori. Therefore, the observed dynamical phenomena appear capable of producing *transient structures which are SIMILAR in SIZE, DURATION and DENSITY to the SiO maser clumps*, and hence are the suggested pre-cursors (Stencel, Carpenter and Hagen 1986; Heske 1989; McIntosh et al. 1989; Bujarrabal and Alcolea 1991).

2. Molecular Catastrophes and SiO Masers: Associative Pumping? Clues from IRAS Low Resolution Spectra and Dust Chemistry

I assert that circumstellar SiO maser clumps, and possibly the analogous H_2O ones, are a consequence of underlying dynamical and chemical processes occuring in the stellar atmosphere. This does not replace infrared pumping as a sustaining mechanism for the masers, but provides a means for initiation and evolution of the phenomenon. These maser structures are a consequence of non-thermal energy input to atmosphere from underlying stellar oscillations and/or waves. The evolutionary-state dependent spectrum of stellar oscillation modes controls the complexity of masers and dust ultimately produced.

The fundamental mechanism relies on thermal bifurcation theory (Kneer 1983; Muchmore 1985; Muchmore, Nuth and Stencel 1987; Cuntz and Muchmore 1989). The essential idea is that simple molecules like CO, SiO, H_2O, etc., can cool an atmospheric region by absorbing UV photons and reradiating the energy in the low opacity infrared, and that this local cooling permits more molecules to form, which promote faster cooling, etc. in a runaway process limited only by component abundances and the global thermal balance of the cloud and atmospheric background. This process has been described by various authors in analogous ways as either the inverse greenhouse effect, or a thermo-chemical instability, "molecular catastrophe", cooling instability or plasma instability, often with application to the multiphase interstellar medium.

The key idea is that molecules associate from component atoms in the rapidly cooling plasma, into highly excited states, and internal rotation-vibrational energy coupling channels selectively into states observed as masers. As they are bathed in IR radiation, further pumping of the masers is possible as a way of sustaining the emission. The reported decreases in radio continuum flux of α Ori by Drake et al. are potentially manifestations of this conversion of chromospheric plasma into molecules. A characteristic timescale can be estimated from the ratio of thermal energy of the plasma to the radiative loss rate. Based on the model chromosphere for α Ori by Hartmann and Avrett (1984), with $n = 10^9$ cm^{-3}, $T = 6000$K and $R_L = 10^{-10}$ ergs cm^{-3} sec^{-1}, the thermal instability timescale is 10^7 seconds. Stochastic wave models simulating mass loss for α Ori with periodicities centered around a few x 10^6 seconds produce stochastic chromospheric velocity fields and changes between episodic atmospheric inflow and outflow events on timescales between 10^7 and 10^8 sec (Cuntz 1992). This is in excellent qualitative agreement with observations reviewed by Querci and Querci (1986). Similar use of thermal instability has been invoked by Murray and Lin (1989) in studies of the formation of stars in proto-globular clusters.

This thermal instability scenario requires more quantitative assessment, but it suggests that the masers clumps might be stationary peaks of discrete waves or columns (at phase velocity) in the outflowing extended atmosphere, rather than discrete, ballistic clouds (at group velocity) being ejected from the star. As soon as the technology permits interferometric time series to monitor the development of maser regions continuously for periods of 10^6 seconds, this scenario can be directly tested observationally. Although this type of associative or chemical pumping has been discounted for OH masers (Litvak 1969), its relevance to more tightly bound SiO molecules in the dynamically active inner circumstellar envelope deserves additional study.

As discussed by Lockett (these Proceedings), remaining problems with the collisional pump mechanism in predicting relative line strengths for SiO maser action could suggest some additional physics plays a role. Two possibilities include state selectivity in vibrational energy redistribution among associating molecules (Krajnovich et al. 1987) and above threshhold ionization bound states that appear in intense laser fields (ATI: Bucksbaum et al. 1990).

There is further evidence of the role of dynamical phenomena in governing the chemistry of circumstellar environments. Supernova 1987A showed the development of strong infrared emission bands of CO and SiO 410 days after the explosion (Arnett et al. 1991). Novae frequently show the development of infrared dust features weeks or months after outburst. In either case, the outflowing matter cools and reaches condensation temperature permitting solids to form. The IRAS Low Resolution Spectrometer [LRS], which obtained 10 micron region spectra for thousands of stars, has allowed us to dissect this process of dust formation chemistry, and in principle provide a connection with the plasma and molecular phases that preceed it in stellar atmospheres.

Stencel et al. (1990) reported on a sequence of LRS silicate features among Miras, with spectral breadths and complexities that correlate with broadband infrared colors and type of circumstellar masers in an apparently evolutionary sequence. Specifically, the bluest Miras examined show only SiO masers and have an LRS silicate profile which is broad and featureless and decidely unlike the cannonical narrowly peaked 10 micron silicate dust feature. The reddest Miras in the sample show this cannonical dust feature and a full complement of SiO, H_2O and OH masers, suggesting they are the most evolved, with longer time involved in filling the circumstellar environment with gas, dust and masers. The next step involves correlating atmospheric dynamics with the degree of dust formation and annealing, in a manner analogous to that begun for slightly less evolved red giant stars by Judge and Stencel (1991).

I am pleased to acknowledge useful conversations with Manfred Cuntz, Carl Lineberger and Charles Parmenter, as well as other members of the JILA "Cool Star Mafiosi". This work has been supported in part by NASA awards NAS5-29337, NAG5-1214 and NAG5-1832 to the University of Colorado.

References

Arnett, D. et al. 1989 Ann.Rev.Astron.Astrophys. 27:677.

Bloemhoef, E., Townes, C. and Vanderwyck, A. 1984 ApJ 276:L21.

Bucksbaum, P. et al. 1990 Phys.Rev.Letters 64:1883.

Buscher, D., Haniff, C., Baldwin, J. and Warner, P. 1990 MNRAS 245:7P.

Chapman, J. and Cohen, R.J. 1986 MNRAS 220:513.

Cuntz, M. and Muchmore, D. 1989 A&A 209:305.

Cuntz, M. 1992 in *Proceedings of Cool Stars VII*, eds. J.Bookbinder and M.Giampapa, ASP Conf. Proc., Vol. 26, p.383.

Cuntz, M. and Stencel, R. 1991 in *Proceedings of Mechanisms of Chromospheric and Coronal Heating*, ed. P.Ulmschneider et al. (Springer-Verlag; Berlin), p.206.

Drake, S. et al. 1992 in *Proceedings of Cool Stars VII*, J.Bookbinder and M.Giampapa, eds., ASP Conf. Proc., Vo. 26, p. 455.

Dupree, A. et al. 1987 Ap.J. 317:L85.

Elitzur, M. 1980 ApJ 240: 553.

Hartmann, L. and Avrett, E. 1984 Ap.J. 284:238.

Hayes, D. 1984 Ap.J.Suppl. 55:179.

Heske, A. 1989 A&A 208:77.

Jewell, P. et al. A&A 242:211.

Judge, P. and Stencel, R. 1991 ApJ 371:357.

Kneer, F. 1983 A&A 128:311.

Krajnovich, D. and Parmenter, C. 1987 Chem. Rev. 87:237.

Lane, A. et al. 1987 ApJ 323:756.

Langer, S. and Watson, W. 1984 ApJ 284: 751.

Litvak, M. 1969 Science 165:855.

McIntosh, G., et al. 1989 ApJ 337:934.

Muchmore, D. 1986 A&A 155:172.

Muchmore, D., Nuth, J. and Stencel, R. 1987 ApJ 315:L141.

Murray, S. and Lin, D. 1991 ApJ 367:149.

Querci, M. and Querci, F. 1986 in *Proceedings of Cool Stars IV*, eds. M.Zeilik and D.Gibson (Springer-Verlag; Berlin), p.492.

Schwarz, and Clarke, F. 1984 A.&A. 132:375.

Smith, M., Patten, B. and Goldberg, L. 1989 A.J. 98:2233.

Stencel, R., Carpenter, K. and Hagen, W. 1986 Ap.J. 308:859.

Stencel, R., Nuth, J., Little-Marenin, I. and Little, S. 1990 Ap.J. 350:L45.

Wilson, R., Baldwin, J., Buscher, D. and Warner, P. 1992 MNRAS, in press.

16. SOLAR SYSTEM MASERS

NATURAL LASERS AND MASERS IN THE SOLAR SYSTEM

Michael J. Mumma
Laboratory for Extraterrestrial Physics
NASA Goddard Space Flight Center
Greenbelt, Md 20771

ABSTRACT

Population inversions have been found in the atmospheres of planets and comets, and amplification has been inferred for several cases. In this paper, I review the molecular systems that exhibit lasing and masing action, review the properties of atmospheres that permit these natural lasers and masers to exist, and give examples of their use as probes of remote regions. One potential future application is the possible communication over interstellar distances at GHz rates.

1. INTRODUCTION

Much of our knowledge of the structure and composition of planetary atmospheres has been gained through infrared spectroscopy of their emergent emissions. The great majority of these emissions originate at pressure levels where molecular populations can be satisfactorily explained by invoking thermal equilibrium among the vibrational and rotational distributions. The principal reason for this is that most planetary infrared spectroscopy was conducted at spectral resolutions of ~0.1-1 cm^{-1}, for reasons of instrumental sensitivity. With typical pressure broadening coefficients of ~0.1 cm^{-1} atm^{-1}, the shapes of spectral lines could only be measured if the pressure exceeded ~ 1 atmosphere. At such high pressure levels, collisional rates (~10^9 sec^{-1}) greatly exceed typical rates for dipole-allowed radiative relaxation of either vibrational (~10-10^3 sec^{-1}) or rotational (1-10^{-3} sec^{-1}) transitions. Collisions have ample opportunity to control molecular populations, leading to thermally equilibrated distributions.

The ability to perform sub-doppler measurements by heterodyne spectroscopy at infrared and radio wavelengths enabled the investigation of atmospheric regions where pressure broadening was negligible. In such regions, the collisional rates can be substantially smaller than radiative rates, and external pumping can lead to non-thermal population distributions and even to population inversions. It was only in 1974, when doppler-limited infrared spectroscopy was first applied to the atmospheres of Mars and Venus that evidence of non-thermal distributions in atmospheric CO_2 was obtained. These and other examples of vibrational laser emission and of rotational maser emission were explored over the ensuing two decades (Table 1).

The most thoroughly studied examples of vibrational (laser) emission are the super-radiant 10 μm CO_2 bands emitted from the mesospheres of Mars and Venus. The vertical gains in individual lines are modest, but gain narrowing and intensity amplification along tangential paths are comparable to single pass gains in terrestrial lasers and these effects would be observable from orbiting spacecraft. These lines have provided a unique probe of kinetic temperatures in this region of the atmosphere, and have permitted measurement of absolute winds with 2 m/sec accuracy. They thus have proven useful in studies of heating and circulation in the middle atmosphere. Population inversions have also been found for several other vibrational systems, including NO in the terrestrial ionosphere and possibly NH_3 (v_2, near 10 μm) in the Jovian polar regions. Regarding electronic states, O^1S is predicted to be inverted relative to O^1D in the terrestrial atmosphere.

The best documented solar system maser is the OH hyperfine rotational transition near 1667 MHz, which is often inverted in cometary atmospheres. It has recently been used to study magnetic field strengths and collisional transfer between hyperfine states in the coma of comet Levy 1990 XX (Gerard et al., this volume). Other candidate cometary masers include rotational transitions of CN and of parent volatiles such as water and CO. Cometary CN is pumped very strongly by solar fluorescence in its electronic transitions (A-X and B-X) producing non-thermal rotational populations. Strong pumping of cometary parent volatiles at infrared wavelengths coupled with low collisional relaxation rates in the outer coma permit non-thermal distributions in their rotational manifolds, which can lead to population inversions. For example, Chin and Weaver (1984) explored the excitation of cometary CO and showed that inverted populations should occur for that molecule. However, no cometary masers have yet been identified in parent volatiles.

The most significant aspect of masers and lasers in the solar system derives from their use as probes of the media in which they are found.

2. SPECTRAL LINE FORMATION IN PLANETARY ATMOSPHERES

The appearance of planetary infrared spectra is closely related to the temperature structure of the atmosphere and to the nature of its molecular constituents. It is also influenced by the presence of condensed phase matter, either as air-borne material (aerosols, ice, dust) or as a solid surface below the atmosphere. The emergent flux ($I_{Total}(v)$) at each wavelength is governed by radiative transfer and Kirchoff's law (emissivity equals absorptivity).

$$I_{Total}(v) \quad = \quad B_v(T_S)\, \tau_S(v) \quad + \quad \int\ B_v(T_A)\ d\tau\ (P, T_A) \qquad\qquad 1)$$

T_S is the temperature of the surface, T_A (z) is the local atmospheric temperature, and $d\tau = \sum k_v du$ = $\sum k_v n dz$ is the element of optical depth for the layer in question. The sum proceeds over all significant atmospheric constituents. k_v contains all information on molecular spectral lines, including broadening. The exact solution proceeds iteratively, but an overly simplified view is to state that emergent atmospheric radiation at a given wavelength originates at the atmospheric level for which the optical depth is unity (measured from the top to the level in question). Thus the emergent intensity provides a measure of the temperature in that level.

Since the opacity near a spectral line is strongly frequency dependent, the line will appear in absorption (emission) if the atmospheric temperature decreases (increases) with increasing altitude above the continuum-forming level. The emergent line shape can be more complicated if the core is formed in or above a temperature inversion while the wings are formed below it, for example. Examples of temperature profiles measured for four planets are given in Fig. 1, and the spectra measured for Mars and Venus are shown in Fig. 2 and Fig. 3. The appearance of the spectra changes greatly with spectral resolution for the reasons discussed above. The super-radiant CO_2 emission bands are not apparent in moderate resolution spectra (4 cm^{-1} for Mars, 0.25 cm^{-1} for Venus) because the doppler core emission is only ~10^{-3} cm^{-1} FWHM, much narrower than the width of the 'absorption' line underlying the core (~10^{-2} cm^{-1} FWHM for Mars, ~0.1 cm^{-1} FWHM for Venus).

Table 1. Lasers and Masers in the Solar System.

Object	Transition	T_{line} (K)	T_{local}	T_{exc}	ΔN (cm^{-2})	Gain	Comment
Venus	$CO_2\ v_3-2v_2$	170	170	>300	10^{14}	10%	Solar pumped
Mars	$CO_2\ v_3-2v_2$	120	120	>300	10^{14}	10%	Solar pumped
Earth	$CO_2\ v_3-2v_2$	NA	200	NA	5×10^8	Small	Calculated-day
	$NO\ v=3-2$	NA	200	NA	10^{10}	Small	Observed-Aurora
	$O\ (^1S-^1D)$	NA	200	NA	10^9	Small	Calculated-night
Jupiter	$NH_3\ v_2$	63	>103	>400	?	Large	Gain Narrowed; Variable
Comets	OH 1667 MHz	NA	NA	200	?	Variable	Fluorescence Pump; Many comets.

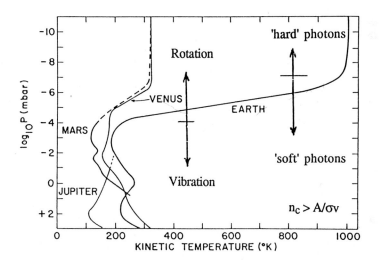

Figure. 1. Vertical temperature profiles (typical) for four planets, arranged on a common pressure scale. Temperature inversions are seen in the mesospheres of the Earth, Mars, and Venus. The range of pressure levels for which rotational and vibrational states may depart from thermal equilibrium are shown schematically. Photo-deposited energy can affect populations in the altitude ranges indicated for 'hard' and 'soft' photons.

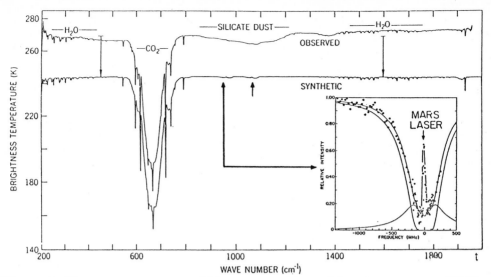

Figure. 2. The emergent spectrum of Mars measured by an infrared spectrometer (IRIS) on Mariner 9, compared with infrared heterodyne spectroscopy of a narrow interval (~0.1 cm^{-1}) near 10.33 μm. The single line (R8 of $^{12}C^{16}O_2$) is fully resolved in the IRHS spectrum (10^{-4} cm^{-1} resolution), but is unrecognizable in the IRIS spectrum (4 cm^{-1} resolution; Hanel and Kunde 1975). A super-radiant emission line is seen at line center. Its emission intensity is 10^8 times brighter than expected from equilibrium considerations. The continuum is formed by thermal emission from the planetary surface (~6 mbar). After Mumma et al. 1981.

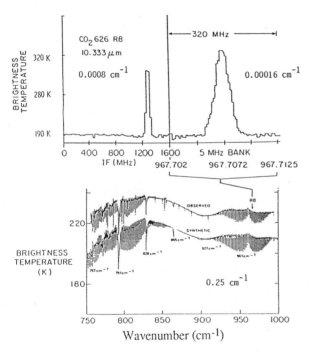

Figure. 3. The emergent spectrum of Venus measured with a resolution of 0.25 cm^{-1} (middle panel, Kunde et al. 1977, and with a resolution of 1.6x10^{-4} cm^{-1} using heterodyne techniques). The super-radiant mesospheric emission line at the core of the R8 line of CO_2 occurs at the center of a very wide 'absorption' line formed at lower altitudes, and is only seen at high resolution. The continuum is formed by thermal radiation from the cloud tops (~60 mbar).

458

The discovery of non-thermal CO_2 emission on Mars and Venus by Johnson et al. (1976) was quickly followed by their use for making the first measurements of Doppler winds (Betz et al. 1976). However, physically unrealistic flow velocities were obtained at disk center on Venus, possibly related to uncertainty in the operating frequency of the CO_2 laser local oscillator used in their infrared heterodyne spectrometer (IRHS). The first absolute wind measurements were made by Goldstein (1989) using a CO_2 laser local oscillator that was stabilized to a saturation resonance of CO_2 in a reference cell (Goldstein et al. 1991).

Betz (1976) and Betz et al. (1977) examined the excitation processes for these emissions. They found that the emission was present only on the planet's dayside and that its intensity was roughly proportional to the local insolation. They examined several excitation mechanisms and showed that the emission was pumped by absorption of combination band quanta (e.g. $00^{\circ}0$-$10^{\circ}1$), followed by resonant collisional transfer of a v_3 quantum to another CO_2 molecule (Fig. 4). Radiative decay of the v_3 level follows promptly, but the optical depth of the v_3 band is much larger than that of the original pump and this prevents escape-to-space of the v_3 photon. Collisional conversion to heat (V-T process) is also very slow. The v_3 quantum is thereby trapped in the medium, permitting a buildup in the population of the $00^{\circ}1$ vibrational level. Betz et al (1977) showed that the emergent intensity was roughly consistent with their pumping model. The relevant physics is reviewed in Fig. 5.

Mumma et al. (1981) studied the kinetic and excitation temperatures on Mars and showed that the population of the upper vibrational level ($00^{\circ}1$) was inverted relative to that of the lower level ($10^{\circ}0$, $02^{\circ}0$)$_I$. They demonstrated that the vertical gain exceeded 0.3%, and termed the emission a 'natural laser'. Deming and Mumma (1983) constructed detailed atmospheric models for Mars and Venus and evaluated all significant pumping bands. They confirmed the combination band solar pump proposed by Betz et al. (1977), and identified a second pump mechanism: absorption in the far wings of v_3 lines followed by collisional frequency re-distribution towards line center. They showed that it provided about 30% of the total pump intensity and that about 1/30,000 of the solar constant at Mars was converted to super-radiant emission at 10 μm. Their models confirmed the presence of lasing action, and showed in detail how the population inversion varied with altitude on Mars (Fig. 6) and Venus (Fig. 7). Independent confirmation of 10 μm lasing action on Venus and Mars was provided by theoretical studies of Gordiets and Panchenko (1983) and Stepanova and Shved (1985). Dickinson and Bougher (1986) provided independent confirmation based on studies of the temperature profile and heating of the Venusian atmosphere.

Deming and Mumma evaluated amplification along horizontal tangent paths through the mesosphere and found that the maximum gain would be ~10% on Mars, slightly less on Venus (Figs. 6,7). They commented that it would be possible to make an oscillator by using the atmosphere as a gain medium and using mirrors in aereosynchronous orbit for the resonator. They suggested that the directed output power could be used for communication over interstellar distances. Sherwood (1988) developed this idea further and showed that although a two mirror system would be subject to low duty cycle as the mirrors orbited the planet, a ring resonator could provide continuous power output. A concept for one such ring laser is shown in Fig. 8.

Sherwood and Mumma (1989) found that sufficient laser power could be obtained to permit interstellar communications at high bandwidth if a 3-meter telescope with currently available 10 μm heterodyne receiver were used as the receiver. Townes (1983) argued that a search for interstellar communications at optical frequencies was warranted, and perhaps preferred to one at radio frequencies, because technical factors favored the development of interstellar communications at optical frequencies (e.g. improved antenna gain). Betz (1986) is currently conducting a search for such signals.

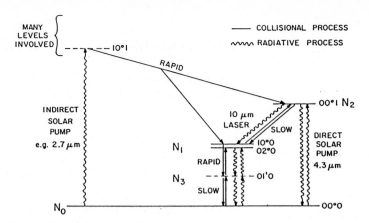

Figure. 4. Molecular physics of natural CO_2 laser. Radiative and collisional processes affecting the pumping of this laser in the mesospheres of Mars and Venus are identified. After Deming and Mumma 1983.

RATE EQUATION FOR 00°1 LEVEL CO_2

$$\frac{dn_u}{dt} = \underbrace{\left[n_0^2 k_\uparrow + n_0 \int I_\nu B_\uparrow d\nu \right]}_{\substack{\text{COLL.} \quad\quad \text{ABS}}} - \underbrace{\left[n_u A + n_u \int I_\nu B_\downarrow d\nu + n_u n_0 k_\downarrow + n_u n_x k_x \right]}_{\substack{\text{SPON.} \quad \text{STIM.} \quad CO_2\text{COLL.} \;\; \text{X-COLL.}}} = 0$$

$$\underbrace{}_{\substack{\text{POPULATION OF}\\ \text{UPPER STATE (00°1)}}} \quad\quad \underbrace{}_{\substack{\text{DE-POPULATION OF}\\ \text{UPPER STATE (00°1)}}}$$

I_ν = Local Radiation Field (λ = 4.26 μm)

$k_{\uparrow\downarrow}$ = Collisional Rate Coefficient for CO_2 (00°1) (cm^3 sec^{-1})

A = Einstein Coefficient for Spontaneous Emission for (00°1) (sec^{-1})

k_x = Collisional Rate Coefficient for Trace Species X.

RELATIVE POPULATION

$$\frac{n_u}{n_0} = \frac{k_\uparrow}{k_\downarrow} \left\{ \frac{1}{\left(1 + \frac{A_{eff}}{n_0 k_\downarrow}\right)} \right\}$$

$$A_{eff} = A \left\{ \frac{1 + \frac{1}{A}\left[\int I_\nu B_\downarrow d\nu - \frac{k_\downarrow}{k_\uparrow} \int I_\nu B_\uparrow d\nu + n_x k_x \right]}{\left(1 + \frac{\int I_\nu B_\uparrow d\nu}{n_0 k_\uparrow}\right)} \right\}$$

IF $A_{eff} > n_0 k_\downarrow$ LTE FAILURE IN 00°1

FROM LABORATORY DATA

$$\frac{A}{n_0 k_\downarrow} = 1 \quad \text{at} \sim 3\,\text{mbar}$$

Figure 5. Simplified rate equation for the upper laser level, and an expression for the relative population of the upper and lower laser states in planetary atmospheres. A_{eff} is the effective transition probability for loss from the medium by spontaneous emission in the ν_3 band followed by escape to space.

460

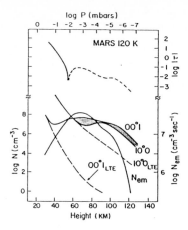

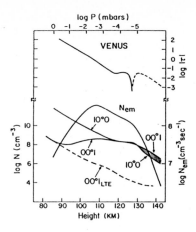

Figure 6. Left. Results of the model calculations for Mars and Venus with the sun at zenith. The local state populations, volume emission rate, and optical depth along a horizontal path are shown vs. altitude above the planet's surface. The region of inverted populations is darkened, and the amplification is shown as negative absorption (dashed line, top). The maximum gain approaches 10%.for tangent altitudes of ~65 km. **Figure 7.** Right. Similar model calculations for Venus. After Deming and Mumma 1983.

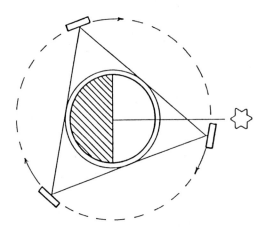

Figure 8. Three satellites defining a triangular ring resonator, in planetary orbit. If used as a communication transmitter, an additional external steering mirror is needed to modulate and direct the emergent laser beam. After Sherwood (1988).

Most measurements of the Doppler core of the Mars and Venus lasers were made at 5 MHz resolution, but a search for gain narrowed emission from regions of enhanced gain on Mars was carried out at 2 MHz resolution. None was found (see Fig. 1 of Deming et al. 1986). A typical measured profile of the R8 line on Mars is shown in Fig. 9. The three principal contributions to the emergent radiation are separately identified: a) thermal emission from the warm surface with atmospheric absorption, b) self-emission from the atmosphere, and c) the super-radiant emission core. Deming et al. (1983) searched for excess limb brightening and verified that the center to limb intensity varied in accord with the pumping models, without recourse to amplification (the maximum air mass was ~3 on Mars, so the amplification was only ~2% in the line of sight).

3. APPLICATIONS

Recent work has emphasized the use of mesospheric lasers to probe atmospheric temperatures and absolute winds. Because the amplification is small along the line-of-sight , gain narrowing can be neglected to a good approximation. The pressure broadening for CO_2 on itself is ~3.5 MHz/mbar, so can also be neglected at mesospheric pressures (~10^{-4} mbar). The measured line width (~35 MHz on Mars) is then a direct measure of the atmospheric doppler width, blended with the instrumental resolution (5 MHz) and drift, if any. The mesospheric temperature can easily be obtained from the measured line width.

Kaufl and his co-workers (Kaufl 1984; Kaufl et al. 1984a, 1984b) studied center-to-pole variations in intensity and argued that an additional pump was needed to explain an observed intensity increase near the poles. They suggested an increased abundance of mesospheric water near the poles. However, from a detailed study of mesospheric temperatures Deming et al. (1986) found a systematic increase of temperatures toward the poles and showed that dynamical warming induced by seasonal transport could account for the observed temperature increase. Their models predicted an increase in super-radiant intensity with increasing atmospheric temperature. The measured intensity increase was consistent with that predicted, hence they argued that no excess mesospheric water was needed. The emission intensity is expected to increase with increasing temperature because the broader rotational distribution permits more solar pump photons to be absorbed in the emitting region.

Accurate wind measurements can only be made with infrared heterodyne spectroscopy if the operating frequency of the local oscillator is known and is stable (cf. Kostiuk and Mumma 1983, Goldstein et al. 1991). In the absence of active frequency stabilization of the laser local oscillator, its actual operating frequency is poorly known. Systematic offsets (pressure shifts) from the rest frequency of the laser transition are introduced by the individual gases found in a typical five-gas mixture. A net offset of ~10-20 m/sec is quite possible, and this can be either towards the red or the blue depending on the exact nature of the mixture. Further, the laser operating frequency is affected by thermal changes in the oscillator length which 'pull' the frequency of the laser unpredictably. The operating frequency may oscillate about some mean value, but with long term drifts. For all of these reasons, absolute velocities are difficult to determine unless the laser is actively locked to a known frequency. The apparent line-of-sight offset (40 m/sec) at disk center on Venus (Betz et al. 1976) may be related to these factors, and the offset of 10 m/sec shown in Fig. 10 is probably also related to these effects (unpublished data, Mumma et al.). The relative velocities in Fig. 10 are probably reliable, and show a retrograde rotation relative to the solid body (heavy dash-dot line) after removing the 10 m/sec offset (compare heavy cross and measurements at disk center).

Goldstein (1989) actively stabilized a CO_2 laser local oscillator to the rest frequency using saturation resonance techniques, and measured absolute winds on Venus and Mars for the first time. His wind measurements are in complete agreement with current models of Venusian

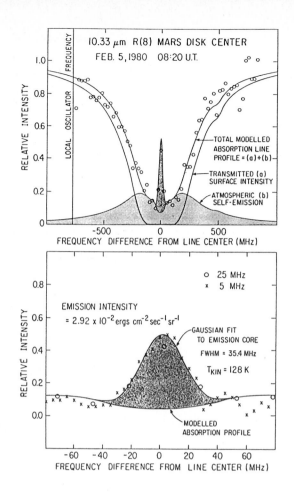

Figure 9. Fully resolved shape of the R8 line of the 10.4 μm CO_2 band [$00°1$—($10°0, 02°0$)ɪ] on Mars. The three components which contribute to the emergent intensity are shown in the upper panel. The super-radiant mesospheric laser emission is darkened to distinguish it from the atmospheric thermal self-emission (grey). Measurements of the shape of the doppler core are shown in the lower panel. The line shape corresponds to a kinetic temperature of 128 K.

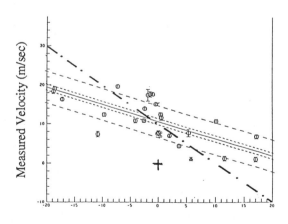

Local Co-rotation Velocity (m/sec)

Figure. 10. Measurements of apparent line of sight doppler winds on Mars using a passively stabilized CO_2 laser local oscillator. The solid body rotation is shown as a heavy dash-dot line, and the expected disk center velocity (i.e. zero) is shown as a heavy cross. The apparent vertical wind at disk center is most easily explained as an offset from rest frequency in the local oscillator. The solid body curve has been shifted upward by the offset frequency for ease of comparison with the relative atmospheric velocities. It is evident that the atmosphere is rotating in a retrograde sense with a velocity of ~10m/sec, relative to the solid body.

circulation patterns (Goldstein et al. 1991, Bougher et al. 1988). An example of his absolute wind measurements is shown in Fig. 11.

4. OTHER SYSTEMS

The search for other solar system lasers can be guided by limits imposed by energy deposition in planetary atmospheres, and considerations of relaxation processes. The solar spectrum is conveniently divided into several wavelength regimes, depending on the effect on atmospheric gases.

First, are the 'hard' photons or particles (e.g. electrons) with energies in the KeV range. When entering the upper atmosphere they produce mainly photoionization, losing ~35 eV per ion-electron pair produced. Typically, ~15 eV is needed for ionization, and the remaining ~20 eV appears as kinetic energy. The 'hot' secondary electrons (e_s) cool by collisions leading to molecular dissociation and/or molecular excitation, with the result that ~15% of the secondary electron energy can be partitioned into vibrational excitation. For example,

$$e_s + N_2\,(X^1\Sigma^+_g) \quad \longrightarrow\ N(^2D) + N(^2D)$$

$$\longrightarrow\ N_2\,(a^1\Pi_g) \tag{2}$$

$$\longrightarrow\ N_2(X^1\Sigma^+_g,\,v)$$

Chemistry can increase the yield of vibrational quanta, e.g.

$$N\,(^2D) \quad + \quad O_2 \quad \longrightarrow \quad NO\,(v=4) \quad + \quad O$$

A rough estimate is that 'hard' photons (or particles) produce ~500-1000 vibrational quanta per KeV.

Photons with insufficient energy to ionize the principal atmospheric gases ('soft' photons) lose energy by dissociating molecules,

$$h\nu \quad + \quad O_3 \quad \longrightarrow \quad O \quad + \quad O_2$$
$$h\nu \quad + \quad XH \quad \longrightarrow \quad X \quad + \quad H$$

and these processes are often followed by chemistry-related vibrational excitation

$$O + O_2 + M \quad \longrightarrow \quad O_3\,(v_3=7) \quad + \quad M$$
$$H + O_3 \quad \longrightarrow \quad OH(v=9) \quad + \quad O_2$$

At infrared wavelengths, direct photo-excitation of vibration occurs

$$h\nu + CO_2(00^00) \quad \longrightarrow \quad CO_2(10^01) \tag{3}$$

and collisional redistribution can follow

$$CO_2(10^01) + CO_2(00^00) \quad \longrightarrow \quad CO_2(10^00) + CO_2(00^01) \tag{4}$$

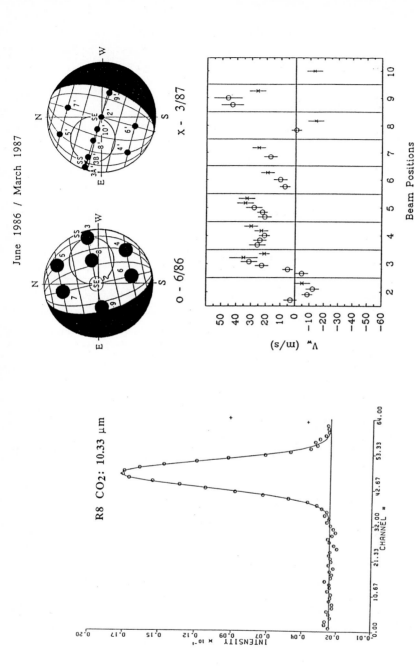

Figure 11. Measurements of absolute Doppler winds in the Venusian atmosphere. A typical measurement of the Doppler core of the R8 line of the 10.4 μm CO_2 band [00°1—(10°0, 02°0)1], is shown at left. Each channel is 5 MHz wide (50 m/sec). The line center is measured to an absolute accuracy ~2 m/sec. The instrumental beam sizes and pointing locations are shown for the McMath Solar Telescope at Kitt Peak (top, left) and for the NASA Infrared Telescope Facility on Mauna Kea (top, right). After Goldstein et al. 1991.

The latter two processes (3,4) are active in the atmospheres of Mars and Venus, which are entirely composed of CO_2 to first order.

Finally, 'soft' photons can produce electronic excitation followed by fluorescence leading to rotational excitation, e.g. the CN $(X^2\Sigma^+ - A^2\Pi)$ bands, or to vibrational excitation through a Franck-Condon process, e.g.

$$h\nu + CO(X^1\Sigma^+, v=0) \longrightarrow CO(A^1\Pi, v=2) \longrightarrow CO(X^1\Sigma^+, v=4).$$

A special case is found for H Lyman α, which pumps the $CO(A^1\Pi, v=14)$ level, and is followed by cascade to $CO(X^1\Sigma^+, v=4)$.

In cometary atmospheres, the OH molecule experiences solar pumping through the strong A-X transition near 306.1 nm, and weaker vibrational pumping in the 1-0 band near 3 μm. The details of the Franck-Condon pump and subsequent cascade depend on the heliocentric velocity of the comet because of the overlap of many solar Fraunhofer lines with the absorption spectrum of OH. This causes populations in the hyperfine rotational levels to switch from inverted to non-inverted and back again as the comet moves along its orbit (Despois et al. 1981). The cometary OH line (1667 MHz) is observed to switch from emission to absorption and back again to emission in concert with the changes in doppler shift. Electron collisions equilibrate the hyperfine levels at distances less than about 5×10^4 km from the cometary nucleus, and the maser is said to be 'quenched' within that region. The reader is referred to the paper by Gerard et al (this volume), which deals specifically with the OH cometary maser.

It is likely that the CN molecule undergoes similar inversions as a result of the very strong fluorescence of the violet bands in the solar radiation field, and indeed the observed rotational distribution deviates significantly from a thermal distribution. The CN radical has not been detected at radio wavelengths since it is much less abundant ($\sim 10^{-3}$) than OH.

Other possible targets for cometary lasers and masers include the water molecule and other parent volatiles. It is now recognized that solar infrared fluorescence is the dominant excitation mechanism for infrared cometary emissions, and many of the pump bands can produce branching into vibrationally excited levels (e.g. H_2O 000—>101—>100). However, recent research has emphasized the detection of fundamental band emission, not potential laser emissions (for a recent review, see Mumma et al. 1992).

Observations of a 'possible' planetary laser in the Jovian atmosphere were reported in 1977 (Kostiuk et al. 1977). Searches for Jovian auroral infrared emissions in NH_3 were begun by our group, although at that time the only evidence supporting the possibility of aurorae on Jupiter was the existence of a strong non-axisymmetric magnetic field, and the known variability in the decametric radio emission. Using our infrared heterodyne spectrometer, we searched for non-thermal emission in NH_3, a known constituent of the Jovian atmosphere, but one which we now know is not normally found above the clouds. To our great surprise, we found evidence for line emission exactly at the expected frequency but with a frequency width that was less than 1/2 as wide as the minimum expected for NH_3. Further, had the emitting region filled our beam, planetary rotation would have required another 2-fold broadening. An accidental overlap with a line from another heavier molecule could explain the observations, but the line frequencies would have to overlap to within about 10 MHz (3 parts in 10^7), and the source region would still need to be smaller than our beam. If it were NH_3 emission, the emitting region could not lie at altitudes below the temperature minimum because precipitating charged particles (the source of auroral energy) cannot penetrate to that level. Also, the rotational and vibrational levels should be equilibrated there, and the population of 0100 can not be easily inverted relative to the ground vibrational level.

The only feasible model is chemical production of NH_3 in vibrationally excited states at high altitudes, followed by rapid destruction of NH_3 after its initial creation, such that a population inversion could be created (e.g. $H_2^+ + N_2 \longrightarrow NH_2 + H^+$; $NH_2 + H_2^+ \longrightarrow NH_3 + H^+$. Other possibilities include $H_3^+ + N \longrightarrow NH_2^+ + H$; $NH_2^+ + H_2 \longrightarrow NH_3 + H^+$). The chemical destruction rate of NH_3 in the ground state must be faster than decay from 0100 into the ground state. These suggestions have not been quantitatively explored, and heterodyne observational studies of emissions in the auroral region now emphasize hydrocarbons at lower altitudes. At this point, the observed emission is not understood.

The known examples of lasers and masers in our solar system are summarized in Table 1.

5. REFERENCES AND ACKNOWLEDGMENT

Betz, A. L. 1976. Ph. D. Thesis, University of California, Berkeley.
Betz, A. L. 1986. Acta Astronautica 13, 623-629.
Betz, A., McLaren, R., Sutton, E., Johnson, M.1977. Icarus 30, 650-662.
Betz, A.L., Johnson, M. A., McLaren, R. A., Sutton, E. C. 1976. Ap. J. 208, L141-L144.
Bougher, S. W., Dickinson, R. E., Ridley, E. C., and Roble, R. G., 1988. Icarus 73, 545-573.
Chin, G., and Weaver, H. A. 1984. Ap. J. 285, 858-869.
Deming, D., and Mumma, M. J. 1983. Icarus 55, 356-368.
Deming, D., Espenak, F., Jennings, D., Kostiuk, T., Mumma, M. J., and Zipoy, D. 1983. Icarus 55, 347-355.
Deming, D., Mumma, M. J., Espenak, F., Kostiuk, T., and Zipoy, D. 1986. Icarus 66, 366-379.
Dickinson, R. E., and Bougher, S. W. 1986. J. Geophys. Res. 91, 70-80.
Despois, D., Gerard, E., Crovisier, J., and Kazes, I. 1981. Astron. Astrophys. 99, 320.
Goldstein, J. 1989. Ph. D. Thesis, Univ of Pennsylvania.
Goldstein, J., Mumma, M. J., Kostiuk, T., Deming, D., Espenak, F., and Zipoy, D. 1991. Icarus 94, 45-63.
Gordiets, B. F., and Panchenko, V. Ya. 1983. Cosmic Res. (USA) 21, 725-734. Translation of Kosm. Issled. (USSR) 21, 929-939 (1983).
Hanel, R. A., and Kunde, V. G. 1975. Spa. Sci. Rev. 18, 201-256.
Johnson, M. A., Betz, A. L., McLaren, R. A., Sutton, E. C., and Townes, C. H. 1976. Ap. J. 208, L145-L148.
Kaufl, H. U. 1984. Ph. D. Thesis, Max-Planck Society.
Kaufl, H. U., Rothermal, H., and Drapatz, S. 1984a. Astron. Astrophys. 136, 319-325.
Kaufl, H. U., Rothermal, H., and Drapatz, S. 1984b. Astron. Astrophys. 141, 430-432.
Kostiuk, T., and Mumma, M. J. 1983. Appl. Optics 22, 2644-2654.
Kostiuk, T., Mumma, M. J., Buhl, D., Brown, L., Faris, J., and Spears, D. 1977. Infrared Physics 17, 431-.
Kunde, V. G., Hanel, R. A., and Herath, L. W. 1977. Icarus 32, 210.
Mumma, M. J., Buhl, D., Chin, G., Deming, D., Espenak, F., Kostiuk, T., and Zipoy, D. 1981. Science 212, 45-49.
Mumma, M. J., Weissman, P. R., and Stern, S. A. 1992. In Protostars and Planets. III. , eds J. Lunine, E. Levy, and M. Matthews (Tucson, Univ. of Arizona Press), in press.
Sherwood, B. 1988. M. S.. Thesis, University of Maryland, College Park.
Sherwood, B., and Mumma, M. J. 1989. Personal communication.
Stepanova, G. I. and Shved, G. M. 1985. Sov. Astron. Lett. 11 (3), 390-394.
Townes, C. H. 1983. Proc. Natl. Acad. Sci. U. S. A. 80, 1147-1151.

This work was supported by the NASA Planetary Astronomy Program under RTOP 196-41-54.

THE OH RADIO EMISSION OF COMET LEVY 1990 XX

E. Gérard, D. Bockelée-Morvan, P. Colom, J. Crovisier
Observatoire de Paris-Meudon
F-92195 Meudon France

ABSTRACT

As part of our monitoring program of the OH radio lines at 18-cm wavelength in bright comets at the Nançay radio telescope, comet Levy 1990 XX was observed from mid-June to September 1990. The cometary maser was strong, due to a favourable excitation of the OH radical, a gas production rate in excess of 10^{29} mol.s^{-1} and a small earth-comet distance. The high signal-to-noise ratio provides useful information on the kinematics of the cometary atmosphere (expansion velocity and anisotropic outgassing). It also allows a sensitive measurement of the average line-of-sight magnetic field by the Zeeman effect: its mean value on Aug. 31 to Sept. 3 is -22^{+}_{-} 3 nT (1 nT=10μG). It finally reveals hyperfine anomalies among the 4 transitions at 1612 MHz, 1665 MHz, 1667 MHz and 1720 MHz. The radiative transfer within the coma cannot explain, alone, all the anomalies and a redistribution of the hyperfine levels, possibly by collisions, is suggested.

1. INTRODUCTION

The 18-cm OH lines were observed in 38 comets with the Nançay radio telescope since 1972. One faint comet, 1991g1, was marginally detected in early 1992 but the most exciting recent comet was undoubtedly Levy 1990 XX, which was monitored almost daily, from June through September 1990 (Bockelée-Morvan et al., 1992). The signal was most of the time larger than that recorded in comet Halley at Nançay and exhibited a strong enhancement in early September when the comet crossed the galactic background. Cometary OH masers are certainly the best understood astrophysical masers. The pump mechanism is fluorescence by UV solar radiation. The Fraunhofer spectrum is Doppler shifted and causes inversion or anti-inversion (depending upon the heliocentric radial velocity of the comet) of the ground-state Λ-doublet where 99% of the OH molecules reside. However, in the inner coma, the UV pump rate is overrun by the collisional transition rates and the maser is quenched. (Despois et al., 1981; Schloerb and Gérard, 1985; Gérard, 1990). The outer coma merely amplifies or attenuates the background radiation according to $\Delta T_F = K\, T_{BG}\, i\, N_{OH}$ (optically thin case), with ΔT_F the line brightness temperature integrated over frequency, K a constant that depends on the hyperfine transition, T_{BG} the background brightness temperature, i the ground state population inversion and N_{OH} the OH column density. In most cases, the OH production rate can be calculated to better than 50% and is in good agreement with the values deduced from UV measurements (Bockelée-Morvan et al., 1990). The exceptionnally strong signal-to-noise ratio in comet Levy 1990 XX permits an analysis of the following topics:
(a) OH production rate versus time (nucleus rotation and/or precession), heliocentric distance and visual magnitude;

(b) mapping of the OH coma: spatial variation of the line shapes and intensities in connection with anisotropic outgassing (jets), scalelength of the OH radical and size of the collisionally quenched inner region;

(c) line shapes and study of the kinematics of the cometary atmosphere with two interesting applications. First, the H_2O outflow velocity (obtained by deconvolution of the OH profile) as a function of heliocentric distance and OH production rate (Bockelée-Morvan et al., 1990). Second, the anisotropic outgassing of the nucleus which generally produces a non-gravitational force which perturbs the cometary orbit (Colom et al., 1990);

(d) Zeeman effect in the OH main lines and corresponding values or limits on the line-of-sight magnetic field averaged over the cometary atmosphere;

(e) relative intensities of the four hyperfine transitions as a test of the excitation model (competition between fluorescence and collisions).

With comet Levy 1990 XX, new effects have been revealed, as it is often the case when high signal-to-noise ratio is achieved. Our strategy is then to attempt to explain such effects and model them in order to make predictions and verify the underlying physics, not only for the geometrical conditions encountered for this comet but also for other comets in our data base. This method has proved to be very useful in the past as the physical and geometric conditions vary greatly from comet to comet and often also for the same comet along its orbit.

The analysis of items (a) and (b) is still under way because it requires careful measurements of the galactic background to deduce precise column densities. We will only present a brief progress report on items (d) and (e).

Figure 1: Average 1667-1665 MHz spectrum of Levy 1990 XX from Aug. 31 to Sept. 3 in right-handed (thick line) and left-handed (dotted line) circular polarizations. The frequency shift from left to right-handed polarization corresponds to a line-of-sight magnetic field of -22^{+3}_{-3} nT.

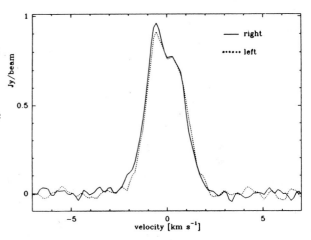

2. THE ZEEMAN SPLITTING IN LEVY AND THE MAGNETIC FIELD

The Zeeman effect (i.e. the frequency difference between the left and right circular polarization components) of the OH 18-cm lines is proportional to the average magnetic field projected on the line-of-sight $< B_p >$. Gérard (1985) measured the Zeeman effect in the

OH lines of comet Austin 1982 VI and obtained a tentative detection $< B_p >= 50^{+}_{-}21\,\mathrm{nT}$ (1 nT=10μG) in early August 1982 (a '+' sign indicates a magnetic field pointing away from the observer). Fig.1 shows the average 1667-1665 MHz spectrum of comet Levy in early September where the frequency shift from left to right circular polarization yields $< B_p >= -22^{+}_{-}3\,\mathrm{nT}$.

More interesting, $< B_p >$ is not constant with time, as illustrated in Fig. 2, and most likely changed sign twice between 8 Aug. (day 220) and 17 Sept. (day 260). As discussed by Gérard (1985), $< B_p >$ depends on several parameters including the heliocentric distance, the OH production rate, the phase angle θ (Sun-Comet-Earth angle) and the heliomagnetic latitude of the comet ϕ. While it is beyond the scope of this presentation to make any quantitative prediction of $< B_p >$, it may be interesting to verify that its sign is correct. In Fig. 2, we have also plotted $sin\theta.sin\phi$ and the agreement appears encouraging between 8 Aug. and 17 Sept.

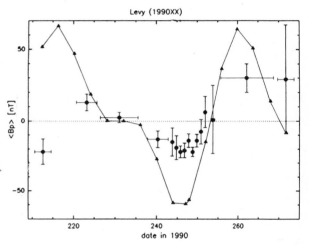

Figure 2: The line-of-sight projected magnetic field $< B_p >$ as a function of time. Circles with error bars: observed values. Triangles: calculated values (see text). Dates are counted from 0 Jan. 1990.

3. THE HYPERFINE LINE RATIOS

Following Despois et al., (1981), the 1720:1667:1665:1612 line area intensities were expected in the LTE ratios of 0.111:1.0:0.555:0.111 when referred to 1667 MHz. Significant departures from the LTE ratios are present in comet Levy: in early Sept., the 1612/1667 and 1720/1667 ratios are respectively $0.069^{+}0.012$ and $0.137^{+}0.015$ instead of 0.111. The 1665/1667 ratio is below normal before 3 Sept. ($0.506^{+}0.01$ in early Aug.) and above normal after 3 Sept. ($0.582^{+}0.008$ on Sept. 5-6). Anomalous 1665/1667 line ratios were searched and indeed found in other comets of our data base with a systematic trend for enhanced ratios at negative inversions of the OH Λ-doublet. Departures from normal ratios should occur when the line optical depths are not negligible, since the line intensity is then proportional to $(e^{-\tau}-1)$, the optical depth τ having the opposite sign of the inversion. Therefore, for positive inversions, the 1667 MHz line should be enhanced compared to the 1665 MHz line and both main lines should be enhanced with respect to the satellite lines. This is in contradiction with what is observed in comet Levy. On the other hand, the

470

relative line intensities could well be explained by a net population transfer from the F=2 to F=1 hyperfine levels within both the '+' and '-' parity states of the Λ-doublet (Fig. 3). Sub-normal values of the 1665/1667 ratio suggest a more efficient transfer inside the lower than the upper level, while values above the normal one indicate the opposite. For other comets observed in absorption (negative inversion), this mechanism predicts 1665/1667 ratios well above the LTE value, in agreement with the observations. A possible explanation for this redistribution of the hyperfine populations might be inelastic collisions.

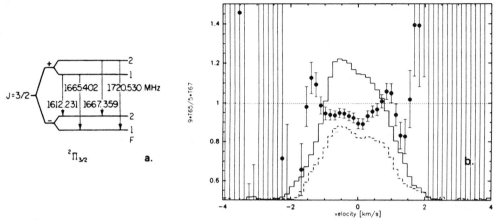

Figure 3:a) The Λ-doublet of the OH ground state. **b)** The 1667 MHz (thick line) and 1665 MHz (dotted line) profiles in Levy 1990 XX from Aug. 27 to Sept. 2 (intensity scale 0 to 1 K) and the 1665/1667 line ratio (circles with error bars) normalized to the LTE ratio of 5/9.

Useful information on this mechanism might also be obtained by measuring the 1665/1667 line ratio in different parts of the coma i.e. in space or frequency. Fig. 4 clearly shows that this ratio varies with velocity and tends to be higher towards the wings of the profiles of comet Levy. Again, by looking back in our data base, we also find anomalous line ratios in comet Halley in Jan. 1986. Clearly much more work remains to be done in order to unravel such effects and model the collisional mechanism involved.

REFERENCES

Bockelée-Morvan,D., Colom,P., Crovisier,J., Gérard,E., Bourgois,G.: 1992 "Asteroids, Comets, Meteors 1991" Edts A. Harris and E. Bowell (in press)

Bockelée-Morvan,D., Crovisier,J., Gérard,E.: 1990, Astron. Astrophys. 238, 382

Colom,P., Gérard,E., Crovisier,J.: 1990 "Asteroids, Comets, Meteors III", Edts C.-I. Lagerkvist, H. Rickman, B.A. Lindblad and M.Lindgren Uppsala University, p.293

Despois,D., Gérard,E., Crovisier,J., Kazès,I.: 1981, Astron.Astrophys. 99, 320

Gérard,E.: 1985, Astron.Astrophys. 146, 1

Gérard,E.: 1990, Astron.Astrophys. 230, 489

Schloerb,F.P., Gérard,E: 1985, Astron.J. 90, 1117

17. CONCLUDING REMARKS

ASTROPHYSICAL MASERS FROM A HISTORICAL PERSPECTIVE

P. R. Schwartz

Naval Research Laboratory

Code 4210, Washington, D.C. 20375

I think that it is appropriate for the concluding remarks for this meeting on Astrophysical Masers be given by a purely historical figure - namely me. My credentials for this task can be established by noting that the most recent maser paper I authored was published in 1982. Incidentally, its abstract states that "These observations allow us to present a snapshot of...maser phenomena to aid in the construction of models..." At least I was on the right track when I left this field. I will review how far we've come in nearly 30 years, what we now agree about concerning Astrophysical Masers, and what are, in my opinion, the most exciting recent results.

We have come a long way in maser research but, before I try to give you my impressions of just how far we have progressed, let's look at where we've been. Everyone knows the story of how, in attempting to follow up the detection of OH interstellar absorption, the University of California at Berkeley group found the OH masers. They, of course, didn't really call them masers but, rather, attributed their results to emission by OH and "Mysterium". One question is then: Who really first called astrophysical masers "masers"? After looking through almost all of the early OH papers, I have concluded that the maser idea first appeared in print in "Observations of Polarized OH Emission" (Weinreb et al. 1965). The authors wrote that "Possible polarization mechanisms are: (1) Zeeman effect; (2) Stark effect; (3) resonance scattering; (4) amplification, through a maser type population inversion, of polarized background radiation." They certainly weren't sure (and were covering their bets).

It is interesting to me that, the polarization paper like the OH detection and "Mysterium" papers, appeared in Nature. Apparently, in 1963-65, OH was more a physical than an astronomical phenomenon. Even more interesting is the fact that the first clear connection of OH masers to conventional astronomy, their discovery in stellar sources, also did not appear in the astronomical literature but rather in <u>Science</u> (Wilson and Barrett, 1968). Similarly, the detection of the H_2O maser appeared in <u>Nature</u> (Cheung et al. 1969). Somewhere around 1969-70, however, masers obviously became respectable astronomy since the CH_3OH maser detection paper made <u>Astrophysical Journal Letters</u> (Barrett, Schwartz and Waters, 1971).

Clearly, today we view astrophysical masers as both an interesting physical phenomenon and as an important astrophysical tool. The greatest measure of the progress of the field is that we can list all of the ways we use masers in astrophysics. We use masers as:

• Signposts of various stages of stellar evolution,
• Probes of star forming regions,
• Probes of the kinematics and structure of mass outflows,
• Probes of YSO environments,
• Distance indicators,
• Galactic structure tracers,
• Probes of the ISM.

We can make the above list because we agree about a number of characteristics of astrophysical masers:

• Astrophysical masers are a ubiquitous phenomenon - that is, there are many masers, many maser lines and many, many sources.
• They are found in YSOs, late-type stars, and (possibly) large galactic features.
• There are strong masers - OH, H_2O, SiO, CH_3OH, and
• Weaker maser lines - CH, NH_3, CHOH.
• We generally agree concerning the physical state of maser regions - i.e. temperatures, densities, magnetic field strengths and velocity dispersions.
• We are converging upon a view involving outflows as a key element in the maser phenomenon - most of our "cartoons" look like wind blown cavities with a disk channeled flow.

At this point, what we all agree about and what most of us agree about begins to become fuzzy but, I would assert the following. The late star scenario and the YSO maser scenarios are beginning to, at least structurally, converge. Perhaps some sort of evolutionary connection is becoming evident.

Finally, I would like to conclude these concluding remarks by pointing out what I think were the most interesting current results discussed at this meeting. I am fascinated by the observations of extra-galactic masers. There appears to be a choice between a "many-maser" and "mega-maser" view that I don't think we understand. But, the truly exciting prospect here is that we may be seeing galactic scale coherent processes. Being a physicist, however, it's explanations rather than puzzles that excite me the most. The discussions of both theory and experiment concerning CH_3OH have been impressive. A real predictive capability for maser characteristics has been demonstrated for this molecule. Similarly, convincing structural and kinematic models for masers associated with whole classes of objects, namely YSOs and late type stars, were presented. These two comments are really part of a third point. Particularly in the last decade, both observers and theoreticians have learned to apply multiple observational tools to masers and the associated astronomical objects. In some sense we've come full circle and are acting like physicists again. Perhaps maser papers should again appear in Nature and Science.

References:

Barrett, A.H., Schwartz, P.R., Waters, J.W. (1971) Ap.J.(Lett) 168, L101.
Cheung, A.C., Rank, D.M., Townes, C.H., Thornton, D.D., Welch, W.J. (1969) Nature, 221, 626.
Weinreb, S., Meeks, M.L., Carter, J.C., Barrett, A.H., Rogers, A.E.E. (1965) Nature, 206, 440.
Wilson, W.J., Barrett, A.H. (1968) Science, 161, 778.

INDEX OF AUTHORS

Printing: Druckhaus Beltz, Hemsbach
Binding: Buchbinderei Schäffer, Grünstadt

RETURN TO: PHYSICS-ASTRONOMY LIBRARY
351 LeConte Hall

LOAN PERIOD 1	2	3
1-MONTH		
4	5	6

ALL BOOKS MAY BE RECALLED AFTER 7 DAYS
Books may be renewed by calling 510-642-3122

DUE AS STAMPED BELOW

FORM NO. DD 22
2M 7-10

UNIVERSITY OF CALIFORNIA, BERKELEY
Berkeley, California 94720–6000